Translational Advances in Gynecologic Cancers

This work is dedicated to the memory of Prof. Helga B. Salvesen who was an outstanding scientist and clinician. She was a leader in the translational science in gynecologic oncology which benefited many of our patients.

Translational Advances in Gynecologic Cancers

Edited by

Michael J. Birrer
Harvard Medical School, Boston, MA, United States
Massachusetts General Hospital, Boston, MA, United States

Lorenzo Ceppi
Massachusetts General Hospital, Boston, MA, United States

ACADEMIC PRESS

An imprint of Elsevier
elsevier.com

Academic Press is an imprint of Elsevier
125 London Wall, London EC2Y 5AS, United Kingdom
525 B Street, Suite 1800, San Diego, CA 92101-4495, United States
50 Hampshire Street, 5th Floor, Cambridge, MA 02139, United States
The Boulevard, Langford Lane, Kidlington, Oxford OX5 1GB, United Kingdom

Notices
Knowledge and best practice in this field are constantly changing. As new research and experience broaden
our understanding, changes in research methods, professional practices, or medical treatment may become
necessary.

Practitioners and researchers must always rely on their own experience and knowledge in evaluating and
using any information, methods, compounds, or experiments described herein. In using such information or
methods they should be mindful of their own safety and the safety of others, including parties for whom they
have a professional responsibility.

To the fullest extent of the law, neither the Publisher nor the authors, contributors, or editors, assume any
liability for any injury and/or damage to persons or property as a matter of products liability, negligence or
otherwise, or from any use or operation of any methods, products, instructions, or ideas contained in the
material herein.

British Library Cataloguing-in-Publication Data
A catalogue record for this book is available from the British Library

Library of Congress Cataloging-in-Publication Data
A catalog record for this book is available from the Library of Congress

ISBN: 978-0-12-803741-6

For Information on all Academic Press publications
visit our website at https://www.elsevier.com/books-and-journals

  Working together
to grow libraries in
developing countries

www.elsevier.com • www.bookaid.org

Publisher: Mica Haley
Acquisition Editor: Rafael Teixeira
Editorial Project Manager: Lisa Eppich
Production Project Manager: Edward Taylor
Designer: Matthew Limbert

Typeset by MPS Limited, Chennai, India

Contents

CHAPTER 6   Homologous Recombination and BRCA Genes in Ovarian
Cancer: Clinical Perspective of Novel Therapeutics 111
J.F. Liu and P.A. Konstantinopoulos

CHAPTER 7   Molecular Basis of PARP Inhibition and Future
Opportunities in Ovarian Cancer Therapy........................ 129
B.L. Collins, A.N. Gonzalez, A. Hanbury,
L. Ceppi and R.T. Penson

List of Contributors

Daphne W. Bell   National Institutes of Health, Bethesda, MD, United States
Jennifer Bergstrom   Johns Hopkins Medicine, Baltimore, MD, United States
Michael J. Birrer   Harvard Medical School, Boston, MA, United States;
Massachusetts General Hospital, Boston, MA, United States
David M. Boruta II   Harvard Medical School, Boston, MA, United States;
Massachusetts General Hospital, Boston, MA, United States
Horacio Cardenas   Northwestern University Feinberg School of Medicine,
Chicago, IL, United States
Lorenzo Ceppi   Massachusetts General Hospital, Boston, MA, United States
Dennis S. Chi   Memorial Sloan Kettering Cancer Center, New York, NY,
United States
Bradley L. Collins   Massachusetts General Hospital, Boston, MA,
United States
Don S. Dizon   Harvard Medical School, Boston, MA, United States
Louis Dubeau   University of Southern California, Los Angeles, CA,
United States
Amanda N. Fader   Johns Hopkins Medicine, Baltimore, MD, United States
Amalia N. Gonzalez   Massachusetts General Hospital, Boston, MA,
United States
Ashley Hanbury   Massachusetts General Hospital, Boston, MA,
United States
Fady Khoury-Collado   Maimonides Medical Center, Brooklyn, NY,
United States
Panagiotis A. Konstantinopoulos   Dana-Farber Cancer Institute, Boston,
MA, United States
Joyce F. Liu   Dana-Farber Cancer Institute, Boston, MA, United States
Lainie P. Martin   Fox Chase Cancer Center, Philadelphia, PA, United States
Daniela Matei   Northwestern University Feinberg School of Medicine,
Chicago, IL, United States; Robert H. Lurie Comprehensive Cancer Center,
Chicago, IL, United States
Kenneth P. Nephew   Indiana University Simon Cancer Center, Indianapolis,
IN, United States; Indiana University, Bloomington, IN, United States

M.H.M. Oonk University of Groningen, Groningen, The Netherlands

Richard T. Penson Massachusetts General Hospital, Boston, MA, United States

Jessica Thomes Pepin Indiana University School of Medicine, Indianapolis, IN, United States

Andrea L. Russo Harvard Medical School, Boston, MA, United States; Massachusetts General Hospital, Boston, MA, United States

Mary E. Sabatini Harvard Medical School, Boston, MA, United States; Massachusetts General Hospital, Boston, MA, United States

Russell J. Schilder Thomas Jefferson University, Philadelphia, PA, United States

Angeles Alvarez Secord Duke University Medical Center, Durham, NC, United States

Ie-Ming Shih Johns Hopkins Medicine, Baltimore, MD, United States

Sharareh Siamakpour-Reihani Duke University Medical Center, Durham, NC, United States

Caryn M. St. Clair Maimonides Medical Center, Brooklyn, NY, United States

Jose Teixeira Michigan State University, Grand Rapids, MI, United States

Krishnansu S. Tewari University of California, Irvine, CA, United States

Mary Ellen Urick National Institutes of Health, Bethesda, MD, United States

A.G.J. van der Zee University of Groningen, Groningen, The Netherlands

Translational Advances in Gynecologic Cancers

INTRODUCTION

In the last few years we have witnessed a dramatic increase in the number of translational discoveries in oncology, taking us from the bench to the bedside more quickly than ever before.

Precision medicine that implies the use of targeted and "patient-based" therapies is turning out to be the most cost-effective approach, and this approach is likely to completely change cancer treatment in the next few years. Regulatory agency approvals for the specific treatment of cancer types, such as homologous recombination deficiency carriers, are brilliant examples of the paradigm shift that is occurring in the treatment of these diseases. Novelties in exploiting targeted approaches are on the cusp of setting a higher bar for better treatments and surgical approaches.

Separate research areas are getting closer to the clinical side as the translation is getting faster. As such, the work of clinicians and researchers has to cover new and transversal areas of knowledge to better guide patients' treatment and to better focus efforts in preclinical investigation. This book starts from the need to create a comprehensive and updated single platform capable of providing a broad outline of the current translational research in gynecologic oncology. The focus starts from the biologic origins of gynecologic cancers, their genomic profiles, and their implications in the therapy. Also included are surgical hot topics, from the role of debulking surgery in ovarian cancer to the sentinel lymph node sentinel mapping in endometrial and vulvar cancer. Moreover, a broad overview of the latest findings regarding target therapies in gynecologic cancers is provided from outstanding and cutting-edge investigators. As an interesting recent topic, we also added fertility preservation in gynecologic cancer patients. We aimed to provide a broad overview of the scientific scenario and are also aware that there is more yet to come.

If this book serves as a tool to better link and speed up the process of transforming promising discoveries from our laboratories into clinical practice, creating improved oncologic outcomes, investigation designs, and patients care, our common goal will be met.

M.J. Birrer and L. Ceppi
Massachusetts General Hospital, Boston, MA, United States

Ovarian Cancer

CHAPTER 1

Origins of Epithelial Ovarian Cancer

L. Dubeau[1] and J. Teixeira[2]
[1]University of Southern California, Los Angeles, CA, United States
[2]Michigan State University, Grand Rapids, MI, United States

CONTENTS

INTRODUCTION

The mechanisms driving both histological subtype differentiation and the early events in progression of epithelial ovarian cancer (EOC) are not clear. Also not well understood are the mechanisms that drive differentiation of epithelial EOC into its common histotypes: serous, endometrioid, mucinous, and clear cell. These deficiencies are largely due to uncertainties about their exact cell of origin, which greatly hampered studies of the biology of their normal counterpart, including how they respond to EOC risk factors.

Accurate knowledge of the cell of origin of EOCs would also lead to a better understanding of the interrelationship between the common histological subtypes of these tumors, which in turn should lead to more effective, histo-type-specific therapeutic approaches. Indeed, the standard of care for all EOC histotypes is essentially the same, characterized by tumor resection with or without prior neoadjuvant therapy and followed by platinum and taxane com-bination chemotherapy. Patients who initially present with advanced stages of EOC typically die from recurrent metastatic disease, but again, the mechanisms of progression to peritoneal/pelvic organ involvement are not well understood. While management of breast cancer patients has significantly improved, in part based on the molecular characteristics and classification of individual tumor types driving the choice of therapy, similar progress for EOC is still on the horizon.

Historically, all EOCs were thought to originate from the mesoepithelial cells covering the ovary, known as the ovarian surface epithelium (OSE) [1], by rupture/repair mechanisms resulting from ovulation, or else, by the forma-tion of inclusion cysts during menopause and ovarian aging and concomitant shrinking. However, the common EOC histotypes, serous, endometrioid, and

Translational Advances in Gynecologic Cancers. DOI: http://dx.doi.org/10.1016/B978-0-12-803741-6.00001-X

mucinous, are not mesotheliomas, but resemble the Müllerian duct-derived Fallopian tubes, endometrium, and endocervix, respectively [2]. This led to the search for evidence of extraovarian sites of origin for these tumors [3]. Here we will present the data supporting various theories on the cell of origin for these most common ovarian cancers and describe the implications on their clinical management.

EMBRYONIC DEVELOPMENT

During embryonic development of vertebrates, the urogenital ridges, made up of the gonadal primordia and mesonephros, form from longitudinal swellings of the coelomic epithelium and intermediate mesoderm, which later differentiates into the kidneys and ureters and into the reproductive ducts and gonads. The same primitive coelomic epithelium also invaginates from stoma near the anterior aspect of the fetal kidney (mesonephros) and expands caudally to form the Müllerian ducts, which are the anlagen of the female reproductive tract [4–6]. Although current evidence suggests that these epithelial ducts are derived from coelomic epithelium, there is little support for the idea that extensive coelomic invagination occurs along the entire Müllerian duct length.

The embryonic developmental pathways driving differentiation of the female reproductive tract from the embryonic Müllerian ducts are controlled, in part, by a subset of homeobox (HOX) genes, HOXA9, 10, 11, the segmental expression of which is required for the proper differentiation of the Müllerian ducts into the oviduct, uterus, and endocervix, respectively (Fig. 1.1) [7]. HOXA9, 10, 11 expression has been observed in serous, endometrioid, and mucinous ovarian cancer subtypes, respectively, and in direct correlation with their Müllerian duct-derived phenotypes [8]. Proponents of the idea that the major EOC subtypes develop from the coelomic epithelium have argued that this illustrates reawakening and metaplasia of OSE cells to their alleged common origin in the multipotent coelomic epithelium of the urogenital ridge, a process that has been debated for many years [9]. Others have argued that this instead merely reflects an origin of the common EOCs not from the OSE, but from tissues embryologically derived from the Müllerian ducts. These arguments will now be discussed in the context of each major subtype of ovarian carcinoma.

SEROUS EOC

The high-grade serous carcinoma (HGSC) histotype is the most common and most deadly form of EOC. Already in the 19th century pathologists recognized that these tumors were histologically similar to those arising in

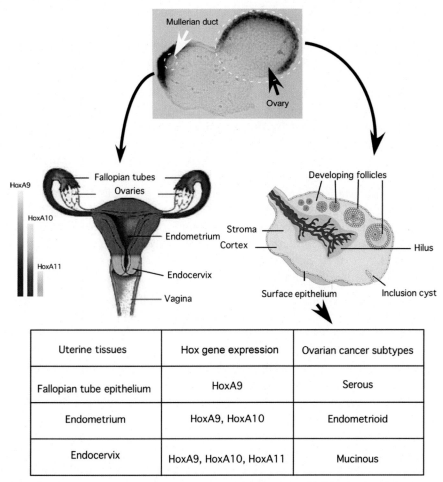

FIGURE 1.1 Correlation of epithelial ovarian cancer (EOC) histotypes with Müllerian duct derivatives. *Amhr2* mRNA expression was detected in mouse embryonic day 15 urogenital ridges by in situ hybridization with an antisense probe (*blue staining*) in the coelomic epithelium of the ovary (*black arrow*) and the distal mesenchyme (*white arrow*) of the Müllerian duct epithelium (*circled with a dotted line*). The EOC subtypes, mucinous, endometrioid, and serous, are histologically similar to and express HOXA9, 10, and 11 genes in the appropriate pattern required for differentiation into the female reproductive tract tissues.

the fallopian tube fimbria [10,11], which are in close proximity to the ovaries. These pathologists argued that a fimbrial origin could not account for all pelvic HGSCs because not only are fallopian tubes not always involved, but also the benign counterparts of serous tumors (serous cystadenomas) as well as serous borderline tumors rarely involve the fallopian tubes [12]. These pathologists instead adopted the view that these tumors develop in OSE cells that have undergone metaplastic change into cells resembling fallopian tube fimbrial epithelium due to changes in their microenvironment after invagination into the ovarian parenchyma. Although a possible origin from the fallopian tube fimbria was acknowledged, it was agreed, by convention, that all serous tumors affecting both the fimbria and the ovary should be categorized as of OSE origin unless strict histopathological criteria were met [10]. This notion is supported by the common presence of intraovarian cortical cysts (CICs) lined by cells similar to fimbrial epithelium based on their ciliated nature and on the fact that they express PAX8, a paired box transcription factor that is often used as a biomarker for diagnosing HGSCs [13]; neither of these characteristics are generally observed in the OSE. The rationale for suggesting the OSE as the source of HGSC has been recently described [1], including the suggested common embryonic origin of both the ovarian surface epithelium and Müllerian duct-derived epithelium and the fact that an estimated 40% of HGSC ovarian tumors are detected without any evidence of Fallopian tube involvement [14–16].

Further support for a role for the OSE in cancer development comes from genetic or induced modification models of OSE in experimental animals and in xenotransplants that are associated with tumors histologically compatible with high-grade serous ovarian cancer. For example, a study from the Teixeira laboratory showed that conditional deletion of the *Stk11* and *Pten* genes in the mouse OSE driven by the Amhr2 promoter, regarded as driving expression of Cre recombinase in the OSE but not in any Müllerian duct-derived epithelium, results in tumors compatible with low-grade or borderline serous ovarian tumors in adolescent and young adult mice that can with time progress to HGSC in older mice [17]. Indirect evidence suggesting that the OSE could be a source of EOC has been demonstrated by its plasticity or stem cell characteristics in mouse [18,19] and in humans [20,21].

The absence of preneoplastic regions within the OSE led Dubeau [22] to consider alternatives to the hypothesis that ovarian tumors arise from metaplastic foci in this epithelium. Reports that microscopic serous tubal intraepithelial carcinomas (STICs) are often observed in secretory cells of the infundibular Fallopian tube epithelium (FTE) collected from patients with BRCA1/2 mutations undergoing prophylactic salpingo-oophorectomies and also from EOC patients without BRCA1/2 mutations [23–26] led to the realization that the fallopian tube epithelium is a much more common site of origin of HGSCs than initially hypothesized [27–29]. Other evidence for the FTE as a site of cancer development has been provided by genetically modified mouse models that develop HGSC after conditional deletion of tumor suppressor genes [30,31] and by observations of stem cell qualities in OSE that are also present in the FTE [18,20,32].

The recognition that the fallopian tube is a more important source of HGSC than initially appreciated needs to be discussed in the context of the arguments used by early pathologists that fallopian tube cancers cannot account for all observations with extrauterine serous tumors, which are as valid today as they were in the early 20th century. The most proximal portion of the Müllerian ducts do not only give rise to the fimbrial end of the fallopian tube but also to microscopic structures lined by fimbrial-like epithelium that are abundant in the peritubal regions as well as in lymph nodes and peritoneal fat away from the tubes and ovaries. These structures, known as endosalpingiosis, also readily account for the aforementioned intraovarian cortical cysts. This led Dubeau to hypothesize that extrauterine HGSCs not originating in the fallopian tube fimbriae could instead originate in foci of endosalpingiosis. This would account for the fact that the benign counterparts of serous carcinomas, known as serous cystadenomas, are frequently found in periovarian and peritubal tissues and that serous borderline tumors have been observed to originate from endosalpingiosis [33]. This theory also readily accounts for the entity known as serous primary peritoneal carcinoma [34–37], as it implies that these tumors do not arise from the peritoneal surface, but from foci of endosalpingiosis away from the ovaries or Fallopian tubes.

ENDOMETRIOID EOC

The endometrioid histotype is the second most common form of EOC. It is often said that these tumors are usually better differentiated than the serous tumors and therefore associated with a better prognosis, although lesser differentiated endometrioid tumors may be underdiagnosed due to difficulties in distinguishing them from poorly differentiated serous tumors using histopathological criteria. Originally described by Sampson in conjunction with his seminal studies of endometriosis [38], the endometrioid histotype has a characteristic glandular morphology similar to that of the human endometrium and is morphologically undistinguishable from the endometrioid subtype of endometrial carcinomas. Sampson raised the possibility, in his original report [38], that retrograde menstruation and subsequent development of endometrioid cysts (endometriomas) could be the origin of this histotype. Several subsequent studies have indeed described a strong association between endometriosis, which is clinically defined by the presence of painful but benign ectopic lesions of endometrial tissue in reproductive age women, and increased risk for developing the endometrioid histotype of EOC (reviewed in [39,40]), but direct evidence in support of this theory has yet to be shown. Tubal ligation, which should theoretically block retrograde menstruation, does appear to significantly correlate negatively with risk for developing the disease later in life [41], supporting the Sampson hypothesis, although this procedure is also protective against serous carcinomas [41–50], implying that the mechanism may be more complex than simple blockage of retrograde menstrual flow. In addition, the idea that all endometrioid ovarian carcinomas arise from endometriosis does not readily account for the occasional presence of both serous and endometrioid differentiation coexisting within the same tumor [51]. The frequency of such tumors showing mixed differentiation is now considered to be low given the number of biomarkers currently available to help pathologists distinguishing between the various histological subtypes, but is still acknowledged as a true entity [52]. Nevertheless, an association between endometrioid ovarian carcinomas and endometriosis is strongly supported by epidemiological, histopathological, and molecular studies [53–55]. Studies comparing the gene expression profile of endometriosis-associated endometrioid ovarian cancer to that of endometrioid ovarian cancer not associated with endometriosis revealed a subset of genes expressed in both cancer types but not in endometriosis [56]. However, in the same study, expression of another subset of genes was concordant between endometriotic lesions and adjacent endometrioid cancer underscoring similarities in differentiation lineages of the malignant and contiguous morphologically benign lesions.

Endometrioid carcinomas are not always associated with endometriosis [57]. The nonendometriosis-associated endometrioid EOCs tend to develop in older women [58], to be of higher histological grade [58], and to harbor genetic changes usually not observed in the endometriotic lesions [59]. Although this may be regarded as evidence for different histogeneses, another interpretation is that higher grade endometrioid carcinomas, because of their increased biological aggressiveness, have often infiltrated into and replaced any adjacent focus of endometriosis, thus masking their site of origin. This idea is similar to the notion that an origin from endosalpingiosis for serous tumors can only be seen in low-grade carcinomas or in tumors of borderline malignancy because these foci are invariably infiltrated by more aggressive high-grade lesions. However, the possibility that some endometrioid carcinomas can arise from metaplasia of the ovarian surface is supported by experimental models showing that specific genetic manipulations of the OSE can lead to tumors with features compatible with endometrioid differentiation, including studies from the Teixeira laboratory [60–63].

Unlike the serous histotype, there is little evidence that the endometrioid histotype involves the FTE in any significant way. However, a mouse model in which APC is lost in the oviductal cells but not in the OSE develop endometrioid ovarian cancer [64], suggesting that disruptions in β-catenin signaling can drive endometrioid differentiation regardless of the cell of origin.

Another feature of extrauterine endometrioid carcinomas is their frequent association with synchronous uterine endometrioid carcinomas, which is seen in up to 25% in younger women [65]. Pathologists have long debated whether these lesions developed in a single primary site that later metastasized or represent two independent primaries. Historically, such synchronous lesions have been regarded as representing two independent primary tumors because these patients typically show clinical courses that are more favorable than what would be expected of metastatic lesions. However, recent massively parallel sequencing analyses showed that synchronous uterine and extrauterine endometrioid carcinomas are clonally related [66]. The presence of such synchronous tumors supports the idea that the different histotypes of cancers of Müllerian duct derivation can originate from either intra- or extrauterine tissues [3]. Indeed, HGSCs and clear cell carcinomas commonly are not only seen in extrauterine tumors but also in endometrial lesions while mucinous carcinomas can develop from the endocervix. This raises the issue of whether or not Müllerian-derived cancers should be classified solely based on their differentiation lineage and be grouped, for example, under the umbrella of cancers of the upper reproductive tract without specifying whether they are of intra- or extrauterine origin. The authors discourage such practice because in spite of the similarities between such tumors, histopathological parameters used in the clinical assessment and management of intrauterine tumors, such as depth of invasion, intravascular invasion, and several others, apply differently to extrauterine cancers of the same differentiation lineages.

MUCINOUS EOC

The least common and least studied of the Müllerian histotypes of EOC is the mucinous variety, traditionally regarded as developing from endocervical metaplasia of the OSE. The incidence of these tumors may have been overestimated in the past because of difficulties in distinguishing them from metastatic colorectal carcinomas. However, it is clear that some mucinous tumors are of primary ovarian or periovarian origin. Intraovarian cysts lined by mucinous epithelium are common [51]. Benign, borderline, and malignant lesions can coexist within the same lesion [51]. In addition, benign mucinous cysts are frequently present in the periovarian and peritubal regions, so much so that pathologists often do not report them unless they are large enough to be clinically significant. The term endocervicosis has been used for such extrauterine mucinous foci, which led Dubeau to suggest that mucinous EOCs arise from endocervicosis, similarly to the notion that serous carcinomas can arise from endosalpingiosis and endometrioid carcinomas can arise from endometriosis [22]. Other proposed potential sites of origin include the idea that these tumors represent a variants of Brenner tumors [67], or that they are derived from mature cystic teratomas [68].

CLEAR CELL CARCINOMA

Clear cell carcinomas have traditionally been considered a variant of endometrioid carcinomas because they are often a component of otherwise typical endometrioid tumors, of either intra- or extrauterine origin. This is also consistent with the fact that mutations that are typically found in endometrioid tumors, such as mutations in PIK3CA and ARID1A, have also been described in clear cell tumors [69,70]. However, epidemiological studies have shown that clear cell differentiation can also be associated with serous tumors [71]. A better understanding of their histogenesis would provide clues as to their biology, which could lead to better clinical management of these cancers that typically are poorly responsive to standard ovarian carcinoma treatment.

Recently, a transgenic mouse was developed in which the *Arid1a* gene was deleted along with expression of an activated allele of *Pik3ca* in the OSE. The mutant mice developed ovarian tumors with complete penetrance and histological similarities to human ovarian clear cell carcinoma [72], suggesting that the OSE could be a cell of origin for this allegedly non-Müllerian histotype. The loss of ARID1A, a component of the SWI/SNF chromatin remodeling complex, is significant because *ARID1A* mutations are highly correlated with both the endometriosis-associated endometrioid ovarian cancer and with clear cell histotypes in humans [69,70]. Another group, using a mouse with a different *Arid1a* mutation and with a superimposed *Pten* deletion, showed instead that some of the tumors had histological similarities to the endometrioid histotype [73], perhaps underscoring the confounding nature of ovarian clear cell carcinoma when compared to the endometrioid histotype discussed in the previous paragraph.

The differentiation lineage of all major subtypes of tumors traditionally classified as ovarian carcinomas, except that of clear cell tumors, is similar to a segment of the normal upper reproductive tract. For example, serous, endometrioid, and mucinous tumors have a differentiation lineage similar to that of the fimbriae, endometrium, and endocervix, respectively, all of which are derived from the Müllerian ducts (Fig. 1.1). There is no known normal component of the upper reproductive tract that shows clear cell differentiation, hence the question of the exact nature of clear cell carcinomas. Potential insights into a source from outside the reproductive tract and gonads came from recent work in mice from the Dubeau laboratory [74] in which a transgenic construct was introduced into a reporter mouse to determine which adult tissues have, at any point during development, expressed the Müllerian specific Amhr2 gene. As expected, all tissues currently known to be derived from the Müllerian ducts showed evidence of such expression. Unexpectedly, a segment of the renal tubules, at the boundary between the renal cortex and medulla, also showed evidence of such prior expression in female, but not in male mice [74]. This strongly suggests a connection between the mesonephric and Müllerian ducts during female development. These findings can account for the facts that tumors belonging to the clear cell subtype share morphological resemblances to clear cell carcinomas of the kidney, have a gene expression profile reported to share similarities with that of renal tumors [75], and also share similarities with renal tumors in their response to chemotherapy [76]. Thus, mesonephric remnants, which are abundant in the paraovarian and paratubal areas, may be integral components of extrauterine Müllerian epithelium and may play a role in the histogenesis of clear cell carcinomas.

CONCLUDING REMARKS

The main differences between the traditional coelomic versus Müllerian hypotheses about the site of origin of cancers historically categorized as ovarian carcinomas are illustrated in Fig. 1.2. The coelomic hypothesis states that the majority of these tumors arise primarily in the ovarian coelomic epithelium, but accounts for their Müllerian appearance by postulating an intermediate step involving Müllerian metaplasia, which is triggered by the hormonal milieu of the ovary. The Müllerian hypothesis states that these tumors come instead from extrauterine derivatives of the Müllerian ducts, which include the fallopian tube fimbriae, endosalpingiosis, endometriosis, endocervicosis, and mesonephric remnants, each one accounting for the specific differentiation lineages of these tumors.

Differences between the two theories have profound implications not only on our understanding of the biology of the different histological subtypes of ovarian carcinomas but also on strategies for risk-reducing surgical intervention in individuals carrying elevated risks, e.g., carriers of germline *BRCA1/2* mutations or individuals with Lynch syndrome. These interventions used to be primarily focused on the ovary. More recently, following the realization of the importance of the fimbriae in the histogenesis of serous carcinomas, salpingectomies became an important component of these operations. The idea that not only the fimbriae but also endosalpingiosis is an important site of origin of serous carcinomas while endometriosis is important for the histogenesis of endometrioid carcinomas implies that risk-reducing surgical interventions should target wide segments of peritubal/ovarian soft tissues. It is the authors' opinion that oophorectomies should continue to be included in risk-reducing surgeries in spite of the fact that most tumors previously categorized as ovarian carcinoma may arise outside this organ because

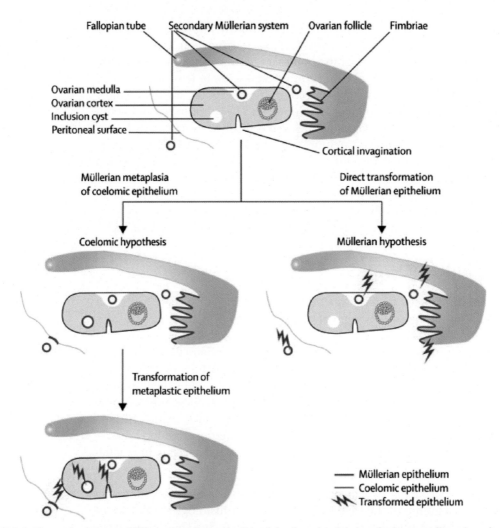

FIGURE 1.2 The coelomic versus Müllerian hypotheses for the origin of ovarian, tubal, and primary peritoneal carcinomas.
According to the coelomic hypothesis, cortical invaginations and cortical inclusion cysts, which are initially lined by coelomic epithelium (*thin black line*), undergo metaplasia and change to Müllerian-like epithelium (*thicker blue lines*) before undergoing malignant transformation (*lightning signs*). The coelomic epithelium covering peritoneal surfaces outside the ovary can give rise to primary peritoneal tumors only after undergoing metaplasia to acquire characteristics of Müllerian epithelium. No intermediary metaplastic step is necessary with the Müllerian hypothesis, which stipulates that Müllerian-like tumors arise directly and exclusively from Müllerian epithelium that is already present, either in the fimbriae or in components of the secondary Müllerian system. *From Dubeau L. The cell of origin of ovarian epithelial tumors. Lancet Oncol 2008;9:1191–7, with permission.*

(1) endosalpingiosis, endometriosis, and endocervicosis may be found within the ovary and (2) regardless of the importance of the ovary as a cancer site of origin, the fact remains that this organ is an important driver of these cancers. Indeed, ovarian hormones play a central role in controlling the menstrual cycle, which is the most important risk factor for all extrauterine Müllerian carcinoma subtypes except mucinous [77–79]. The importance of the ovary as a driver of these cancers is further underscored by observations, from both mouse models and human studies, that *BRCA1* mutations lead to alterations in the circulating levels of sex steroid hormones and in the dynamics of progression of the menstrual cycle [80,81].

The notion that most cancers historically categorized as ovarian carcinomas arise from outside the ovary also underscores the need for a new terminology that more accurately reflects their true site of origin. The term extrauterine Müllerian carcinomas, further subdivided into serous, endometrioid, mucinous, and clear cell subtypes, has been proposed [3].

References

[1] Auersperg N. Ovarian surface epithelium as a source of ovarian cancers: unwarranted speculation or evidence-based hypothesis? Gynecol Oncol 2013;130:246–51.

[2] Scully RE. Recent progress in ovarian cancer. Hum Pathol 1970;1:73–98.

[3] Dubeau L. The cell of origin of ovarian epithelial tumours. Lancet Oncol 2008;9:1191–7.

[4] Guioli S, Sekido R, Lovell-Badge R. The origin of the Mullerian duct in chick and mouse. Dev Biol 2007;302:389–98.

[5] Jacob M, Konrad K, Jacob HJ. Early development of the Mullerian duct in avian embryos with reference to the human. An ultrastructural and immunohistochemical study. Cells Tissues Organs 1999;164:63–81.

[6] Orvis GD, Behringer RR. Cellular mechanisms of Mullerian duct formation in the mouse. Dev Biol 2007;306:493–504.

[7] Kobayashi A, Behringer RR. Developmental genetics of the female reproductive tract in mammals. Nat Rev Genet 2003;4:969–80.

[8] Cheng W, Liu J, Yoshida H, Rosen D, Naora H. Lineage infidelity of epithelial ovarian cancers is controlled by HOX genes that specify regional identity in the reproductive tract. Nat Med 2005;11:531–7.

[9] von Numers C. Observations on metaplastic changes in the germinal epithelium of the ovary and on the aetiology of ovarian endometriosis. Acta Obstet Gynecol Scand 1965;44:107–16.

[10] Finn WF, Javert CT. Primary and metastatic cancer of the fallopian tube. Cancer 1949;2:803–14.

[11] Orthmann EG. Ein primäres carcinoma papillare tubae dextrae, verbunden mit ovarial-abscess. Centrabl f Gynäk 1886;10:816–18.

[12] Dubeau L, Drapkin R. Coming into focus: the non-ovarian origins of ovarian cancer. Ann Oncol 2013;24:viii28–viii35.

[13] Banet N, Kurman RJ. Two types of ovarian cortical inclusion cysts: proposed origin and possible role in ovarian serous carcinogenesis. Int J Gynecol Pathol 2015;34:3–8.

[14] Carlson J, Roh MH, Chang MC, Crum CP. Recent advances in the understanding of the pathogenesis of serous carcinoma: the concept of low- and high-grade disease and the role of the fallopian tube. Diagn Histopathol 2008;14:352–65.

[15] Kindelberger DW, Lee Y, Miron A, Hirsch MS, Feltmate C, Medeiros F, et al. Intraepithelial carcinoma of the fimbria and pelvic serous carcinoma: evidence for a causal relationship. Am J Surg Pathol 2007;31:161–9.

[16] Przybycin CG, Kurman RJ, Ronnett BM, Shih Ie M, Vang R. Are all pelvic (nonuterine) serous carcinomas of tubal origin? Am J Surg Pathol 2010;34:1407–16.

[17] Tanwar PS, Mohapatra G, Chiang S, Engler DA, Zhang L, Kaneko-Tarui T, et al. Loss of LKB1 and PTEN tumor suppressor genes in the ovarian surface epithelium induces papillary serous ovarian cancer. Carcinogenesis 2014;35:546–53.

[18] Flesken-Nikitin A, Hwang CI, Cheng CY, Michurina TV, Enikolopov G, Nikitin AY. Ovarian surface epithelium at the junction area contains a cancer-prone stem cell niche. Nature 2013;495:241–5.

[19] Szotek PP, Chang HL, Brennand K, Fujino A, Pieretti-Vanmarcke R, Lo Celso C, et al. Normal ovarian surface epithelial label-retaining cells exhibit stem/progenitor cell characteristics. Proc Natl Acad Sci USA 2008;105:12469–73.

[20] Auersperg N. The stem-cell profile of ovarian surface epithelium is reproduced in the oviductal fimbriae, with increased stem-cell marker density in distal parts of the fimbriae. Int J Gynecol Pathol 2013;32:444–53.

[21] Viswanathan SR, Powers JT, Einhorn W, Hoshida Y, Ng TL, Toffanin S, et al. Lin28 promotes transformation and is associated with advanced human malignancies. Nat Genet 2009;41:843–8.

[22] Dubeau L. The cell of origin of ovarian epithelial tumors and the ovarian surface epithelium dogma: does the emperor have no clothes? Gynecol Oncol 1999;72:437–42.

[23] Colgan TJ, Murphy J, Cole DE, Narod S, Rosen B. Occult carcinoma in prophylactic oophorectomy specimens: prevalence and association with BRCA germline mutation status. Am J Surg Pathol 2001;25:1283–9.

[24] Lee Y, Miron A, Drapkin R, Nucci MR, Medeiros F, Saleemuddin A, et al. A candidate precursor to serous carcinoma that originates in the distal fallopian tube. J Pathol 2007;211:26–35.

[25] Medeiros F, Muto MG, Lee Y, Elvin JA, Callahan MJ, Feltmate C, et al. The tubal fimbria is a preferred site for early adenocarcinoma in women with familial ovarian cancer syndrome. Am J Surg Pathol 2006;30:230–6.

[26] Piek JM, van Diest PJ, Zweemer RP, Jansen JW, Poort-Keesom RJ, Menko FH, et al. Dysplastic changes in prophylactically removed fallopian tubes of women predisposed to developing ovarian cancer. J Pathol 2001;195:451–6.

[27] Crum CP, Drapkin R, Kindelberger D, Medeiros F, Miron A, Lee Y. Lessons from BRCA: the tubal fimbria emerges as an origin for pelvic serous cancer. Clin Med Res 2007;5:35–44.

[28] Karst AM, Drapkin R. Ovarian cancer pathogenesis: a model in evolution. J Oncol 2010;2010:932371.

[29] Perets R, Drapkin R. It's totally tubular...riding the new wave of ovarian cancer research. Cancer Res 2016;76:10–17.

[30] Kim J, Coffey DM, Creighton CJ, Yu Z, Hawkins SM, Matzuk MM. High-grade serous ovarian cancer arises from fallopian tube in a mouse model. Proc Natl Acad Sci USA 2012;109:3921–6.

[31] Perets R, Wyant GA, Muto KW, Bijron JG, Poole BB, Chin KT, et al. Transformation of the fallopian tube secretory epithelium leads to high-grade serous ovarian cancer in Brca;Tp53;Pten models. Cancer Cell 2013;24:751–65.

[32] Brenton JD, Stingl J. Anatomy of an ovarian cancer. Nature 2013;495:183–4.

[33] Kadar N, Krumerman M. Possible metaplastic origin of lymph node "metastases" in serous ovarian tumor of low malignant potential (borderline serous tumor). Gynecol Oncol 1995;59:394–7.

[34] Altaras MM, Aviram R, Cohen I, Cordoba M, Weiss E, Beyth Y. Primary peritoneal papillary serous adenocarcinoma: clinical and management aspects. Gynecol Oncol 1991;40:230–6.

[35] August CZ, Murad TM, Newton M. Multiple focal extraovarian serous carcinoma. Int J Gynecol Pathol 1985;4:11–23.

[36] Dalrymple JC, Bannatyne P, Russell P, Solomon HJ, Tattersall MHN, Atkinson K, et al. Extraovarian peritoneal serous papillary carcinoma. Cancer 1989;64:110–15.

[37] Fromm G-L, Gershenson DM, Silva EG. Papillary serous carcinoma of the peritoneum. Obstet Gynecol 1990;75:89–95.

[38] Sampson J. Endometrial carcinoma of the ovary, arising in endometrial tissue of that organ. Arch Surg 1925;10:1–75.

[39] Heidemann LN, Hartwell D, Heidemann CH, Jochumsen KM. The relation between endometriosis and ovarian cancer—a review. Acta Obstet Gynecol Scand 2014;93:20–31.

[40] Ness RB. Endometriosis and ovarian cancer: thoughts on shared pathophysiology. Am J Obstet Gynecol 2003;189:280–94.

[41] Rosenblatt KA, Thomas DB. Reduced risk of ovarian cancer in women with a tubal ligation or hysterectomy. The World Health Organization Collaborative Study of Neoplasia and Steroid Contraceptives. Cancer Epidemiol Biomarkers Prev 1996;5:933–5.

[42] Cornelison TL, Natarajan N, Piver MS, Mettlin CJ. Tubal ligation and the risk of ovarian carcinoma. Cancer Detect Prev 1997;21:1–6.

[43] Cramer DW, Xu H. Epidemiologic evidence for uterine growth factors in the pathogenesis of ovarian cancer. Ann Epidemiol 1995;5:310–14.

[44] Green AC, Purdie DM, Bain CJ, Siskind V, Russell P, Quinn M, et al. Tubal sterilization, hysterectomy and decreased risk of ovarian cancer. Survey of Women's Health Group. Int J Cancer 1997;71:948–51.

[45] Hankinson SE, Hunter DJ, Colditz GA, Willet WC, Stampfer MJ, Rosner B, et al. Tubal ligation, hysterectomy, and risk of ovarian cancer. A prospective study. JAMA 1993;270:2813–18.

[46] Irwin KL, Weiss NS, Lee NC, Peterson HB. Tubal sterilization, hysteractomy, and the subsequent occurence of epithelial ovarian cancer. Am J Epidemiol 1991;134:362–9.

[47] Kreiger N, Sloan M, Cotterchio M, Parsons P. Surgical procedures associated with risk of ovarian cancer. Int J Epidemiol 1997;26:710–15.

[48] Loft A, Lidegaard O, Tabor A. Incidence of ovarian cancer after hysterectomy: a nationwide controlled follow up. Br J Obstet Gynaecol 1997;104:1296–301.

[49] Miracle-McMahill HL, Calle EE, Kosinski AS, Rodriguez C, Wingo PA, Thun MJ, et al. Tubal ligation and fatal ovarian cancer in a large prospective cohort study. Am J Epidemiol 1997;145:349–57.

[50] Weiss NS, Harlow BL. Why does hysterectomy without bilateral oophorectomy influence the subsequent incidence of ovarian cancer? Am J Epidemiol 1986;124:856–8.

[51] Scully RE. Tumors of the ovary and maldeveloped gonads. Armed Forces Institute of Pathology 1978; Fascicle 16:145.

[52] Mackenzie R, Talhouk A, Eshragh S, Lau S, Cheung D, Chow C, et al. Morphologic and molecular characteristics of mixed epithelial ovarian cancers. Am J Surg Pathol 2015;39:1548–57.

[53] Ali-Fehmi R, Khalifeh I, Bandyopadhyay S, Lawrence WD, Silva E, Liao D, et al. Patterns of loss of heterozygosity at 10q23.3 and microsatellite instability in endometriosis, atypical

endometriosis, and ovarian carcinoma arising in association with endometriosis. Int J Gynecol Pathol 2006;25:223–9.

[54] Fuseya C, Horiuchi A, Hayashi A, Suzuki A, Miyamoto T, Hayashi T, et al. Involvement of pelvic inflammation-related mismatch repair abnormalities and microsatellite instability in the malignant transformation of ovarian endometriosis. Hum Pathol 2012;43:1964–72.

[55] Wei JJ, William J, Bulun S. Endometriosis and ovarian cancer: a review of clinical, pathologic, and molecular aspects. Int J Gynecol Pathol 2011;30:553–68.

[56] Banz C, Ungethuem U, Kuban RJ, Diedrich K, Lengyel E, Hornung D. The molecular signature of endometriosis-associated endometrioid ovarian cancer differs significantly from endometriosis-independent endometrioid ovarian cancer. Fertil Steril 2010;94:1212–17.

[57] Mangili G, Bergamini A, Taccagni G, Gentile C, Panina P, Vigano P, et al. Unraveling the two entities of endometrioid ovarian cancer: a single center clinical experience. Gynecol Oncol 2012;126:403–7.

[58] Erzen M, Rakar S, Klancnik B, Syrjanen K. Endometriosis-associated ovarian carcinoma (EAOC): an entity distinct from other ovarian carcinomas as suggested by a nested case–control study. Gynecol Oncol 2001;83:100–8.

[59] Vestergaard AL, Thorup K, Knudsen UB, Munk T, Rosbach H, Poulsen JB, et al. Oncogenic events associated with endometrial and ovarian cancers are rare in endometriosis. Mol Hum Reprod 2011;17:758–61.

[60] Dinulescu DM, Ince TA, Quade BJ, Shafer SA, Crowley D, Jacks T. Role of K-ras and Pten in the development of mouse models of endometriosis and endometrioid ovarian cancer. Nat Med 2005;11:63–70.

[61] Tanwar PS, Kaneko-Tarui T, Lee HJ, Zhang L, Teixeira JM. PTEN loss and HOXA10 expression are associated with ovarian endometrioid adenocarcinoma differentiation and progression. Carcinogenesis 2013;34:893–901.

[62] Tanwar PS, Zhang L, Kaneko-Tarui T, Curley MD, Taketo MM, Rani P, et al. Mammalian target of rapamycin is a therapeutic target for murine ovarian endometrioid adenocarcinomas with dysregulated Wnt/beta-catenin and PTEN. PLoS One 2011;6:e20715.

[63] Wu R, Hendrix-Lucas N, Kuick R, Zhai Y, Schwartz DR, Akyol A, et al. Mouse model of human ovarian endometrioid adenocarcinoma based on somatic defects in the Wnt/beta-catenin and PI3K/Pten signaling pathways. Cancer Cell 2007;11:321–33.

[64] van der Horst PH, van der Zee M, Heijmans-Antonissen C, Jia Y, DeMayo FJ, Lydon JP, et al. A mouse model for endometrioid ovarian cancer arising from the distal oviduct. Int J Cancer 2014;135:1028–37.

[65] Walsh C, Holschneider C, Hoang Y, Tieu K, Karlan B, Cass I. Coexisting ovarian malignancy in young women with endometrial cancer. Obstet Gynecol 2005;106:693–9.

[66] Schultheis AM, Ng CK, De Filippo MR, Piscuoglio S, Macedo GS, Gatius S, et al. Massively parallel sequencing-based clonality analysis of synchronous endometrioid endometrial and ovarian carcinomas. J Natl Cancer Inst 2016;108.

[67] Seidman JD, Khedmati F. Exploring the histogenesis of ovarian mucinous and transitional cell (Brenner) neoplasms and their relationship with Walthard cell nests: a study of 120 tumors. Arch Pathol Lab Med 2008;132:1753–60.

[68] Vang R, Gown AM, Zhao C, Barry TS, Isacson C, Richardson MS, et al. Ovarian mucinous tumors associated with mature cystic teratomas: morphologic and immunohistochemical analysis identifies a subset of potential teratomatous origin that shares features of lower gastrointestinal tract mucinous tumors more commonly encountered as secondary tumors in the ovary. Am J Surg Pathol 2007;31:854–69.

[69] Jones S, Wang TL, Shih Ie M, Mao TL, Nakayama K, Roden R, et al. Frequent mutations of chromatin remodeling gene ARID1A in ovarian clear cell carcinoma. Science 2010;330:228–31.

[70] Wiegand KC, Shah SP, Al-Agha OM, Zhao Y, Tse K, Zeng T, et al. ARID1A mutations in endometriosis-associated ovarian carcinomas. N Engl J Med 2010;363:1532–43.

[71] Pearce CL, Templeman C, Rossing MA, Lee A, Near AM, Webb PM, et al. Association between endometriosis and risk of histological subtypes of ovarian cancer: a pooled analysis of case–control studies. Lancet Oncol 2012;13:385–94.

[72] Chandler RL, Damrauer JS, Raab JR, Schisler JC, Wilkerson MD, Didion JP, et al. Coexistent ARID1A-PIK3CA mutations promote ovarian clear-cell tumorigenesis through pro-tumorigenic inflammatory cytokine signalling. Nat Commun 2015;6:6118.

[73] Guan B, Rahmanto YS, Wu RC, Wang Y, Wang Z, Wang TL, et al. Roles of deletion of Arid1a, a tumor suppressor, in mouse ovarian tumorigenesis. J Natl Cancer Inst 2014;106.

[74] Liu Y, Yen HY, Austria T, Pettersson J, Peti-Peterdi J, Maxson R, et al. A mouse model that reproduces the developmental pathways and site specificity of the cancers associated with the human BRCA1 mutation carrier state. EBioMedicine 2015;2:1318–30.

[75] Zorn KK, Bonome T, Gangi L, Chandramouli GV, Awtrey CS, Gardner GJ, et al. Gene expression profiles of serous, endometrioid, and clear cell subtypes of ovarian and endometrial cancer. Clin Cancer Res 2005;11:6422–30.

[76] Anglesio MS, George J, Kulbe H, Friedlander M, Rischin D, Lemech C, et al. IL6-STAT3-HIF signaling and therapeutic response to the angiogenesis inhibitor sunitinib in ovarian clear cell cancer. Clin Cancer Res 2011;17:2538–48.

[77] Brose MS, Rebbeck TR, Calzone KA, Stopfer JE, Nathanson KL, Weber BL. Cancer risk estimates for BRCA1 mutation carriers identified in a risk evaluation program. J Natl Cancer Inst 2002;94:1365–72.

[78] Pike MC, Pearce CL, Peters R, Cozen W, Wan P, Wu AH. Hormonal factors and the risk of invasive ovarian cancer: a population-based case–control study. Fertil Steril 2004;82:186–95.

[79] Whittemore AS, Harris R, Itnyre J. Characteristics relating to ovarian cancer risk: collaborative analysis of 12 US case–control studies. II. Invasive epithelial ovarian cancers in white women. Collaborative Cancer Group. Am J Epidemiol 1992;136:1184–203.

[80] Hong H, Yen HY, Brockmeyer A, Liu Y, Chodankar R, Pike MC, et al. Changes in the mouse estrus cycle in response to BRCA1 inactivation suggest a potential link between risk factors for familial and sporadic ovarian cancer. Cancer Res 2010;70:221–8.

[81] Widschwendter M, Rosenthal AN, Philpott S, Rizzuto I, Fraser L, Hayward J, et al. The sex hormone system in carriers of BRCA1/2 mutations: a case–control study. Lancet Oncol 2013;14:1226–32.

CHAPTER 2

Ovarian Cancer Genomics

L. Ceppi[1] and M.J. Birrer[1,2]
[1]Massachusetts General Hospital, Boston, MA, United States
[2]Harvard Medical School, Boston, MA, United States

CONTENTS

INTRODUCTION

Epithelial ovarian cancer (OC) is a clinically important disease [1]. It affects 22,000 women each year in the United States and produces 15,000 fatalities annually, making one of the most lethal forms of cancer in women [2]. The high mortality rate of OC is primarily due to its advanced stage at diagnosis with the clinical presentation of widespread abdominal dissemination. OC therapy includes debulking surgery followed by taxol/platinum chemotherapy [3]. It has been demonstrated that the residual disease after primary debulking surgery has a crucial impact on survival [4]. However, the limited overall survival is mainly due to the high rate of tumor relapse with the development of chemoresistant disease [5].

A paradigm shift has occurred in the definition of OC disease. OC is no longer considered a single disease but rather a composite number of unique cancers, characterized by different precursor lesions with completely different patterns of genomic alterations.

LOW-GRADE SEROUS OVARIAN CANCER

Existing clinical [6] and genomic data demonstrate critical differences between low-grade serous carcinoma (LGSC) and high-grade serous carcinoma (HGSC) in pathologic features, molecular changes, and clinical outcome [7,8], as well as some remarkable similarities between borderline-LMP (low malignant potential) and LGSC tumors [9–11].

Expression profiling of HGSC and LGSC revealed no enhancement of expression of genes involved in chromosomal instability and cell proliferation in LGSC and LMP tumors as opposed to HGSC [12,13]. Pathology series

19

Translational Advances in Gynecologic Cancers. DOI: http://dx.doi.org/10.1016/B978-0-12-803741-6.00002-1

analyzing LMP tumors found mutations in KRAS or BRAF to be present in 25–50% [14]. Similarly LGSC has been reported to have mutations in KRAS and BRAF in one-third of the cases [15] whereas lower mutational rates were reported in more recent [16] and prospective series [17]. GOG-0239 observed a lower occurrence of BRAF (6%) rather than KRAS (41%) mutation, even though a fraction of these LGS tumors were primary (82%). Signature mutations of HGSC such as TP53 and BRCA1/2 are rarely present in LGSC [18,19].

Targeting MAPK hyperactivation is a rational therapeutic approach for the treatment of LGSC, which has been shown to have a dismal response to conventional platinum-based therapy [20]. GOG-239 phase II clinical study using the small molecule inhibitor of MEK1/2 selumetinib (AZD6244) in women with recurrent LGSC [17] exhibited substantial activity together with minimal toxicity. It demonstrated a fivefold increase in the response rate (15% vs 3.7%) in comparison to conventional chemotherapy and over 65% of the patients had stable disease with an overall median progression-free survival of 11 months. No clear correlations between response or PFS and tumor mutation were identified. However, recently a meticulous investigation over selumetinib in an extreme responder identified a novel mutation related to MAPK pathway activation [21]. Overall, these findings support MAPK inhibition as an effective therapy in LGSC and warrant further evaluations of this approach in treating LGSC with single agent regimens as well as with combinatorial therapies that leverage on different therapeutic mechanisms, e.g., antiangiogenesis drugs [22].

CLEAR CELL OVARIAN CANCER

Numerous studies have characterized the mutational and gene expression differences between serous and clear cell ovarian cancer (CCOC). CCOC is characterized by recurrent mutations in ARID1A in 46–57% of cases [23,24] with the subsequent loss of BAF250. Interestingly, preneoplastic lesions such as endometriosis nodules in proximity to CCOC harbor the ARID1A mutation. PIK3CA mutations are present in 33% of cases of CCOC [25], constituting a possible driving mutation in this disease. Microarray gene expression analyses have shown that CCOC carries a uniform expression pattern regardless of the organ of origin between ovarian and endometrial cancers. This is in contrast to serous and endometrioid subtypes of both ovarian and endometrial cancers whose expression profiles are clearly distinct. Interestingly, the similarity among clear cell cancers from different organs extends to clear cell renal cell carcinoma (CCRCC) and suggests that targeting the mTOR pathway or angiogenesis (targets in CCRCC) as potential targets may be rationale [26]. In addition, a distinct molecular signature of CCOC includes HIF-1α and HIF-2α differentially overexpressed genes (compared to serous subtypes [27,28]).

Indeed, preclinical studies comparing serous and clear-cell OC cell lines revealed that the former are less resilient to hypoxic conditions than clear-cell cell lines. This was confirmed by knockdown assays of critical genes in the hypoxia pathway (ENO-1 and HIF-1α) that showed sensitization of clear-cell cell lines to hypoxia and glucose deprivation [28]. Consequently, disruption of the hypoxia response or targeting of angiogenesis pathways leads to a recent phase II clinical trials involving an inhibitor of vascular endothelial growth factor receptor (VEGFR) and platelet-derived growth factor receptor (PDGFR) for recurrent or persistent clear cell ovarian cancer (GOG 254, NCT00979992) [29]. Additionally, PIK3CA mutations [25,30] supported the basis to test PI3K/PTEN/mTOR pathway inhibition through downstream effectors blockade, such as mTOR [31]. Phase II trials are ongoing and results are expected for newly diagnosed CCOC (Gynecologic Oncology Group (GOG) 268, NCT01196429).

MUCINOUS OVARIAN CANCER

Mucinous ovarian cancer (MOC) represents a unique entity. Most of the morphologic and genomic aspects of MOC differ from other OC histotypes, and MOC appears to be more closely related to colorectal cancer. The evidence of many MOCs being diagnosed as metastatic colorectal cancers raised questions concerning the existence of primary origin of MOC. It has been reported that just a small subgroup (3%) of MOC should be classified as of Mullerian origin [32] and a well-established pathology workflow appears to be crucial to rule out the metastatic origin of MOC [33].

MOC is known to carry a similar genomic profile to colorectal cancers with hallmark mutations occurring in the KRAS gene [34] in more than half of the cases, regardless of stage and grade. Kurman and Shih summarized the etiology of mucinous cancers as a mutational relationship exists among benign mucinous cystoadenomas, LMP mucinous tumors, and mucinous ovarian cancer [14], suggesting that KRAS represents an early event in this cascade. Other molecular events include HER2 overexpression which occurs in 6.2% of mucinous LMP tumors and 18.8% of MOC [35]. A more recent study [36] analyzed the mutational status [35]. Targeted exome sequencing described pattern of mutations in known important oncogenes: RAS was mutated in 54% in benign lesions, 62% in LMP and in 45% of the cases of MOC, respectively; BRAF in 10.3% and 22.6% in LMP and MOC, respectively. Interestingly, TP53 mutations were also present in 8% of benign, 13% of LMP and 51% of MOC, with increased frequency related to increasing degree of tumor malignancy. Additionally, also RRAS2 mutation and cell cycle regulatory CDKN2A homozygous deletion have been reported in low frequency.

Recent reports observed mutations in an MOC dataset through hotspot exome sequencing [37] and confirmed the mutational burden in KRAS, TP53, and CDKN2A. Additional mutated genes were identified in MOC samples: PIK3CA was mutated in 13.5%, PTEN in 2.7%, CTNNB1 in 5.4% of the samples [37]. These alterations may represent additional actionable targets in this rare disease.

ADVANCED STAGE SEROUS OVARIAN CANCER

The Cancer Genome Atlas (TCGA) Project provided meaningful insight into advanced stage high-grade serous ovarian cancer (ASOC) genomic data. A national network of research teams was able to generate an integrated analysis and provide a full publicly available dataset for further data analysis, validation, and development (https://tcga-data.nci.nih.gov/tcga/). For OC, almost 489 patient samples with complete clinical data were analyzed for DNA copy number variation analysis, whole genome sequencing, mRNA and microRNA expression, and DNA methylation.

DNA Copy Number Analysis

DNA analysis revealed extensive gene copy number alterations in multiple chromosomal regions, representing one of the hallmarks of OC genomic alterations. 8 genome regions with copy number gain and 22 with losses were identified as occurring in up to 50% of the tumors, and 63 additional regions had focal amplifications. Throughout these altered regions, approximately 30 growth-stimulatory genes amplifications (e.g., CCNE1, MYC, and MECOM) were consistently altered in up to 20% of the cases, and the expression of these genes is known to have prognostic significance [38]. Other genes are located focal amplifications and include ZMYND8, IRF2BP2, ID4, PAX8, and TERT. Among deleted regions, known regulators (such as PTEN, RB1, NF1) were affected in 6–8% of cases, with increasing alteration status when coupled with mutational events.

Mutational Status

TP53 gene mutation is the most common event in serous OC, present in 96% tumors of TCGA dataset. It is believed to be one of the first driving events in OC carcinogenesis since it was found in in situ serous tubal carcinoma lesions, likely a candidate for the precursor form of high-grade serous ovarian cancer [39]. The early loss of TP53 function, key regulator of genomic stability, contributes to the multifocal genomic alteration as discussed before. Recently, an expert pathologist group performed a revision of the TCGA TP53 wild-type samples and questioned their description as high-grade serous tumors [40]: 13/14 could be reclassified as different histotypes or some rare high-grade serous cancers that may develop from low-grade diseases. Other genes are found with high frequency mutation in OC. BRCA 1 and 2 genes are altered in up to 32% of the cases of OC: germline (14.8%) and somatic (6.1%) mutations account for 20.9%. With the exception of 2 cases, all BRCA germline and sporadic mutations were mutually exclusive and 81% of BRCA1 mutations and 72.4% BRCA2 were accompanied by heterozygous loss as predicted by Knudson's two-hit hypothesis. EMSY alteration, that inactivates BRCA2 by specific binding, was present in 7.9% of cases. PTEN, an important tumor suppressor with a key regulatory role in several cancers, is deficient in 6.7% of cases due to homozygous deletion or mutation. Additional low frequency mutations (<5%) to the core set of RAD51, ATM, PALB2, Fanconi Anemia genes represent an interesting group of alterations. These mutations resemble the clinical representation of homologous recombination deficiency tumors, and patients with them are described to carry a "BRCAness phenotype" [41]. This subgroup of patients carries a defined profile of better response rates to first and second line platinum-based chemotherapy, longer tumor free intervals, ending in a significant improvement in overall survival [42]. Also they have specific clinical features compared to HR proficient patients [43]. Taken together, alterations in BRCA genes and other members of the HR pathway (Table 2.1) account for up to half of ASOC cases.

Table 2.1 Frequent Mutations in HGSOC

	Germline	Somatic	Total
BRCA1 mutation	8.5%	3.2%	11.7%
BRCA2 mutation	6.3%	2.9%	9.2%
BRCA1 epigenetic silencing	–	–	10.8%
EMSY alteration (amplification–mutation)			7.9%
PTEN			6.7%
RAD51			0.3%
ATM			1.3%
ATR			0.6%
PALB2			1.3%
FANCA			1.0%
FANCC			0.6%
FANCI			0.6%
FANCL			0.6%
FANCD2			0.3%
FANCE			0.3%
FANCG			0.3%
FANCM			0.3%
FANCF promoter hypermethylation			0.0%

Modified from TCGA Bell D, et al. Integrated genomic analyses of ovarian carcinoma. Nature 2011;474:609–15 [48].

Recent achievement in exploiting the intrinsic genomic instability and HR deficiency of the disease [44] is an important example of the clinical application of genomic discoveries. However, numerous molecular targets remain unexplored, and understanding their biology may allow for the development of novel therapies for larger patient population.

Low prevalence (<6%) somatic mutations are present in six genes: RB, NF1, FAT3, CSMD3, GABRA6, and CDK12. RB and NF1 are tumor suppressor genes, CDK12 is implicated in regulation of RNA splicing, and other oncogenes such as KRAS, NRAS, PIK3CA, BRAF were found to be mutated in less than 1% of the cases.

Gene Expression

There is extensive microarray data containing transcriptional data from OCs. Gene overexpression or downregulation can be due to processes such as gene amplification, activating mutation, or epigenetic activation. Expression profiles in OC have been demonstrated to be related to ovarian cancer oncogenic development [45], to distinguish between histologic subtypes [46], and to carry prognostic value [47]. TCGA reported a gene expression profile [48] where four different gene expression subtypes were identified and characterized by specific activation pathways. These subtypes were identified as mesenchymal, immunoreactive, differentiated, or proliferative subtype. However, these subtypes were not associated with clinical outcome. Subsequently, a validation study using the TCGA gene expression signature was reported [49], further identifying better signatures to predict patients' survival, but lacking a clear demonstration of the clinical impact.

Methylation Analysis

TCGA analysis also provided methylation patterns. One hundred sixty-eight epigenetically silenced genes were observed to contribute to gene expression modulation. The epigenetic-based signature segregated subgroups with different prognosis. However, the observed survival differences were small with no clinical relevance (Fig. 2.1).

GENERATION OF CLINICALLY RELEVANT SIGNATURES IN OC

OC does not harbor commonly mutated oncogenes, making it unique compared to other epithelial cancers. Instead, a small number of recurrent mutations are present, coupled with larger alterations in gene copy numbers and methylation of oncogene promoters. For this reason, OC has an intrinsic

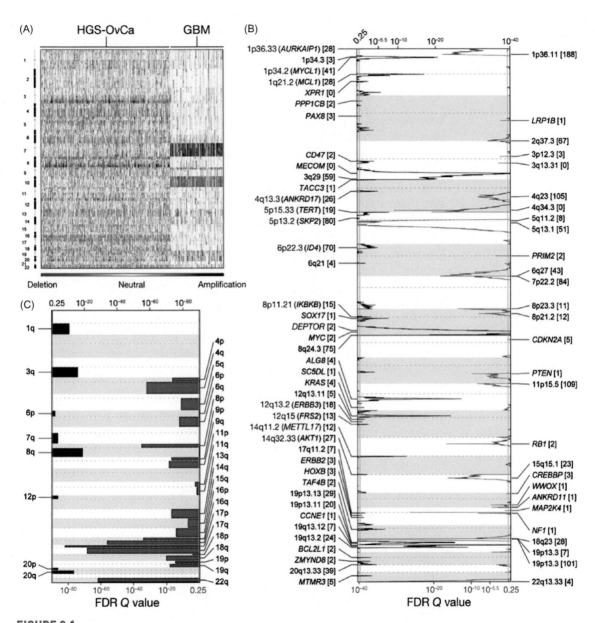

FIGURE 2.1

Genome copy number abnormalities. (A) Copy number profiles of 489 HGS-OvCa, compared with profiles of 197 glioblastoma multiforme (GBM) tumors [47]. Copy number increases (red) and decreases (blue) are plotted as a function of distance along the normal genome (vertical axis, divided into chromosomes). (B) Significant, focally amplified (red) and deleted (blue) regions are plotted along the genome. Annotations include the 20 most significant amplified and deleted regions, well-localized regions with 8 or fewer genes, and regions with known cancer genes or genes identified by genome-wide loss-of-function screens. The number of genes included in each region is given in brackets (*FDR*, false-discovery rate). (C) Significantly amplified (red) and deleted (blue) chromosome arms. *From TCGA Bell D, et al. Integrated genomic analyses of ovarian carcinoma. Nature 2011;474:609–15 [48].*

mutational plasticity, harboring multiple and heterogeneous mutations, driver and passenger events occur within the tumor at the same time. In this scenario, it is clearly difficult to sort the initiating alterations from subsequent events and, consequently, to identify discrete gene expression subgroups in OC.

Unsupervised clustering of high-throughput gene expression data in OC has shown to be capable of identifying subgroups with different expression patterns that may underline differentially expressed tumor-driving activated pathways. This method is per se capable of generating signatures [50], however, most of the patients' tumors may express overlapping features that don't allow for discrete separation of patients genotypes. This pitfall results in undermining the clinical prognostic significance of the signature [51]. In our perspective, the effort of establishing and validating gene expression signatures for prediction of survival and other meaningful predictors of patient outcome remains valid. To the purpose of unraveling the composite data from OC gene expression data, our group created a centralized curated database [52] comprised of datasets of Affymetrix gene expression profiles of women with HGSC treated with primary surgery and chemotherapy. This database required complete annotated survival information and a sample size of ≥40 patients. We identified 13 publicly available datasets corresponding to 1525 samples' microarrays [48,53–63].

In comparison to existing prognostic factors and gene signatures, we generated a meta-analysis signature that performed better than all previous established models, with the highest capability of patient stratification into low- and high-risk groups of overall survival. While encouraging, the signature requires prospective validation (Fig. 2.2A).

Through the same meta-analytic approach, we aimed to establish a gene expression signature for the prediction of patient outcome following primary debulking surgery. Our hypothesis assumed the existence of activated pathway profiles that underlined intrinsic aggressive biology which affected surgical outcome [4]. Interestingly, the signature demonstrated 94% accuracy on predicting debulking status. The debulking signature included genes such as POSTN, CXCL14, FAP, NUAK1, PTCH1, and TGFBR2, and demonstrated those genes as independent predictors of cytoreductive surgery (Fig. 2.2B). These results suggest TGFβ signaling activation has a relevant role in tumor dissemination (Fig. 2.2C). The signature could be used for patient stratification for primary or secondary debulking surgery as well as targeted therapies. The above-mentioned meta-analysis [64] is one example of how genomic data, multiplexed statistical analysis, and external validation can generate meaningful predictive signatures.

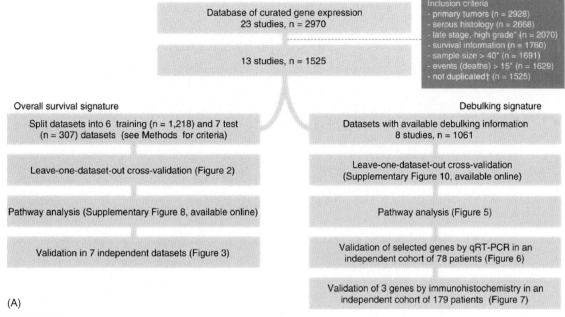

FIGURE 2.2

Meta-analysis of 1525 late-stage ovarian cancer samples. (A) Flowchart of the study outlining the steps for training and validating the prognostic models presented in this meta-analysis study. (B) Validation of POSTN, pSmad2/3, and CXCL14 in an independent cohort by immunohistochemistry and validation of selected genes associated with debulking status by quantitative reverse-transcription polymerase chain reaction (qRT-PCR) in the Bonome et al. validation data. (C) Pathway analysis of the debulking signature. Using the Pathway Studio 7.1 (Ariadne Genomics) software and a novel signature of 200 debulking-associated genes with suboptimal debulking surgery. Genes are labeled in *red* when overexpressed in tumors that were subsequently suboptimally debulked. Conversely, genes overexpressed in tumors with optimal cytoreduction are labeled in *blue*. Genes with predictive power toward poor prognosis based on the meta-analysis are highlighted with *pink borders*. *Red broken arrows* indicate direct stimulatory modification. *Green arrows* indicate EGR-1-based transcriptional regulations. *Orange arrows* indicate TGF-β/Smad-based transcriptional regulations. *Blue solid arrows* indicate other direct regulations. *Blue broken arrows* indicate other indirect regulations. *Purple sticks* indicate binding. *Adapted from Riester M, et al. Risk prediction for late-stage ovarian cancer by meta-analysis of 1525 patient samples. J Natl Cancer Inst 2014;106 [64].*

GENE EXPRESSION SIGNATURES FOR DRUG SENSITIVITY IN OC

Recent efforts have focused on the identification of gene expression signatures to predict OC chemosensitivity. Gourley [65] interrogated microarray platforms based data from an independent Scottish dataset. Unsupervised hierarchical clustering identified three major subgroups, two with diverse angiogenic pathways upregulation and the last with angiogenic repression and immune hallmarks pathways upregulation. It was observed a better prognosis for OS in the immune subgroup compared to the combined angiogenic

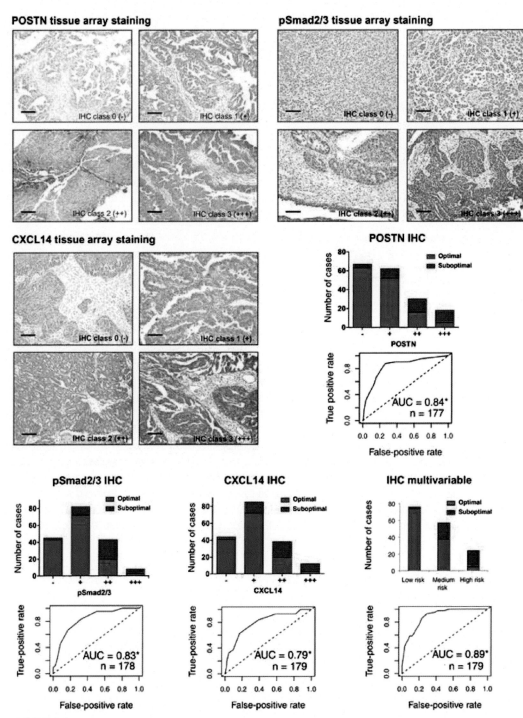

FIGURE 2.2

(Continued)

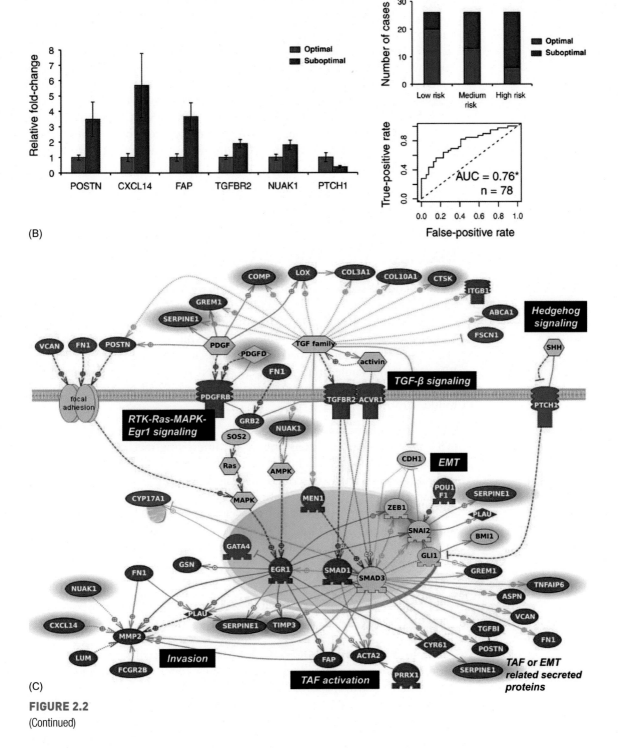

(B)

(C)

FIGURE 2.2

(Continued)

subgroups [HR = 0.66 (0.46–0.94)]. The identified gene expression signature was independently validated and further, a different response rate to bevacizumab first line additional therapy was observed between the groups in the ICON7 dataset (Phase III clinical trial where 3 first-line chemotherapy regimens were compared: first line paclitaxel/carboplatin ± concomitant ± maintenance bevacizumab for 12 months): the addition of bevacizumab displayed a worse interaction for progression-free survival (PFS) [HR = 1.73 (95% CI 1.12–2.68)] and overall survival (OS) [HR = 2.00 (95% CI 1.11–3.61)] compared to chemotherapy alone. In the proangiogenic group, nonsignificant improvement in PFS or OS with the addition of Bevacizumab was observed. However, these findings need to be validated in an independent dataset.

To find prediction of drug response in an exposed population represents an interesting alternative approach in genomic interrogation.

RECURRENT OVARIAN CANCER

Whole genome sequencing data from primary and paired recurrent OC disease has been performed to provide insight into the mechanisms of chemoresistance [66]. In the primary tumors, high prevalence genomic alterations included mutational inactivation of TP53, BRCA1-2, and HR associated genes. Also low frequency inactivation events included mutations of NF1 and RB1. As a remarkable oncogenic event, CCNE1 amplification was present in 19% of primary cases being a predictive marker of bad prognosis compared to BRCA carriers.

On the other hand, mutational status in the relapsing sample counterparts (15 paired samples) revealed only a few interesting mechanisms of chemoresistance: (1) BRCA1/2 deficient gene expression reversions were observed, according to previously described investigations [67]; (2) mutated multidrug resistance gene 1 (MDR1) was observed in 8% of recurrent samples; and (3) desmoplastic reactions in chemorefractory tumor nodules subsequent to inflammatory response found in necropsies of patients deceased for the disease. Larger datasets with paired biopsies performed prior and postchemotherapy treatment to observe insights in tumor heterogeneity and de novo alterations are an urgent need for the scientific community.

CONCLUSION

Genomic landscaping of OC subtypes has led to a revolution in our understanding of these tumors, demonstrating their unique profiles, and identifying targetable pathways. These discoveries have already led to subtype specific clinical trials in an attempt to develop specific therapies for these OC subtypes. Future research will focus on further identification and refinement of the activated pathways and the development of relevant clinical trials. Understanding feedback pathways and the mechanisms of resistance will be extremely important for this effort. In addition, defining tumor heterogeneity and its role in the above processes will be needed. These efforts which should be accomplished over the next several years will undoubtedly lead to the establishment of more effective and less toxic therapies for OC. These molecular discoveries will lead to personalized medicine for OC.

LIST OF ACRONYMS AND ABBREVIATIONS

ATM ataxia telangiectasia mutated gene
ARID1A AT-rich interactive domain 1A (SWI-like) gene
BRCA 1-2 Breast Cancer 1-2, Early Onset gene
CCNE1 cyclin E1 gene
CDK12 cyclin-dependent kinase 12 gene
CSMD3 CUB and sushi multiple domains 3 gene
CTNNB1 catenin (cadherin-associated protein), beta 1 gene
EMSY BRCA2-interacting transcriptional repressor gene
ERBB2 erb-b2 receptor tyrosine kinase 2 gene
Fanconi Anemia complementation genes FANCA, FANCB, FANCC, FANCD2, FANCE, FANCF, FANCG, FANCI, BRIP1 or FANCJ, FANCL, FANCM, FANCN or PALB2, RAD51C or FANCO, FANCP or SLX4, and also BRCA1 and BRCA2 genes
FAT3 FAT Atypical Cadherin 3 gene
GABRA6 gamma-aminobutyric acid (GABA) α receptor, alpha 6
ID4 DNA-binding protein inhibitor 4 gene
IRF2BP2 interferon regulatory factor 2 binding protein 2
KRAS Kirsten rat sarcoma viral oncogene homolog gene
MECOM MDS1 and EVI1 complex locus gene
MYC v-myc avian myelocytomatosis viral oncogene homolog gene
NF1 neurofibromin 1
PALB2 partner and localizer of BRCA2 gene
PAX8 embryonic development gene
PIK3CA phosphatidylinositol-4,5-bisphosphate 3-kinase, catalytic subunit alpha
PTEN phosphatase and tensin homolog gene
RAD51 RAD51 recombinase gene
RB1 retinoblastoma 1 gene
TERT telomerase catalytic subunit gene
TP53 tumor protein p53 gene
ZMYND8 receptor for activated C-kinase

References

[1] Blagden SP. Harnessing pandemonium: the clinical implications of tumor heterogeneity in ovarian cancer. Front Oncol 2015;5.

[2] Siegel RL, Miller KD, Jemal A. Cancer statistics, 2015. CA Cancer J Clin 2015;65:5–29.

[3] Jayson GC, Kohn EC, Kitchener HC, Ledermann JA. Ovarian cancer. The Lancet 2014;384:1376–88.

[4] Chang S-J, Bristow RE, Ryu H-S. Impact of complete cytoreduction leaving no gross residual disease associated with radical cytoreductive surgical procedures on survival in advanced ovarian cancer. Ann Surg Oncol 2012;19:4059–67.

[5] Chornokur G, Amankwah EK, Schildkraut JM, Phelan CM. Global ovarian cancer health disparities. Gynecol Oncol 2013;129:258–64.

[6] Bodurka DC, et al. Reclassification of serous ovarian carcinoma by a 2-tier system: a Gynecologic Oncology Group study. Cancer 2012;118:3087–94.

[7] Diaz-Padilla I, et al. Ovarian low-grade serous carcinoma: a comprehensive update. Gynecol Oncol 2012;126:279–85.

[8] Gourley C, et al. Gynecologic Cancer InterGroup (GCIG) consensus review for ovarian and primary peritoneal low-grade serous carcinomas. Int J Gynecol Cancer 2014;24:S9–S13.

[9] Gershenson DM, et al. Clinical behavior of stage II–IV low-grade serous carcinoma of the ovary. Obstet Gynecol 2006;108:361–8.

[10] Anglesio MS, et al. Mutation of ERBB2 provides a novel alternative mechanism for the ubiquitous activation of RAS-MAPK in ovarian serous low malignant potential tumors. Mol Cancer Res 2008;6:1678–90.

[11] Meinhold-Heerlein I, et al. Molecular and prognostic distinction between serous ovarian carcinomas of varying grade and malignant potential. Oncogene 2005;24:1053–65.

[12] Bonome T, et al. Expression profiling of serous low malignant potential, low-grade, and high-grade tumors of the ovary. Cancer Res 2005;65:10602–12.

[13] May T, et al. Low-grade and high-grade serous Mullerian carcinoma: review and analysis of publicly available gene expression profiles. Gynecol Oncol 2013;128:488–92.

[14] Mok SC, et al. Mutation of K-ras protooncogene in human ovarian epithelial tumors of borderline malignancy. Cancer Res 1993;53:1489–92.

[15] Singer G, et al. Mutations in BRAF and KRAS characterize the development of low-grade ovarian serous carcinoma. JNCI J Natl Cancer Inst 2003;95:484–6.

[16] Wong K-K, et al. BRAF mutation is rare in advanced-stage low-grade ovarian serous carcinomas. Am J Pathol 2010;177:1611–17.

[17] Farley J, et al. Selumetinib in women with recurrent low-grade serous carcinoma of the ovary or peritoneum: an open-label, single-arm, phase 2 study. Lancet Oncol 2013;14:134–40.

[18] Singer G, et al. Patterns of p53 mutations separate ovarian serous borderline tumors and low- and high-grade carcinomas and provide support for a new model of ovarian carcinogenesis: a mutational analysis with immunohistochemical correlation. Am J Surg Pathol 2005;29:218–24.

[19] Risch HA, et al. Prevalence and penetrance of germline BRCA1 and BRCA2 mutations in a population series of 649 women with ovarian cancer. Am J Hum Genet 2001;68:700–10.

[20] Gershenson DM, et al. Recurrent low-grade serous ovarian carcinoma is relatively chemoresistant. Gynecol Oncol 2009;114:48–52.

[21] Grisham RN, et al. Extreme outlier analysis identifies occult mitogen-activated protein kinase pathway mutations in patients with low-grade serous ovarian cancer. J Clin Oncol 2015. http://dx.doi.org/10.1200/JCO.2015.62.4726.

[22] Grisham RN, et al. Bevacizumab shows activity in patients with low-grade serous ovarian and primary peritoneal cancer. Int J Gynecol Cancer 2014;24:1010–14.

[23] Wiegand KC, et al. ARID1A mutations in endometriosis-associated ovarian carcinomas. N Engl J Med 2010;363:1532–43.

[24] Jones S, et al. Frequent mutations of chromatin remodeling gene ARID1A in ovarian clear cell carcinoma. Science 2010;330:228–31.

[25] Kuo K-T, et al. Frequent activating mutations of PIK3CA in ovarian clear cell carcinoma. Am J Pathol 2009;174:1597–601.

[26] Zorn KK, et al. Gene expression profiles of serous, endometrioid, and clear cell sub-types of ovarian and endometrial cancer. Clin Cancer Res Off J Am Assoc Cancer Res 2005;11:6422–30.

[27] Anglesio MS, et al. Clear cell carcinoma of the ovary: a report from the first Ovarian Clear Cell Symposium, June 24th, 2010. Gynecol Oncol 2011;121:407–15.

[28] Stany MP, et al. Identification of novel therapeutic targets in microdissected clear cell ovarian cancers. PLoS ONE 2011;6:e21121.

[29] Jackson AL, Eisenhauer EL, Herzog TJ. Emerging therapies: angiogenesis inhibitors for ovarian cancer. Expert Opin Emerg Drugs 2015;20:331–46.

[30] Campbell IG, et al. Mutation of the PIK3CA gene in ovarian and breast cancer. Cancer Res 2004;64:7678–81.

[31] Takano M, et al. Weekly administration of temsirolimus for heavily pretreated patients with clear cell carcinoma of the ovary: a report of six cases. Int J Clin Oncol 2011;16:605–9.

[32] Köbel M, et al. Differences in tumor type in low-stage versus high-stage ovarian carcinomas. Int J Gynecol Pathol Off J Int Soc Gynecol Pathol 2010;29:203–11.

[33] Seidman JD, Kurman RJ, Ronnett BM. Primary and metastatic mucinous adenocarcinomas in the ovaries: incidence in routine practice with a new approach to improve intraoperative diagnosis. Am J Surg Pathol 2003;27:985–93.

[34] Auner V, et al. KRAS mutation analysis in ovarian samples using a high sensitivity biochip assay. BMC Cancer 2009;9:111.

[35] Lee KR, Young RH. The distinction between primary and metastatic mucinous carcinomas of the ovary: gross and histologic findings in 50 cases. Am J Surg Pathol 2003;27:281–92.

[36] Ryland GL, et al. Mutational landscape of mucinous ovarian carcinoma and its neoplastic precursors. Genome Med 2015;7.

[37] Mackenzie R, et al. Targeted deep sequencing of mucinous ovarian tumors reveals multiple overlapping RAS-pathway activating mutations in borderline and cancerous neoplasms. BMC Cancer 2015;15.

[38] Farley J, et al. Cyclin E expression is a significant predictor of survival in advanced, suboptimally debulked ovarian epithelial cancers: a Gynecologic Oncology Group study. Cancer Res 2003;63:1235–41.

[39] Mehra K, et al. STICS, SCOUTs and p53 signatures; a new language for pelvic serous carcinogenesis. Front Biosci Elite Ed 2011;3:625–34.

[40] Vang R, et al. Molecular alterations of TP53 are a defining feature of ovarian high-grade serous carcinoma: a rereview of cases lacking TP53 mutations in The Cancer Genome Atlas Ovarian Study. Int J Gynecol Pathol 2015;1.

[41] Turner N, Tutt A, Ashworth A. Opinion: hallmarks of "BRCAness" in sporadic cancers. Nat Rev Cancer 2004;4:814–19.

[42] Tan DSP, et al. "BRCAness" syndrome in ovarian cancer: a case–control study describing the clinical features and outcome of patients with epithelial ovarian cancer associated with BRCA1 and BRCA2 mutations. J Clin Oncol Off J Am Soc Clin Oncol 2008;26:5530–6.

[43] Gourley C, et al. Increased incidence of visceral metastases in Scottish patients with BRCA1/2-defective ovarian cancer: an extension of the ovarian BRCAness phenotype. J Clin Oncol 2010;28:2505–11.

[44] Ledermann J, et al. Olaparib maintenance therapy in platinum-sensitive relapsed ovarian cancer. N Engl J Med 2012;366:1382–92.

[45] Welsh JB, et al. Analysis of gene expression profiles in normal and neoplastic ovarian tissue samples identifies candidate molecular markers of epithelial ovarian cancer. Proc Natl Acad Sci USA 2001;98:1176–81.

[46] Schwartz DR, et al. Gene expression in ovarian cancer reflects both morphology and biological behavior, distinguishing clear cell from other poor-prognosis ovarian carcinomas. Cancer Res 2002;62:4722–9.

[47] Berchuck A, et al. Microarray analysis of early stage serous ovarian cancers shows profiles predictive of favorable outcome. Clin Cancer Res Off J Am Assoc Cancer Res 2009;15:2448–55.

[48] Bell D, et al. Integrated genomic analyses of ovarian carcinoma. Nature 2011;474:609–15.

[49] Verhaak RGW, et al. Prognostically relevant gene signatures of high-grade serous ovarian carcinoma. J Clin Invest 2012. http://dx.doi.org/10.1172/JCI65833.

[50] McShane LM, et al. Methods for assessing reproducibility of clustering patterns observed in analyses of microarray data. Bioinforma Oxf Engl 2002;18:1462–9.

[51] Waldron L, Riester M, Birrer M. Molecular subtypes of high-grade serous ovarian cancer: the holy grail? J Natl Cancer Inst 2014;106 dju297.

[52] Ganzfried BF, et al. curatedOvarianData: clinically annotated data for the ovarian cancer transcriptome. Database J Biol Database Curation 2013;2013 bat013.

[53] Mok SC, et al. A gene signature predictive for outcome in advanced ovarian cancer identifies a survival factor: microfibril-associated glycoprotein 2. Cancer Cell 2009;16:521–32.

[54] Bentink S, et al. Angiogenic mRNA and microRNA gene expression signature predicts a novel subtype of serous ovarian cancer. PLoS ONE 2012;7:e30269.

[55] Partheen K, Levan K, Osterberg L, Horvath G. Expression analysis of stage III serous ovarian adenocarcinoma distinguishes a sub-group of survivors. Eur J Cancer Oxf Engl 1990 2006;42:2846–54.

[56] Crijns APG, et al. Survival-related profile, pathways, and transcription factors in ovarian cancer. PLoS Med 2009;6:e24.

[57] Yoshihara K, et al. Gene expression profile for predicting survival in advanced-stage serous ovarian cancer across two independent datasets. PLoS ONE 2010;5:e9615.

[58] Konstantinopoulos PA, et al. Gene expression profile of BRCAness that correlates with responsiveness to chemotherapy and with outcome in patients with epithelial ovarian cancer. J Clin Oncol 2010;28:3555–61.

[59] Bonome T, et al. A gene signature predicting for survival in suboptimally debulked patients with ovarian cancer. Cancer Res 2008;68:5478–86.

[60] Gillet J-P, et al. Multidrug resistance-linked gene signature predicts overall survival of patients with primary ovarian serous carcinoma. Clin Cancer Res 2012;18:3197–206.

[61] Yoshihara K, et al. High-risk ovarian cancer based on 126-gene expression signature is uniquely characterized by downregulation of antigen presentation pathway. Clin Cancer Res 2012;18:1374–85.

[62] Tothill RW, et al. Novel molecular subtypes of serous and endometrioid ovarian cancer linked to clinical outcome. Clin Cancer Res Off J Am Assoc Cancer Res 2008;14:5198–208.

[63] Dressman HK, et al. An integrated genomic-based approach to individualized treatment of patients with advanced-stage ovarian cancer. J Clin Oncol 2007;25:517–25.

[64] Riester M, et al. Risk prediction for late-stage ovarian cancer by meta-analysis of 1525 patient samples. J Natl Cancer Inst 2014;106.

[65] Gourley C. Molecular subgroup of high-grade serous ovarian cancer (HGSOC) as a predictor of outcome following bevacizumab. J Clin Oncol 2014.

[66] Patch A-M, et al. Whole-genome characterization of chemoresistant ovarian cancer. Nature 2015;521:489–94.

[67] Norquist B, et al. Secondary somatic mutations restoring BRCA1/2 predict chemotherapy resistance in hereditary ovarian carcinomas. J Clin Oncol Off J Am Soc Clin Oncol 2011;29:3008–15.

CHAPTER 3

Epigenetics

J.T. Pepin[1], H. Cardenas[2], K.P. Nephew[3,4] and D. Matei[2,5]

[1]Indiana University School of Medicine, Indianapolis, IN, United States [2]Northwestern University Feinberg School of Medicine, Chicago, IL, United States [3]Indiana University Simon Cancer Center, Indianapolis, IN, United States [4]Indiana University, Bloomington, IN, United States [5]Robert H. Lurie Comprehensive Cancer Center, Chicago, IL, United States

CONTENTS

INTRODUCTION

Epithelial ovarian cancer (OC) causes more deaths than any other female reproductive tract cancer [1]. Most patients are diagnosed with advanced disease, and despite progress in surgical and chemotherapy approaches, the 5-year survival rate remains below 25% for this group of patients [2]. Advances in genomic technologies over the past decade have firmly established that genetic and epigenetic changes accompany ovarian tumor initiation and progression. The recently completed Tumor Cancer Genome Atlas (TCGA) described genomic chaos encompassing mutations and chromosomal alterations as representing a hallmark of OC [3]. However, a second layer of transcriptional regulation occurs through epigenetic modifications, which have begun to be characterized in ovarian tumors in concert with genomic alterations.

Epigenetics refers to heritable changes in gene expression not involving alterations of the DNA sequence, including alterations in DNA methylation, histone modifications, nucleosome repositioning, and posttranscriptional gene regulation mediated by microRNAs (miRNAs) [4–7]. Alterations of the epigenome have been broadly described in cancer, including in OC [8,9], and have been functionally linked to tumor initiation, development of chemotherapy resistance, survival of cancer stem cells (CSCs), metastasis, and tumor progression. The transfer of a methyl group to the carbon-5 position of cytosines, almost always within the context of cytosine–guanine (CpG) dinucleotides, is the best-studied epigenetic mark and the only known covalent modification of DNA in mammalian cells. DNA-associated histones undergo extensive

Translational Advances in Gynecologic Cancers. DOI: http://dx.doi.org/10.1016/B978-0-12-803741-6.00003-3

posttranslational modifications (methylation, acetylation) which tightly regulate the assembly of transcriptionally permissive or repressive (i.e., open or closed) chromatin and it is now recognized that DNA methylation and histone modifications are intimately linked [4]. MicroRNAs (miR) are single stranded 16–24 nucleotide regulatory RNAs that repress target gene expression by inhibiting translation or by promoting target *mRNA* degradation. MiRs have been linked to human cancer, where they act either as oncogenic factors (through repression of tumor suppressor genes, TSGs) or as tumor suppressors [10,11]. This chapter will provide a general overview of epigenetic alterations in OC, their functional consequences, and therapeutic strategies to disrupt the epigenome and to restore normal cellular function.

EPIGENETIC CHANGES IN OVARIAN CANCER

Alterations in DNA Methylation

DNA methylation at carbon-5 of cytosines (5-methylcytosine or 5mC) typically occurring in a CpG context plays an important role in the regulation of gene transcription. Both increased methylation and hypomethylation have been documented in the context of cancer genomes and strongly implicated in cancer initiation as well as in tumor progression [12]. In OC, a global loss in DNA methylation mainly due to decreased methylation in noncoding DNA sequences has been described [13]. In addition, the promoters of a number of genes show increased methylation [14]. CpG methylation is regulated by DNA methyltransferases (DNMTs), primarily by DNMT1 which mediates maintenance (one strand) methylation and by DNMT3A and -3B which catalyze *de novo* methylation [12,15]. Demethylases responsible for removing the cytosine methyl group through hydroxylation or glycosylation have been recently characterized [16,17]. The ten–eleven translocation (Tet) proteins catalyze the conversion of 5mC to 5-hydroxymethylcytosine (5hmC) and have been implicated in the survival and maintenance of embryonic stem cells [18], but the role of Tets in OC has yet to be elucidated.

Many tumors, including ovarian, show increased methylation in CpG-rich regions usually, but not exclusively, associated with gene promoters [19–21]. CpG islands aberrantly methylated in ovarian tumors are associated with silencing of genes involved in control of the cell cycle, apoptosis, and drug sensitivity, as well as tumor suppressor genes [22]. Genes whose promoters are known to be hypermethylated in OC, leading to loss of expression, include the classical TSGs *BRCA1* (breast cancer susceptibility gene-1) [23,24], *p16* [25], and *MLH1* [24,26], the putative tumor suppressors (*RASSF1A, OPCML, SPARC, ANGPTL2, CTGF*) [27–31], the imprinted genes (*ARH1* and *PEG3*) [32], and other genes which play a role in apoptosis

(*LOT1*, *DAPK*, *TMS1/ASC*, and *PAR-4*) [33–35], cell adhesion (*ICAM-1*, *CDH1*) [36], cell signaling (*HSulf-1*) [37,38], genome stability *(PALB2)* [39], responsiveness to taxanes (*TUBB3*) [40], embryonic development and differentiation (*HOXA10*, *HOXA11*) [41].

BRCA1 is one of the best-studied OC-associated genes with well-defined hypermethylated promoter CpG sites. BRCA1 silencing through promoter methylation occurs in ~10–15% of sporadic OC cases [42] and has been correlated with clinical outcome [43]. Hypermethylation of BRCA1 in circulating DNA in the serum of women with OC was proposed as a diagnostic biomarker [44]. Interestingly, BRCA1 methylation does not commonly occur as a second hit in patients with familial OC related to loss of function BRCA mutations [45]. The functional significance of BRCA1 silencing due to promoter methylation remains debatable, as highlighted recently by the TCGA dataset analysis suggesting that epigenetic repression of BRCA1 has a lesser effect on platinum-induced DNA repair responses and subsequent clinical outcomes as compared to loss of function mutations [3]. Indeed, the overall survival of patients with BRCA1 methylated tumors is intermediate between those with BRCA1 mutated (best outcomes) and those with wild-type tumors (worst outcomes) [3]. Methylation of other mismatch repair genes (MMR), particularly *MLH1*, has been described in association with microsatellite instability (MSI) and responsiveness to chemotherapy in OC [46]. Up to 25% of patients with relapsed ovarian tumors have detectable *MLH1* methylation in serum compared to baseline, supporting the involvement of this methylation regulated DNA repair mechanism in response to platinum [47].

Global hypomethylation of heterochromatin and specific CpG sites has also been reported in OC [48]. Specifically, chromosome 1 satellite 2 and LINE-1 repetitive elements have been found to be hypomethylated [49,50]. Promoter CpG island hypomethylation correlated with gene overexpression was reported for genes associated with chemoresistance, such as *MCJ* [51], *SNCG* [25], and *BORIS* [52] and other transcripts, such as *IGF2*, an imprinted gene [12], and *claudin-4*, related to formation of tight junctions between epithelial OC cells [53].

Advances in the sequencing technology over the past few years have enabled deeper investigation of the methylome at a single base-pair resolution. The most prominent genome-wide approaches include whole-genome bisulfite sequencing (WGBS), methyl-binding domain capture sequencing (MBDCap-Seq), reduced-representation bisulfite sequencing (RRBS), and Infinium BeadChips arrays (27K, 450K, and Epic from Illumina) [54]. Examination of global DNA methylation of OC cell lines and human tumors suggests that the degree of aberrant methylation (i.e., the total number of methylated genes) directly correlates with ovarian tumor progression and recurrence, and associated specific methylated loci associate with poor progression-free survival (PFS) [55–60]. For instance, our group developed a model to examine DNA methylation changes correlated with the onset of drug resistance in OC [60]. By integrating DNA methylation and gene expression profiles through applied bioinformatic approaches, we identified a specific DNA methylation signature associated with platinum resistance and clinical outcome [61,62]. Another recent study based on a large cohort of ovarian tumors analyzed by gene expression and methylation arrays identified hypermethylation of stroma-related genes overlapping with the poor prognosis stromal transcriptomic signature associated with poor prognosis high-grade serous ovarian cancer (HGSOC) [63]. Thus, newly proposed "methylation signatures" rather than methylation of single gene promoters may be more useful for disease classification, monitoring response to therapies [64], and identifying chemoresistance-associated pathways.

Histone Modifications in Ovarian Cancer

It has been demonstrated that DNA methylation is associated with post-translational histone alterations which tightly regulate the degree of chromatin compaction and transcription factor availability within specific DNA sequences. A cooperative system of histone acetylation, methylation, and monoubiquitination of lysine residues within the N-terminal "tails" emanating from the core nucleosome histone octamer maintains the degree of chromatin compaction in a dynamic fashion. These functions are modulated by histone methyltransferases, acetyltransferases, deacetylases, ubiquitin ligases, and deubiquitinases, whose roles can be studied and manipulated by using newly developed enzymatic inhibitors [65].

Active transcription characterized by "open" chromatin is associated with di- and trimethylation of histone H3 lysine 4 (H3Kme2 and H3K4me3), methylation of histone 3 lysine 79 (H3K79me), methylation of histone H3 lysine 36 (H3K36me), acetylation of histone H3 lysine 9 (H3K9ac), acetylation of histone H3 lysine 14 (H3K14ac), and monoubiquitination of histone H2B lysine 120 (H2Bub1). Transcriptional repression, on the other hand, marking "closed" chromatin, is associated with methylation at histones H3 lysine 9 (H3K9me), H3 lysine 27 (H3K27me), and H3 lysine 20 (H3K20me) [66].

Histone alterations have been studied in OC in association with chemotherapy resistance and cancer stem cell phenotypes. For example, overexpression of the dominant negative mutant H3-K27R led to suppression of H3K27 trimethylation inducing re-expression of the *RASSF1* tumor suppressor, which re-sensitized OC cells to cisplatin, by reversing compacted chromatin at the promoter site [67]. Genome-wide loss of the repressive trimethyl-H3K27me3 mark was associated with reduced global DNA methylation, allowing platinum resensitization of chemoresistant OC cells [67] and loss of H3K27 trimethylation has recently been associated with poor prognosis in OC and other malignancies [36]. Gene silencing by histone modifications, in the absence of DNA methylation, has been also reported for *GATA4* and *GATA6* [68], *cyclinB1* [69], and p21WAF1/CIP1 [70]. Similarly, repressive histone modifications (trimethyl-H3K27 and dimethyl-H3K9) play a role downregulating *ADAM19* in the absence of CpG island methylation in TGF-β1-refractory OC cells [71], demonstrating that aberrant TGF-β1 signaling can result in formation of a repressive chromatin environment, without DNA methylation.

The simultaneous presence of an activating histone mark, H3K4me3, and a repressive histone mark, H3K27me3, marks "poised" or bivalent chromatin commonly associated with cellular differentiation associated genes in embryonic stem (ES) cells. Bivalent marks have been recently described in HGSOC in association with a repressed gene set enriched in PI3K and TGF-β signaling pathways [72]. Interestingly, this gene set, marked by H3K27me3 and bivalent modifications, was expressed at lower levels in chemoresistant compared with chemosensitive OC cells, suggesting that epigenetically repressed genes play a role in responsiveness to chemotherapy. Additionally, stem cell populations known to be chemoresistant are marked by bivalent histone marks [73]. The methyltransferase Enhancer of Zeste Homolog 2 (EZH2), which represents the catalytic unit of Polycomb repressive complex 2 (PRC2) and trimethylates H3K27, is also associated with chemoresistance in OC [33], further supporting the role of histone alterations in the process. Various environmental cues propagating a stem cell-like phenotype of malignant-associated stromal cells under the regulation of EZH2 have been described in OC [33,72].

Histone ubiquitination is considered an "activating" marker controlled by ubiquitin ligases, such as RING finger proteins (RNF20 and RNF40) [74] and deubiquitinases. One of the functions of OC-associated *BRCA1* gene is that of maintaining global heterochromatin integrity accounting for many of its tumor suppressor functions through ubiquitylation of histone H2A at satellite repeats. Ectopic expression of H2A fused to ubiquitin reversed the effects of *BRCA1* loss in murine BRCA1 deficient breast cancer models [75]. Altogether, intricately involved with the development and aggressive nature of OC, histone modifications appear to influence stem cell differentiation, transcription, replication, and DNA repair through concerted regulation of gene expression [76].

EXPRESSION AND FUNCTION OF EPIGENOME MODIFYING ENZYMES IN OC

DNA Methyltransferases (DNMTs)

DNA is methylated de novo and methylation patterns are maintained by the DNA methyltransferases (DNMTs) 1, -3A, and 3-B (Table 3.1). It has

Table 3.1 Enzymes Involved in DNA Methylation and Histones H3 and H4 Methylation and Acetylation

DNA Methylation		
Enzyme	**Action and Function**	**References**
DNA methyltransferase 1	Maintenance methylation, transcriptional repression	[29]
DNA methyltransferase 3A	De novo methylation, transcriptional repression	[29,159]
DNA methyltransferase 3B	De novo methylation, transcriptional repression	[29,159]

Histone Methylation and Demethylation				
Histone	**Site**	**Enzyme**	**Action and Function**	**References**
H3	K4	SET1	Mono-, di-, and trimethylation, transcriptional activation	[79,81]
		Set7/9	Monomethylation, transcriptional activation	[19,160]
		LSD1, LSD2	Demethylation of H3K4me2/me1	[22,150]
		JHDM1B	Demethylation of H3K4me3	[48]
		JARID1A, JARID1B, JARID1C, JARID1D, NO66	Demethylation of H3K4me3/me2, transcriptional repression	[161]
	K9	G9a	Mono- and dimethylation, transcriptional repression	[162]
		SUV39H1	Trimethylation, transcriptional repression	[163]
		JMJD1A, JHDM1D, JHDM1F	Demethylation of H3K9me2/1	[20,79]
		JHDM1E	Demethylation of H3K9me2	[49]
		JMJD2A, JMJD2B, JMJD2C, JMJD2D	Demethylation of H3K9me3/2	[79]
	K27	EZH2	Trimethylation, transcriptional repression	[20]
		G9a	Mono- and dimethylation, transcriptional repression	[162]
		JMJD3, UTX	Demethylation of H3K27me3/2	[20]
		JHDM1D	Demethylation of H3K27me2/1	[20]
	K36	Set2	Transcriptional activation	[20]
		JHDM1A, JHDM1B	Demethylation of H3K36me2/1	[20]
		JMJD2A, JMJD2B, JMJD2C, JMJD2D, NO66	Demethylation of H3K36me3/2	[20]
	K79	Dot1	Transcriptional activation	[20]
H4	K20	PR-Set7 (SET8)	Monomethylation, transcriptional repression	[23,50]
		SUV4-20h1, SUV4-20h2	Trimethylation, transcriptional repression	[24]

Histone Acetylation and Deacetylation				
Histone	**Site**	**Enzyme**	**Action and Function**	**References**
H3	K9	Gcn5	Acetylation, transcriptional activation	[79]
	K14	Gcn5, PCAF, p300/CBP	Acetylation, transcriptional activation	[79]
	K18, K23, K27	Gcn5, p300/CBP	Acetylation, transcriptional activation	[79]
	K9, K14, K18, K23, K27	HDACs, mainly Class I	Deacetylation, transcriptional repression	[25]
H4	K5, K12	p300	Acetylation, transcriptional activation	[79]
	K8	Gcn5, PCAF	Acetylation, transcriptional activation	[79]
	K16	Gcn5	Acetylation, transcriptional activation	[79]
	K5, K12, K8, K16	HDACs, mainly Class I	Deacetylation, transcriptional repression	[25]

been speculated that changes in methylation might be related to alterations in expression and (or) enzymatic activity of the DNMTs, however, no simple association between methylation changes and expression levels of the DNMTs has been clearly established in OC. In a study that included 60 ovarian tumors, both DNA hypermethylation and hypomethylation were described, however, neither satellite nor global DNA hypomethylation were correlated with *DNMT mRNA* levels, and hypermethylation of only 2 out of 55 genes examined was associated with DNMT1 expression [26]. *DNMT mRNA* and protein expression levels differ in OC compared with benign ovarian tumors or ovarian surface epithelial (OSE) cells, however, studies reporting such differences have been inconsistent. Bai et al. observed increased *DNMT3A* levels in HGSOC compared to benign tumors, while *DNMT3B* levels were not different and *DNMT1* expression was only modestly higher in malignant tissue. Another study revealed increased *mRNA* and protein levels of *DNMT1* and *DNMT3B* but no differences in DNMT3A levels between malignant tumors relative to normal ovarian tissues [28]. Comparisons of DNMTs *mRNA* expression levels between OC cell lines and OSE cells showed greater DNMT1 *mRNA* levels only in HeyA8 and HeyC2, greater DNMT3B in SKOV3 and PA-1 cells, and no differences in DNMT3A *mRNA* levels in OC cell lines compared to OSE cells [28,77]. Despite this variability, it has been proposed that protein levels of DNMT1 and -3B are associated with OC progression and patients' overall survival [27]. Overall, a less than straightforward relationship appears to exist between methylation patterns and altered *DNMT mRNA* levels in OC cells [26,77]. However, targeted downregulation of DNMTs results in loss of CpG hypermethylation and OC cell growth inhibition [78], validating the role of promoter methylation in ovarian carcinogenesis. Additionally, significant upregulation of DNMTs was observed in acquired cisplatin resistance [62], further linking DNA methylation to responsiveness to chemotherapy. An association between *DNMT3B* expression levels and DNA methylation and transcriptomic signatures identifying poor prognostic high-grade ovarian cancer has been recently reported [63].

Chromatin Modifiers

Nucleosomal histones are modified posttranslationally through enzymatic addition of different chemical groups resulting in methylation, acetylation, phosphorylation, ubiquitylation, sumoylation, ADP-ribosylation, deimination, or proline isomerization [79]. Histone modifications change chromatin conformation and transcription factor accessibility and play important roles in regulation of gene expression by influencing physiological and pathological processes including tumorigenesis. Most histone modifications are removable by enzyme-catalyzed reactions [79]. Histone methylation and acetylation occur more frequently than other modifications and take place mainly on lysine residues of histones H3 and H4. The enzymes responsible for methylation and acetylation of histones H3 and H4 and those that reverse these modifications are listed in Table 3.1. Only a small number of the enzymes presented in Table 3.1 have been studied to date in OC.

Histone Methyltransferases and Demethylases in OC

As described above, EZH2 is a repressive chromatin modifier, and *EZH2 mRNA* and protein expression levels are increased in OC tumors and cell lines compared with OSE cells [29,30,32,33]. Higher levels of *EZH2 mRNA* have also been detected in stem cell-like populations compared to nonstem-like cells of OC ascites [31]. An association of EZH2 protein levels with OC histological type and grade has been reported [31,33], and EZH2 levels have been correlated with clinical outcome in some studies [34], but not others [33]. Despite the fact that overexpression of *EZH2* has been consistently reported in OC, increased levels of H3K27me3 have not been described [36], illustrating the importance of other factors (i.e., histone demethylases) in regulating the abundance and distribution of this epigenetic mark. Furthermore, the functional significance of the protein is related to the complex transcriptional machinery it regulates *EZH2* target genes including *ALDH1A1*, *SSTR1*, and *DACT3* which regulate OC cell proliferation and metastasis [29,33].

The histone methyltransferase G9a belongs to the SET domain containing Su(var)3-9 family of proteins and performs mono- and dimethylation of histone H3 at lysine 9 [80]. G9a was detected by immunohistochemistry (IHC) in 71.6% of 208 ovarian tumors, with intensity of staining being significantly correlated with stage, grade, and serous type OC. Furthermore, G9a levels were found to be increased in metastases compared with primary tumors suggesting that G9a might regulate genes controlling OC dissemination [35].

Few lysine demethylases have been studied in OC. The lysine-specific demethylase 1 (LSD1) targets the active marks H3K4me2 and H3K4me1 [79,81]. LSD1 is expressed at increased levels in ovarian tumors (72–94%) compared to normal ovarian epithelium. LSD1 expression levels were associated with FIGO surgical stage and correlated with survival [37,38]. Another H3K4 demethylase, JARID1B was found to be overexpressed in OC being also associated with poor prognosis [39]. The functional relevance of these demethylases remains poorly understood.

Histone Acetylases and Deacetylases

Histone acetylation and deacetylation critically regulate gene transcription. Acetylation removes the positive charge from the histone tails rendering them neutral and resulting in a more relaxed chromatin structure permissive for initiation of transcription. Several lysine residues located in histone tails are acetylated by the histone acetyltransferases (HATs) Gcn5, P300, CBP, and

PCAF (Table 3.1), and acetylation is reversed by histone deacetylases (HDAC) resulting in closed, inactive chromatin. The mammalian HDACs have been classified into four groups based on their structural similarities: class I includes HDAC 1, 2, 3, and 8, which are the main deacetylating enzymes, class II are HDACs 4, 5, 7, 9 (subgroup IIa), 6 and 10 (subgroup IIb), class III comprises the Sirtuin family, and class IV includes only HDAC 11 [25,40]. Expression of HDACs is frequently dysregulated in cancer, including in OC [25], and overexpression of HDAC1, HDAC2, and HDAC3 in OC relative to normal ovarian tissue has been reported [28,41] and correlated with stage [28], and acquired resistance to chemotherapy [51]. Similarly, HDAC4 is over-expressed in platinum-resistant OC cells and plays a role in the survival of resistant cells through regulation of STAT1 signaling [52]. Mutations in genes mediating DNA methylation, histone modifications, or nucleosomal positioning have been recently reported in a variety of malignancies, providing additional support to the concept that epigenetic modifications contribute to oncogenesis [82,83].

FUNCTIONAL CONSEQUENCES OF EPIGENETIC ALTERATIONS IN OC

It is now clear that many cancers, including OC, are characterized by epigenetic alterations leading to transcriptional silencing of key TSGs or chemoresponsiveness regulating genes [9,84]. Several studies have suggested that aberrant gene promoter methylation of TSGs acts as an inactivating "hit" during OC initiation and/or progression [4]. TSGs known to be epigenetically silenced in ovarian tumors include *BRCA1* [44,85], *p16* [86], *PALB2* (partner and localizer of BRCA2) [87], *MLH1* [46,47,88], *RASSF1A*, *OPCML* [85,89], and others. Functional studies are needed to elucidate the significance of epigenetic silencing of specific TSGs to OC tumor initiation.

Epigenomic modifications in OC have also been intimately linked to the development of chemotherapy resistance, in particular platinum resistance. Studies performed by us and others showed increased CpG site methylation in isogenic platinum-resistant OC cells compared to sensitive cells [62,90,91]. Proposed DNA methylation drivers in platinum resistance include the tumor suppressors *ARMCX2* and *MLH1*, the extracellular matrix protein *COL1A1*, and the development related genes *MDK* and *MEST* [90]. These genes have been found to be hypermethylated in platinum-resistant OC cells, confirmed to be methylated and silenced in recurrent platinum-resistant human tumors, and are also methylated in the cancer "side population," suggesting an important functional role in the process of emergence of platinum resistance. Hypermethylation-mediated repression of cell adhesion and tight junction pathways has also been implicated in the emergence of platinum resistance [62]. Comprehensive analyses of DNA methylation and histone modifications in OC using high-density microarrays [61,92] described "epigenomic" signatures that occur in conjunction with acquired drug resistance [92].

Recent reports have begun to document the involvement of epigenetic alterations in the process of epithelial to mesenchymal transition (EMT), a process critical to the metastatic dissemination of tumors. For instance, the PRC2 complex has been shown to repress E-cadherin (*CDH1) expression* [60,93] and the demethylase LSD1 was reported to interact with and be required for Snail-1-induced *CDH1* repression [94]. Conversion of the bivalent chromatin configuration bearing both repressive H3K27me3 and active H3K4me3 marks at the *ZEB1* promoter, to an active configuration that lacks the H3K27me3 mark, was reported to lead to increased transcription of the well-characterized EMT inducer Zeb1 promoting transition of noncancer stem cells into cancer stem cells [95]. Our recent work in OC cells demonstrated that TGF-β increases the number of hypermethylated CpG sites impacting transcription of EMT-associated genes, including the key epithelial marker *CDH1* [96]. Focal promoter hypermethylation and global redistribution of genomic histone marks were also observed in a Twist-induced model of EMT in mammary cells [97]. These data in ovarian and other tumor models support that epigenetic changes occur during the plastic process of EMT.

Lastly, CSCs, which are key regulators of tumor initiation and recurrence after chemotherapy, are characterized by unique epigenomic vulnerabilities. CSCs are characterized by expression of specific cell surface markers, the ability to self-renew, differentiate, and generate tumors when injected in small numbers (e.g., 50–1000 cells) in NOD/SCID mice [98,99]. In culture, CSCs grow as spheres, are able to differentiate in cell subtypes with different phenotypes and have been implicated in tumor heterogeneity, tumor dormancy, and recurrence after chemotherapy [100–102]. Several surface markers have been proposed for CSC identification in OC, including CD133/ALDH$^+$, ALDH$^+$, CD44$^+$/CD117$^+$, CD44$^+$/MyoD, or CD133$^+$ [103–107]. ALDH1A1 activity, detectable through the Aldefluor assay, has been validated by several groups and is considered a robust marker [104,108]. ALDH1A1$^+$ cells have tumor initiating capacity, are resistant to cisplatin, and upregulate expression of stem cell transcription factors (Sox2, nanog, Oct4) [104,108,109], fulfilling the CSC criteria. Our group recently demonstrated that ALDH$^+$ OC cells over-express *DNMT1 DNMT3A*, and *DNMT3B* compared to OSE, suggesting that aberrant methylation patterns may be associated with altered DNMT activity in ovarian CSCs [109] and that epigenetically directed therapy induced down-regulation of DNMTs and stemness genes, suppressing OCSC stem-like properties. Recently, a direct association between EZH2 and ALDH1A1 expression has been reported in OC cells [110], consolidating the concept that epigenetic regulatory mechanisms contribute to the maintenance of stem cell characteristics and supporting eradication of CSC by using epigenome targeting strategies.

THERAPEUTIC TARGETING OF THE OC EPIGENOME

Epigenomic alterations associated with cancer provide unique opportunities for therapeutic targeting. Unlike cancer-associated gene mutations, amplifications, and deletions, DNA methylation and other epigenetic modifications are potentially reversible. Inhibitors of DNA methyltransferases (DNMTIs) and of chromatin modifying enzymes are being studied as a means of inducing re-expression of TSGs and reverse malignant phenotypes [111].

DNMT Inhibitors

DNMTIs are analogues of deoxycytosine possessing various substitutions at their 5-carbons. Consequently, upon phosphorylation and incorporation into DNA, DNMTIs irreversibly "trap" the methyltransferase in a transition state complex, which is subsequently eliminated from the cell, effectively preventing methyl group transfer [111]. The first studies focused on hematologic malignancies and myelodyspastic syndromes (MDS) [112], leading to clinical success of 5-azacytidine (5-aza-C) and its deoxyribose analog, 5-aza-2'-deoxycytidine (5-aza-dC, decitabine), both FDA approved for treatment of MDS [113–119]. The effects of these hypomethylating agents have been attributed to induction of cellular differentiation, directly related to reversal of epigenetic alterations [119–122]. Exploration of DNMTI activity in solid tumors was initially limited by toxicity, as early studies followed the traditional model of a drug studied at or near its maximal tolerated dose (MTD) and utilized high doses of DNMTI [123,124]. However, preclinical models showing that low doses of decitabine induce DNA hypomethylation, have fueled the redesign of clinical trials using regimens targeting a *"biologically effective"* dose, not the MTD. Subsequent trials emulating these *in vitro* findings demonstrated that doses as low as 1/10 of MTD preserved clinical effectiveness, while also improving tolerability [125–127]. This experience provided the rationale for using lower doses of DNMTI alone or in combination with chemotherapy. Several trials have been reported to date (Table 3.2). A randomized phase II trial of the UK Cancer Research Group compared the combination decitabine and carboplatin to single agent carboplatin in patients with OC recurring within 6–12 months after first line treatment containing a platinum regimen [128]. The combination regimen was associated with significant myelotoxicity leading to reduction of the dose of decitabine which was administered as a single bolus. Perhaps due to subtherapeutic dosing and long delays in therapy, the combination was less active compared to single agent carboplatin in this patient population with platinum sensitive OC. A phase I–II trial

Table 3.2 DNMT Inhibitors in Ovarian Cancer

Type of Trial	No. Patients	Responses	Toxicity
Phase II [128]	15	RR 20%	G3/4 neutropenia 60%
		PFS 1.9 months	G2/3 carboplatin hypersensitivity 47%
Phase I–II [129,130]	17	RR 35%	G3/4 neutropenia
		PFS 10.2 months	G3/4 thrombocytopenia
Phase Ib–IIa [164]	30	RR 13.8%	G4/4 Neutropenia
		PFS 3.7 months	G4/4 Thrombocytopenia

RR, response rate; PFS, progression-free survival; G, grade.

at Indiana University Simon Cancer Center (IUSCC) investigated the decitabine and carboplatin combination in patients with platinum-resistant OC. To minimize toxicity and enhance the demethylating properties of decitabine, the regimen studied in this trial used low daily doses of decitabine for 5 days prior to carboplatin. The trial demonstrated clinical tolerability of the regimen (phase I) [129], achievement of the desired biological effect at tolerable doses (global DNA) and gene-specific demethylation in PBMCs and tumor biopsies (phase I and II) [129], and impressive clinical activity (phase II) [130]. Among 17 patients with heavily pretreated and platinum-resistant OC treated in the phase II portion of the trial, the objective response rate (RR) was 35% and the PFS was 10.2 months, with 9 patients (53%) being free of disease progression at 6 months. Another phase II trial conducted at M.D. Anderson Cancer Center tested the combination of azacitidine given daily for 5 days and carboplatin [131]. In this cohort, there were 4 objective responses (RR of 14%), of which one was a complete response (CR) among 30 patients treated. Like the previous trial, this study supports the concept that demethylation by decitabine may resensitize platinum-resistant ovarian tumors to platinum. An ongoing randomized phase II (NCT01696032) is testing guadecitabine, a dinucleotide incorporating decitabine (5-aza-2′-deoxycytidine) and deoxyguanosine that act as prodrugs of decitabine, in combination with carboplatin against physician choice chemotherapy in patients with platinum-resistant OC. Guadecitabine is resistant to modification by cytidine deaminase leading to longer half-life that ensures prolonged exposure to the active compound [132]. Results of this trial are expected later this year.

HDAC Inhibitors (HDACi)

Inhibitors of enzymes catalyzing posttranslational histone modifications are being investigated as anticancer agents in a variety of malignancies. HDACi alter gene expression by enabling acetylation of both histones and associated complex proteins [133]. Aside from inhibiting HDACs, these agents also affect nonhistone protein acetylation, which further influences their antitumor activity [134]. Recent studies have reported significant alterations in gene expression profiles, with up to ~7–10% of genes being either up- or downregulated in response to HDACi [133]. HDACi induce apoptosis in platinum-resistant OC cells [135] by downregulating antiapoptotic genes such as caspase inhibitors, oncogenic kinases, DNA synthesis and repair enzymes, transcription factors (E2F-1), subunits of the proteasome, ubiquitin conjugating enzymes, growth factors and their receptors [136]. One of the genes consistently upregulated in response to treatment with HDACi is the cyclin-dependent kinase inhibitor 1 p21, whose expression is regulated by H3K4 acetylation [137]. Induction of p21 by HDACi causes cell cycle arrest at G1/S, being partly responsible for the antitumor effects of these inhibitors.

The HDACi family is comprised of several structural classes including organic hydroxamic acids, cyclic tetrapeptides, short chain fatty acids, sulfonamides, and benzamides. Vorinostat (suberoylanilide hydroxamic acid, SAHA), a hydroxamate, pan-inhibitor of class I and class II HDACs with excellent bioavailability was the first FDA-approved HDACi for clinical use in cutaneous T-cell lymphoma (CTCL) [138]. SAHA promoted cell cycle arrest and apoptosis in OC cells [139], supporting its clinical evaluation in this setting. The Gynecologic Oncology Group (GOG) conducted a phase II clinical trial evaluating vorinostat as a single agent in patients with recurrent OC, relapsing within 12 months after platinum-based therapy [140]. Out of 27 women enrolled on the study, only 2 remained without progression for 6 months, and 1 demonstrated a partial response (PR), deeming the drug insufficiently active as a single agent for further utilization in this setting. Associated grade 3–4 toxicities included leukopenia and neutropenia (2 patients), fatigue (3 patients), and gastrointestinal events (3 patients).

Pretreatment with belinostat (PDX-101), another hydroxamate HDACi, re-sensitized platinum-resistant OC xenografts to platinum [141]. In a phase I trial in solid tumors, single agent belinostat demonstrated efficacy in sarcoma, renal cancer, thymoma and melanoma with dose-limiting toxicities including grade 3 fatigue, diarrhea, and cardiac arrhythmia [142]. In a phase II trial in patients with platinum-resistant OC (n = 21), treatment with single agent belinostat resulted in stable disease in 10 patients and 1 partial response [143].

HDACi have also demonstrated synergistic or additive activity in combination with other antitumor agents, including radiation therapy [65], chemotherapy [144], other epigenetic agents [145], and biologically targeted agents

[146]. For example, vorinostat radiosensitized tumor cell lines, substantially decreasing the surviving cell fraction [65,147]. Pretreatment with HDACi exerted synergistic response with radiotherapy through downregulation of genes/proteins involved with the DNA damage response including Ku70, Ku80, Rad50, and DNA ligase IV [147].

In combination with olaparib, a poly (ADP-ribose) polymerase (PARP) inhibitor, SAHA induced increased apoptosis in BRCA1 wild-type and mutated cells compared to either agent alone [148]. Combination epigenetic agents have demonstrated synergistic upregulation of proapoptotic genes TMS1/ASC in OC cells [149,150]. The combination of decitabine (a DNMT1 inhibitor) and belinostat was more effective in re-sensitizing OC cells to platinum than belinostat alone [151].

The combination of paclitaxel and SAHA decreased OC cell proliferation more prominently than single agent SAHA, inducing G1/G2 cell cycle arrest and causing *CDK1* downregulation [152]. Several studies using in vitro and in vivo OC models demonstrated HDACi in combination with cisplatin increased cancer cell apoptosis induced by platinum [153–155]. Based on these data, the combination of carboplatin and belinostat was evaluated by the GOG in a phase II clinical trial [156]. Among 27 patients, there were only a CR and one PR, deeming the combination not sufficiently active to warrant further development (Table 3.2) [156]. In a phase I study of the combination of carboplatin, gemcitabine, and vorinostat in platinum-resistant OC, all four dose levels tested resulted in hematological dose-limiting toxicities [157]. Seven of the fifteen patients enrolled were eligible for evaluation of RECIST response, of which six demonstrated a PR, suggesting promising clinical activity [157]. However, further development of the combination was halted, because a tolerable dose was not identified. Similarly, a phase I/II clinical trial evaluating the combination of SAHA with carboplatin and paclitaxel was terminated early due to gastrointestinal perforations recorded in three patients [158]. Of 18 patients enrolled in that trial, 7 achieved a CR, while 2 had PR. Grade 3 neutropenia was the most common toxicity (9 of 18 patients), followed by grade 3 thrombocytopenia (2 patients), grade 3 anemia (1 patient), and grade 2 neuropathy (1 patient) [158]. Table 3.3 lists clinical trials that tested HDAC inhibitors in OC, either as single agents or in combination. Based on the clinical experience to date, single agent HDACi have only modest activity in OC. While combination regimens appear more active, they are also associated with greater risk for toxicity. Investigational trials evaluating rational combinations of HDACi and other epigenetic modifiers and other pathway inhibitors (e.g., PARP inhibitors) will be an important step forward.

Table 3.3 Clinical Trials Using HDAC Inhibitors in Ovarian Cancer and Solid Tumors

Principal Investigator	Type of Trial	Site	No. Patients	HDAC Inhibitor	Combination	Response Rate	Grade	Toxicities	Status
Modesitt et al. [140]	Phase II	Ovary	27	Vorinostat 400 mg	None	2 PR 6 months PFS	4 3	Neutropenia and leukopenia constipation, metabolic abnormality, and thrombocytosis	Completed
Mackay et al. [143]	Phase II	Ovary	18 EOC 14 LMP	Belinostat 1000 mg/m^2	None	1 PR (LMP) 10 SD (9 EOC)	3	Thrombosis, hypersensitivity	Completed
Dizon et al. [156]	Phase II	Ovary	27	Belinostat 1000 mg/m^2	Carboplatin (AUC 5)	1 CR 1 PR 12 SD 8 PD	3	Neutropenia, thrombocytopenia, and vomiting	Terminated to lack of activity
Dizon et al. [165]	Phase II	Ovary	35	Belinostat 1000 mg/m^2	Carboplatin (AUC 5) Paclitaxel (175 mg/m^2)	3 CR 12 PR	3	Nausea (83%) Fatigue (74%) Vomiting (63%) Alopecia (57%) Diarrhea (37%)	Completed
Matulonis et al. [157]	Phase I	Ovary	7	Vorinostat DL 1–4	Carboplatin (AUC 4) Gemcitabine 1000 mg/m^2	6 PR	4	Hematologic	Terminated to toxicities
Mendivil et al. [158]	Phase I	Ovary	18	Vorinostat 200 mg	Carboplatin (AUC 6) Paclitaxel (80 mg/m^2)	7 CR 2 PR	3	Neutropenia, thrombocytopenia, anemia, neuropathy	Terminated to (3) bowel anastomotic perforation
Steele et al. [142]	Phase I	Solid tumors	46	Belinostat 1000 mg/m^2	None	SD 18 (39%) MTD group SD (50%)	3 2	Fatigue, diarrhea, and atrial fibrillation nausea and vomiting	Completed
Kummar et al. [166]	Phase I	Solid tumors and lymphoid malignancies	19	Entinostat 6 mg/m^2 (MTD)	None		3	Hypophosphatemia, hyponatremia, hypoalbuminemia	Completed

PR, partial response; CR, complete response; SD, stable disease; PD, progressive disease; PFS, progression-free survival; EOC, epithelial ovarian cancer; LMP, low malignant potential; MTD, maximum tolerated dose.

CONCLUSION

OC is characterized by complex epigenetic modifications modulated by deregulated chromatin modifying enzymes and DNMTs. Functionally, these changes are implicated in tumor progression, development of chemotherapy resistance, and survival of CSCs. As the machinery governing epigenetic changes continues to become deciphered, the OC epigenome will become an important new cancer target that should be exploited for therapeutic benefit.

References

[1] Bukowski RM, Ozols RF, Markman M. The management of recurrent ovarian cancer. Seminars Oncol 2007;34:S1–15.

[2] Vaughan S, Coward JI, Bast Jr. RC, et al. Rethinking ovarian cancer: recommendations for improving outcomes. Nat Rev Cancer 2011;11:719–25.

[3] Integrated genomic analyses of ovarian carcinoma. Nature 2011;474:609–15.

[4] Jones PA, Baylin SB. The epigenomics of cancer. Cell 2007;128:683–92.

[5] Esteller M. Epigenetics in cancer. N Engl J Med 2008;358:1148–59.

[6] Schickel R, Boyerinas B, Park SM, Peter ME. MicroRNAs: key players in the immune system, differentiation, tumorigenesis and cell death. Oncogene 2008;27:5959–74.

[7] Iorio MV, Visone R, Di Leva G, et al. MicroRNA signatures in human ovarian cancer. Cancer Res 2007;67:8699–707.

[8] Balch C, Huang TH, Brown R, Nephew KP. The epigenetics of ovarian cancer drug resistance and resensitization. Am J Obstet Gynecol 2004;191:1552–72.

[9] Barton CA, Hacker NF, Clark SJ, O'Brien PM. DNA methylation changes in ovarian cancer: implications for early diagnosis, prognosis and treatment. Gynecol Oncol 2008;109:129–39.

[10] Calin GA, Croce CM. MicroRNA signatures in human cancers. Nat Rev Cancer 2006;6:857–66.

[11] Foekens JA, Sieuwerts AM, Smid M, et al. Four miRNAs associated with aggressiveness of lymph node-negative, estrogen receptor-positive human breast cancer. Proc Natl Acad Sci USA 2008;105:13021–6.

[12] Das PM, Singal R. DNA methylation and cancer. J Clin Oncol 2004;22:4632–42.

[13] Gama-Sosa MA, Slagel VA, Trewyn RW, et al. The 5-methylcytosine content of DNA from human tumors. Nucleic Acids Res 1983;11:6883–94.

[14] Matei DE, Nephew KP. Epigenetic therapies for chemoresensitization of epithelial ovarian cancer. Gynecol Oncol 2010;116:195–201.

[15] Jones PA, Baylin SB. The fundamental role of epigenetic events in cancer. Nat Rev Genet 2002;3:415–28.

[16] Patra SK, Patra A, Rizzi F, Ghosh TC, Bettuzzi S. Demethylation of (Cytosine-5-C-methyl) DNA and regulation of transcription in the epigenetic pathways of cancer development. Cancer Metastasis Rev 2008;27:315–34.

[17] Zhu JK. Active DNA demethylation mediated by DNA glycosylases. Ann Rev Genet 2009;43:143–66.

[18] Ito S, D'Alessio AC, Taranova OV, Hong K, Sowers LC, Zhang Y. Role of Tet proteins in 5mC to 5hmC conversion, ES-cell self-renewal and inner cell mass specification. Nature 2010;466:1129–33.

[19] Wang H, Cao R, Xia L, et al. Purification and functional characterization of a histone H3-lysine 4-specific methyltransferase. Mol Cell 2001;8:1207–17.

[20] Greer EL, Shi Y. Histone methylation: a dynamic mark in health, disease and inheritance. Nat Rev Genet 2012;13:343–57.

[21] Tsukada Y, Ishitani T, Nakayama KI. KDM7 is a dual demethylase for histone H3 Lys 9 and Lys 27 and functions in brain development. Genes Develop 2010;24:432–7.

[22] Fang R, Barbera AJ, Xu Y, et al. Human LSD2/KDM1b/AOF1 regulates gene transcription by modulating intragenic H3K4me2 methylation. Mol Cell 2010;39:222–33.

[23] Fang J, Feng Q, Ketel CS, et al. Purification and functional characterization of SET8, a nucleosomal histone H4-lysine 20-specific methyltransferase. Curr Biol 2002;12:1086–99.

[24] Schotta G, Lachner M, Sarma K, et al. A silencing pathway to induce H3-K9 and H4-K20 trimethylation at constitutive heterochromatin. Genes Develop 2004;18:1251–62.

[25] Barneda-Zahonero B, Parra M. Histone deacetylases and cancer. Mol Oncol 2012;6:579–89.

[26] Ehrlich M, Woods CB, Yu MC, et al. Quantitative analysis of associations between DNA hypermethylation, hypomethylation, and DNMT RNA levels in ovarian tumors. Oncogene 2006;25:2636–45.

[27] Bai X, Song Z, Fu Y, et al. Clinicopathological significance and prognostic value of DNA methyltransferase 1, 3a, and 3b expressions in sporadic epithelial ovarian cancer. PLoS ONE 2012;7:e40024.

[28] Gu Y, Yang P, Shao Q, et al. Investigation of the expression patterns and correlation of DNA methyltransferases and class I histone deacetylases in ovarian cancer tissues. Oncol Lett 2013;5:452–8.

[29] Li H, Bitler BG, Vathipadiekal V, et al. ALDH1A1 is a novel EZH2 target gene in epithelial ovarian cancer identified by genome-wide approaches. Cancer Prev Res (Philadelphia, PA) 2012;5:484–91.

[30] Emmanuel C, Gava N, Kennedy C, et al. Comparison of expression profiles in ovarian epithelium in vivo and ovarian cancer identifies novel candidate genes involved in disease pathogenesis. PLoS ONE 2011;6:e17617.

[31] Rizzo S, Hersey JM, Mellor P, et al. Ovarian cancer stem cell-like side populations are enriched following chemotherapy and overexpress EZH2. Mol Cancer Ther 2011;10:325–35.

[32] Guo J, Cai J, Yu L, Tang H, Chen C, Wang Z. EZH2 regulates expression of p57 and contributes to progression of ovarian cancer in vitro and in vivo. Cancer Sci 2011;102:530–9.

[33] Li H, Cai Q, Godwin AK, Zhang R. Enhancer of zeste homolog 2 promotes the proliferation and invasion of epithelial ovarian cancer cells. Mol Cancer Res 2010;8:1610–18.

[34] Rao ZY, Cai MY, Yang GF, et al. EZH2 supports ovarian carcinoma cell invasion and/or metastasis via regulation of TGF-beta1 and is a predictor of outcome in ovarian carcinoma patients. Carcinogenesis 2010;31:1576–83.

[35] Hua KT, Wang MY, Chen MW, et al. The H3K9 methyltransferase G9a is a marker of aggressive ovarian cancer that promotes peritoneal metastasis. Mol Cancer 2014;13:189.

[36] Wei Y, Xia W, Zhang Z, et al. Loss of trimethylation at lysine 27 of histone H3 is a predictor of poor outcome in breast, ovarian, and pancreatic cancers. Mol Carcinogen 2008;47:701–6.

[37] Chen C, Ge J, Lu Q, Ping G, Yang C, Fang X. Expression of Lysine-specific demethylase 1 in human epithelial ovarian cancer. J Ovarian Res 2015;8:28.

[38] Konovalov S, Garcia-Bassets I. Analysis of the levels of lysine-specific demethylase 1 (LSD1) mRNA in human ovarian tumors and the effects of chemical LSD1 inhibitors in ovarian cancer cell lines. J Ovarian Res 2013;6:75.

[39] Wang L, Mao Y, Du G, He C, Han S. Overexpression of JARID1B is associated with poor prognosis and chemotherapy resistance in epithelial ovarian cancer. Tumour Biol J Int Soc Oncodevelop Biol Med 2015;36:2465–72.

[40] Li Z, Zhu WG. Targeting histone deacetylases for cancer therapy: from molecular mechanisms to clinical implications. Int J Biol Sci 2014;10:757–70.

[41] Jin KL, Pak JH, Park JY, et al. Expression profile of histone deacetylases 1, 2 and 3 in ovarian cancer tissues. J Gynecol Oncol 2008;19:185–90.

[42] Lynch MA, Nakashima R, Song H, et al. Mutational analysis of the transforming growth factor beta receptor type II gene in human ovarian carcinoma. Cancer Res 1998;58:4227–32.

[43] Chen T, Triplett J, Dehner B, et al. Transforming growth factor-beta receptor type I gene is frequently mutated in ovarian carcinomas. Cancer Res 2001;61:4679–82.

[44] Ibanez de Caceres I, Battagli C, Esteller M, et al. Tumor cell-specific BRCA1 and RASSF1A hypermethylation in serum, plasma, and peritoneal fluid from ovarian cancer patients. Cancer Res 2004;64:6476–81.

[45] Dworkin A, Spearman A, Tseng S, Sweet K, Toland A. Methylation not a frequent "second hit" in tumors with germline BRCA mutations. Familial Cancer 2009;8:339–46.

[46] Geisler JP, Goodheart MJ, Sood AK, Holmes RJ, Hatterman-Zogg MA, Buller RE. Mismatch repair gene expression defects contribute to microsatellite instability in ovarian carcinoma. Cancer 2003;98:2199–206.

[47] Gifford G, Paul J, Vasey PA, Kaye SB, Brown R. The acquisition of hMLH1 methylation in plasma DNA after chemotherapy predicts poor survival for ovarian cancer patients. Clin Cancer Res 2004;10:4420–6.

[48] Frescas D, Guardavaccaro D, Bassermann F, Koyama-Nasu R, Pagano M. JHDM1B/ FBXL10 is a nucleolar protein that represses transcription of ribosomal RNA genes. Nature 2007;450:309–13.

[49] Tsukada Y, Fang J, Erdjument-Bromage H, et al. Histone demethylation by a family of JmjC domain-containing proteins. Nature 2006;439:811–16.

[50] Nishioka K, Rice JC, Sarma K, et al. PR-Set7 is a nucleosome-specific methyltransferase that modifies lysine 20 of histone H4 and is associated with silent chromatin. Mol Cell 2002;9:1201–13.

[51] Kim MG, Pak JH, Choi WH, Park JY, Nam JH, Kim JH. The relationship between cisplatin resistance and histone deacetylase isoform overexpression in epithelial ovarian cancer cell lines. J Gynecol Oncol 2012;23:182–9.

[52] Stronach EA, Alfraidi A, Rama N, et al. HDAC4-regulated STAT1 activation mediates platinum resistance in ovarian cancer. Cancer Res 2011;71:4412–22.

[53] Thiery JP. Epithelial–mesenchymal transitions in tumour progression. Nat Rev Cancer 2002;2:442–54.

[54] Tang J, Fang F, Miller DF, et al. Global DNA methylation profiling technologies and the ovarian cancer methylome. Methods Mol Biol 2015;1238:653–75.

[55] Baldwin RL, Tran H, Karlan BY. Loss of c-myc repression coincides with ovarian cancer resistance to transforming growth factor beta growth arrest independent of transforming growth factor beta/Smad signaling. Cancer Res 2003;63:1413–19.

[56] Dowdy SC, Mariani A, Janknecht R. HER2/Neu- and TAK1-mediated up-regulation of the transforming growth factor beta inhibitor Smad7 via the ETS protein ER81. J Biol Chem 2003;278:44377–84.

[57] Cardillo MR, Yap E, Castagna G. Molecular genetic analysis of TGF-beta1 in ovarian neoplasia. J Exp Clin Cancer Res 1997;16:49–56.

[58] Rodriguez GC, Haisley C, Hurteau J, et al. Regulation of invasion of epithelial ovarian cancer by transforming growth factor-beta. Gynecol Oncol 2001;80:245–53.

[59] Wakahara K, Kobayashi H, Yagyu T, et al. Transforming growth factor-beta1-dependent activation of Smad2/3 and up-regulation of PAI-1 expression is negatively regulated by Src in SKOV-3 human ovarian cancer cells. J Cell Biochem 2004;93:437–53.

[60] Cao Q, Yu J, Dhanasekaran SM, et al. Repression of E-cadherin by the polycomb group protein EZH2 in cancer. Oncogene 2008;27:7274–84.

[61] Wei SH, Balch C, Paik HH, et al. Prognostic DNA methylation biomarkers in ovarian cancer. Clin Cancer Res 2006;12:2788–94.

[62] Li M, Balch C, Montgomery JS, et al. Integrated analysis of DNA methylation and gene expression reveals specific signaling pathways associated with platinum resistance in ovarian cancer. BMC Med Genomics 2009;2:34.

[63] Chen P, Huhtinen K, Kaipio K, et al. Identification of prognostic groups in high-grade serous ovarian cancer treated with platinum-taxane chemotherapy. Cancer Res 2015;75:2987–98.

[64] Abendstein B, Stadlmann S, Knabbe C, et al. Regulation of transforming growth factor-beta secretion by human peritoneal mesothelial and ovarian carcinoma cells. Cytokine 2000;12:1115–19.

[65] Bolden JE, Peart MJ, Johnstone RW. Anticancer activities of histone deacetylase inhibitors. Nature Rev Drug Discov 2006;5:769–84.

[66] Mikkelsen TS, Ku M, Jaffe DB, et al. Genome-wide maps of chromatin state in pluripotent and lineage-committed cells. Nature 2007;448:553–60.

[67] Abbosh PH, Montgomery JS, Starkey JA, et al. Dominant-negative histone H3 lysine 27 mutant derepresses silenced tumor suppressor genes and reverses the drug-resistant phenotype in cancer cells. Cancer Res 2006;66:5582–91.

[68] Caslini C, Capo-chichi CD, Roland IH, Nicolas E, Yeung AT, Xu XX. Histone modifications silence the GATA transcription factor genes in ovarian cancer. Oncogene 2006;25:5446–61.

[69] Valls E, Sanchez-Molina S, Martinez-Balbas MA. Role of histone modifications in marking and activating genes through mitosis. J Biol Chem 2005;280:42592–600.

[70] Richon VM, Sandhoff TW, Rifkind RA, Marks PA. Histone deacetylase inhibitor selectively induces p21WAF1 expression and gene-associated histone acetylation. Proc Natl Acad Sci USA 2000;97:10014–19.

[71] Chan MW, Huang YW, Hartman-Frey C, et al. Aberrant transforming growth factor beta1 signaling and SMAD4 nuclear translocation confer epigenetic repression of ADAM19 in ovarian cancer. Neoplasia (New York, NY) 2008;10:908–19.

[72] Chapman-Rothe N, Curry E, Zeller C, et al. Chromatin H3K27me3/H3K4me3 histone marks define gene sets in high-grade serous ovarian cancer that distinguish malignant, tumour-sustaining and chemo-resistant ovarian tumour cells. Oncogene 2013;32:4586–92.

[73] Ohm JE, McGarvey KM, Yu X, et al. A stem cell-like chromatin pattern may predispose tumor suppressor genes to DNA hypermethylation and heritable silencing. Nat Genet 2007;39:237–42.

[74] Fuchs G, Oren M. Writing and reading H2B monoubiquitylation. Biochim Biophys Acta 2014;1839:694–701.

[75] Zhu Q, Pao GM, Huynh AM, et al. BRCA1 tumour suppression occurs via heterochromatin-mediated silencing. Nature 2011;477:179–84.

[76] Wood A, Schneider J, Shilatifard A. Cross-talking histones: implications for the regulation of gene expression and DNA repair. Biochem Cell Biology 2005;83:460–7.

[77] Ahluwalia A, Hurteau JA, Bigsby RM, Nephew KP. DNA methylation in ovarian cancer. II. Expression of DNA methyltransferases in ovarian cancer cell lines and normal ovarian epithelial cells. Gynecol Oncol 2001;82:299–304.

[78] Leu YW, Rahmatpanah F, Shi H, et al. Double RNA interference of DNMT3b and DNMT1 enhances DNA demethylation and gene reactivation. Cancer Res 2003;63:6110–15.

[79] Kouzarides T. Chromatin modifications and their function. Cell 2007;128:693–705.

[80] Krajewski WA, Nakamura T, Mazo A, Canaani E. A motif within SET-domain proteins binds single-stranded nucleic acids and transcribed and supercoiled DNAs and can interfere with assembly of nucleosomes. Mol Cell Biol 2005;25:1891–9.

[81] Vermeulen M, Timmers HT. Grasping trimethylation of histone H3 at lysine 4. Epigenomics 2010;2:395–406.

[82] Naoe T, Kubo K, Kiyoi H, et al. Involvement of the MLL/ALL-1 gene associated with multiple point mutations of the N-ras gene in acute myeloid leukemia with t(11;17)(q23;q25). Blood 1993;82:2260–1.

[83] Parsons DW, Jones S, Zhang X, et al. An integrated genomic analysis of human glioblastoma multiforme. Science 2008;321:1807–12.

[84] Watts GS, Futscher BW, Holtan N, Degeest K, Domann FE, Rose SL. DNA methylation changes in ovarian cancer are cumulative with disease progression and identify tumor stage. BMC Med Genomics 2008;1:47.

[85] Swisher EM, Gonzalez RM, Taniguchi T, et al. Methylation and protein expression of DNA repair genes: association with chemotherapy exposure and survival in sporadic ovarian and peritoneal carcinomas. Mol Cancer 2009;8:48.

[86] Makarla PB, Saboorian MH, Ashfaq R, et al. Promoter hypermethylation profile of ovarian epithelial neoplasms. Clin Cancer Res 2005;11:5365–9.

[87] Potapova A, Hoffman AM, Godwin AK, Al-Saleem T, Cairns P. Promoter hypermethylation of the PALB2 susceptibility gene in inherited and sporadic breast and ovarian cancer. Cancer Res 2008;68:998–1002.

[88] Helleman J, van Staveren IL, Dinjens WN, et al. Mismatch repair and treatment resistance in ovarian cancer. BMC Cancer 2006;6:201.

[89] Teodoridis JM, Hall J, Marsh S, et al. CpG island methylation of DNA damage response genes in advanced ovarian cancer. Cancer Res 2005;65:8961–7.

[90] Zeller C, Dai W, Steele NL, et al. Candidate DNA methylation drivers of acquired cisplatin resistance in ovarian cancer identified by methylome and expression profiling. Oncogene 2012;31:4567–76.

[91] Yu W, Jin C, Lou X, et al. Global analysis of DNA methylation by methyl-capture sequencing reveals epigenetic control of cisplatin resistance in ovarian cancer cell. PLoS ONE 2011;6:e29450.

[92] Wei SH, Chen CM, Strathdee G, et al. Methylation microarray analysis of late-stage ovarian carcinomas distinguishes progression-free survival in patients and identifies candidate epigenetic markers. Clin Cancer Res 2002;8:2246–52.

[93] Herranz N, Pasini D, Diaz VM, et al. Polycomb complex 2 is required for E-cadherin repression by the Snail1 transcription factor. Mol Cell Biol 2008;28:4772–81.

[94] Lin T, Ponn A, Hu X, Law BK, Lu J. Requirement of the histone demethylase LSD1 in Snail1-mediated transcriptional repression during epithelial–mesenchymal transition. Oncogene 2010;29:4896–904.

[95] Chaffer CL, Marjanovic ND, Lee T, et al. Poised chromatin at the ZEB1 promoter enables breast cancer cell plasticity and enhances tumorigenicity. Cell 2013;154:61–74.

[96] Cardenas H, Vieth E, Lee J, Liu Y, Nephew KP, Matei D. TGF-Beta induces global changes in DNA methylation during the epithelial-to-mesechymal transition in ovarian cancer cells. Epigenetics 2014;9:1461–72.

[97] Malouf GG, Taube JH, Lu Y, et al. Architecture of epigenetic reprogramming following Twist1-mediated epithelial-mesenchymal transition. Genome Biol 2013;14:R144.

[98] Kirk R. Tumour evolution: evidence points to the existence of cancer stem cells. Nat Rev Clin Oncol 2012;9:552.

[99] Nguyen LV, Vanner R, Dirks P, Eaves CJ. Cancer stem cells: an evolving concept. Nat Rev Cancer 2012;12:133–43.

[100] Pattabiraman DR, Weinberg RA. Tackling the cancer stem cells—what challenges do they pose? Nat Rev Drug Discov 2014;13:497–512.

[101] Dean M, Fojo T, Bates S. Tumour stem cells and drug resistance. Nat Rev Cancer 2005;5:275–84.

[102] Bertolini G, Roz L, Perego P, et al. Highly tumorigenic lung cancer CD133+ cells display stem-like features and are spared by cisplatin treatment. Proc Natl Acad Sci USA 2009;106:16281–6.

[103] Zhang S, Balch C, Chan MW, et al. Identification and characterization of ovarian cancer-initiating cells from primary human tumors. Cancer Res 2008;68:4311–20.

[104] Silva IA, Bai S, McLean K, et al. Aldehyde dehydrogenase in combination with CD133 defines angiogenic ovarian cancer stem cells that portend poor patient survival. Cancer Res 2011;71:3991–4001.

[105] Condello S, Morgan CA, Nagdas S, et al. beta-Catenin-regulated ALDH1A1 is a target in ovarian cancer spheroids. Oncogene 2014.

[106] Ayub TH, Keyver-Paik MD, Debald M, et al. Accumulation of ALDH1-positive cells after neoadjuvant chemotherapy predicts treatment resistance and prognosticates poor outcome in ovarian cancer. Oncotarget 2015.

[107] Shank JJ, Yang K, Ghannam J, et al. Metformin targets ovarian cancer stem cells in vitro and in vivo. Gynecol Oncol 2012;127:390–7.

[108] Yasuda K, Torigoe T, Morita R, et al. Ovarian cancer stem cells are enriched in side population and aldehyde dehydrogenase bright overlapping population. PLoS ONE 2013;8:e68187.

[109] Wang Y, Cardenas H, Fang F, et al. Epigenetic targeting of ovarian cancer stem cells. Cancer Res 2014;74:4922–36.

[110] Condello S, Morgan CA, Nagdas S, et al. beta-Catenin-regulated ALDH1A1 is a target in ovarian cancer spheroids. Oncogene 2015;34:2297–308.

[111] Lyko F, Brown R. DNA methyltransferase inhibitors and the development of epigenetic cancer therapies. J Natl Cancer Inst 2005;97:1498–506.

[112] Issa JP, Gharibyan V, Cortes J, et al. Phase II study of low-dose decitabine in patients with chronic myelogenous leukemia resistant to imatinib mesylate. J Clin Oncol 2005;23:3948–56.

[113] Kantarjian HM. Treatment of myelodysplastic syndrome: questions raised by the azacitidine experience. J Clin Oncol 2002;20:2415–16.

[114] Kornblith AB, Herndon 2nd JE, Silverman LR, et al. Impact of azacytidine on the quality of life of patients with myelodysplastic syndrome treated in a randomized phase III trial: a Cancer and Leukemia Group B study. J Clin Oncol 2002;20:2441–52.

[115] Silverman LR, Demakos EP, Peterson BL, et al. Randomized controlled trial of azacitidine in patients with the myelodysplastic syndrome: a study of the Cancer and Leukemia Group B. J Clin Oncol 2002;20:2429–40.

[116] Silverman LR, McKenzie DR, Peterson BL, et al. Further analysis of trials with azacitidine in patients with myelodysplastic syndrome: studies 8421, 8921, and 9221 by the Cancer and Leukemia Group B. J Clin Oncol 2006;24:3895–903.

[117] Kuykendall JR. 5-Azacytidine and decitabine monotherapies of myelodysplastic disorders. Ann Pharmacother 2005;39:1700–9.

[118] de Vos D, van Overveld W. Decitabine: a historical review of the development of an epigenetic drug. Ann Hematol 2005;84(Suppl. 13):3–8.

[119] Kantarjian H, Issa JP, Rosenfeld CS, et al. Decitabine improves patient outcomes in myelodysplastic syndromes: results of a phase III randomized study. Cancer 2006;106:1794–803.

[120] Attadia V. Effects of 5-aza-2'-deoxycytidine on differentiation and oncogene expression in the human monoblastic leukemia cell line U-937. Leukemia 1993;7(Suppl. 1):9–16.

[121] Pinto A, Attadia V, Fusco A, Ferrara F, Spada OA, Di Fiore PP. 5-Aza-2'-deoxycytidine induces terminal differentiation of leukemic blasts from patients with acute myeloid leukemias. Blood 1984;64:922–9.

[122] Jones PA, Taylor SM. Cellular differentiation, cytidine analogs and DNA methylation. Cell 1980;20:85–93.

[123] Willemze R, Archimbaud E, Muus P. Preliminary results with 5-aza-2'-deoxycytidine (DAC)-containing chemotherapy in patients with relapsed or refractory acute leukemia. The EORTC Leukemia Cooperative Group. Leukemia 1993;7(Suppl. 1):49–50.

[124] Kantarjian HM, Issa JP. Decitabine dosing schedules. Semin Hematol 2005;42:S17–22.

[125] O'Brien SM R-KF, Giles S, et al. Decitabine low dose schedule in myelodysplastic syndrome, comparison of three different dose schedules. J Clin Oncol 2005(Suppl. l):16.

[126] Wijermans P, Lubbert M, Verhoef G, et al. Low-dose 5-aza-2'-deoxycytidine, a DNA hypomethylating agent, for the treatment of high-risk myelodysplastic syndrome: a multicenter phase II study in elderly patients. J Clin Oncol 2000;18:956–62.

[127] Samlowski WE, Leachman SA, Wade M, et al. Evaluation of a 7-day continuous intravenous infusion of decitabine: inhibition of promoter-specific and global genomic DNA methylation. J Clin Oncol 2005;23:3897–905.

[128] Glasspool RM, Brown R, Gore ME, et al. A randomised, phase II trial of the DNA-hypomethylating agent 5-aza-2'-deoxycytidine (decitabine) in combination with carboplatin vs carboplatin alone in patients with recurrent, partially platinum-sensitive ovarian cancer. Br J Cancer 2014;110:1923–9.

[129] Fang F, Balch C, Schilder J, et al. A phase 1 and pharmacodynamic study of decitabine in combination with carboplatin in patients with recurrent, platinum-resistant, epithelial ovarian cancer. Cancer 2010;116:4043–53.

[130] Matei D, Fang F, Shen C, et al. Epigenetic resensitization to platinum in ovarian cancer. Cancer Res 2012;72:2197–205.

[131] Bast RC, Iyer RB, Hu W, et al. A phase IIa study of a sequential regimen using azacitidine to reverse platinum resistance to carboplatin in patients with platinum resistant or refractory epithelial ovarian cancer. J Clin Oncol 2008;26 (May 20 suppl; abstr 3500) 2008.

[132] Yoo CB, Jeong S, Egger G, et al. Delivery of 5-aza-2'-deoxycytidine to cells using oligodeoxynucleotides. Cancer Res 2007;67:6400–8.

[133] Gray SG, Qian CN, Furge K, Guo X, Teh BT. Microarray profiling of the effects of histone deacetylase inhibitors on gene expression in cancer cell lines. Int J Oncol 2004;24:773–95.

[134] Minucci S, Pelicci PG. Histone deacetylase inhibitors and the promise of epigenetic (and more) treatments for cancer. Nat Rev Cancer 2006;6:38–51.

[135] Qiu L, Burgess A, Fairlie DP, Leonard H, Parsons PG, Gabrielli BG. Histone deacetylase inhibitors trigger a G2 checkpoint in normal cells that is defective in tumor cells. Mol Biol Cell 2000;11:2069–83.

[136] Mitsiades CS, Mitsiades NS, McMullan CJ, et al. Transcriptional signature of histone deacetylase inhibition in multiple myeloma: biological and clinical implications. Proc Natl Acad Sci USA 2004;101:540–5.

[137] Gui CY, Ngo L, Xu WS, Richon VM, Marks PA. Histone deacetylase (HDAC) inhibitor activation of p21WAF1 involves changes in promoter-associated proteins, including HDAC1. Proc Natl Acad Sci USA 2004;101:1241–6.

[138] Duvic M, Talpur R, Ni X, et al. Phase 2 trial of oral vorinostat (suberoylanilide hydroxamic acid, SAHA) for refractory cutaneous T-cell lymphoma (CTCL). Blood 2007;109:31–9.

[139] Takai N, Kawamata N, Gui D, Said JW, Miyakawa I, Koeffler HP. Human ovarian carcinoma cells: histone deacetylase inhibitors exhibit antiproliferative activity and potently induce apoptosis. Cancer 2004;101:2760–70.

[140] Modesitt SC, Sill M, Hoffman JS, Bender DP. A phase II study of vorinostat in the treatment of persistent or recurrent epithelial ovarian or primary peritoneal carcinoma: a Gynecologic Oncology Group study. Gynecol Oncol 2008;109:182–6.

[141] Qian X, LaRochelle WJ, Ara G, et al. Activity of PXD101, a histone deacetylase inhibitor, in preclinical ovarian cancer studies. Mol Cancer Ther 2006;5:2086–95.

[142] Steele NL, Plumb JA, Vidal L, et al. A phase 1 pharmacokinetic and pharmacodynamic study of the histone deacetylase inhibitor belinostat in patients with advanced solid tumors. Clin Cancer Res 2008;14:804–10.

[143] Mackay HJ, Hirte H, Colgan T, et al. Phase II trial of the histone deacetylase inhibitor belinostat in women with platinum resistant epithelial ovarian cancer and micropapillary (LMP) ovarian tumours. Eur J Cancer 2010;46:1573–9.

[144] Dalgard CL, Van Quill KR, O'Brien JM. Evaluation of the in vitro and in vivo antitumor activity of histone deacetylase inhibitors for the therapy of retinoblastoma. Clin Cancer Res 2008;14:3113–23.

[145] Venturelli S, Armeanu S, Pathil A, et al. Epigenetic combination therapy as a tumor-selective treatment approach for hepatocellular carcinoma. Cancer 2007;109:2132–41.

[146] Dasmahapatra G, Yerram N, Dai Y, Dent P, Grant S. Synergistic interactions between vorinostat and sorafenib in chronic myelogenous leukemia cells involve Mcl-1 and p21CIP1 down-regulation. Clin Cancer Res 2007;13:4280–90.

[147] Munshi A, Tanaka T, Hobbs ML, Tucker SL, Richon VM, Meyn RE. . Vorinostat, a histone deacetylase inhibitor, enhances the response of human tumor cells to ionizing radiation through prolongation of gamma-H2AX foci. Mol Cancer Ther 2006;5:1967–74.

[148] Konstantinopoulos PA, Wilson AJ, Saskowski J, Wass E, Khabele D. Suberoylanilide hydroxamic acid (SAHA) enhances olaparib activity by targeting homologous recombination DNA repair in ovarian cancer. Gynecol Oncol 2014;133:599–606.

[149] Terasawa K, Sagae S, Toyota M, et al. Epigenetic inactivation of TMS1/ASC in ovarian cancer. Clin Cancer Res Off J Am Assoc Cancer Res 2004;10:2000–6.

[150] Shi H, Wei SH, Leu YW, et al. Triple analysis of the cancer epigenome: an integrated microarray system for assessing gene expression, DNA methylation, and histone acetylation. Cancer Res 2003;63:2164–71.

[151] Steele N, Finn P, Brown R, Plumb JA. Combined inhibition of DNA methylation and histone acetylation enhances gene re-expression and drug sensitivity in vivo. Br J Cancer 2009;100:758–63.

[152] Dietrich 3rd CS, Greenberg VL, DeSimone CP, et al. Suberoylanilide hydroxamic acid (SAHA) potentiates paclitaxel-induced apoptosis in ovarian cancer cell lines. Gynecol Oncol 2010;116:126–30.

[153] Ong PS, Wang XQ, Lin HS, Chan SY, Ho PC. Synergistic effects of suberoylanilide hydroxamic acid combined with cisplatin causing cell cycle arrest independent apoptosis in platinum-resistant ovarian cancer cells. Int J Oncol 2012;40:1705–13.

[154] Lin CT, Lai HC, Lee HY, et al. Valproic acid resensitizes cisplatin-resistant ovarian cancer cells. Cancer Sci 2008;99:1218–26.

[155] Wilson AJ, Lalani AS, Wass E, Saskowski J, Khabele D. Romidepsin (FK228) combined with cisplatin stimulates DNA damage-induced cell death in ovarian cancer. Gynecol Oncol 2012;127:579–86.

[156] Dizon DS, Blessing JA, Penson RT, et al. A phase II evaluation of belinostat and carboplatin in the treatment of recurrent or persistent platinum-resistant ovarian, fallopian tube, or primary peritoneal carcinoma: a Gynecologic Oncology Group study. Gynecol Oncol 2012;125:367–71.

[157] Matulonis U, Berlin S, Lee H, et al. Phase I study of combination of vorinostat, carboplatin, and gemcitabine in women with recurrent, platinum-sensitive epithelial ovarian, fallopian tube, or peritoneal cancer. Cancer Chemother Pharmacol 2015;76:417–23.

[158] Mendivil AA, Micha JP, Brown 3rd JV, et al. Increased incidence of severe gastrointestinal events with first-line paclitaxel, carboplatin, and vorinostat chemotherapy for advanced-stage epithelial ovarian, primary peritoneal, and fallopian tube cancer. Int J Gynecol Cancer 2013;23:533–9.

[159] Okano M, Bell DW, Haber DA, Li E. DNA methyltransferases Dnmt3a and Dnmt3b are essential for de novo methylation and mammalian development. Cell 1999;99:247–57.

[160] Keating ST, El-Osta A. Transcriptional regulation by the Set7 lysine methyltransferase. Epigenetics 2013;8:361–72.

[161] Takeuchi T, Watanabe Y, Takano-Shimizu T, Kondo S. Roles of jumonji and jumonji family genes in chromatin regulation and development. Develop Dyn Off Publ Am Assoc Anat 2006;235:2449–59.

[162] Shankar SR, Bahirvani AG, Rao VK, Bharathy N, Ow JR, Taneja R. G9a, a multipotent regulator of gene expression. Epigenetics 2013;8:16–22.

[163] Wang T, Xu C, Liu Y, et al. Crystal structure of the human SUV39H1 chromodomain and its recognition of histone H3K9me2/3. PLoS ONE 2012;7:e52977.

[164] Fu S, Hu W, Iyer R, et al. Phase 1b–2a study to reverse platinum resistance through use of a hypomethylating agent, azacitidine, in patients with platinum-resistant or platinum-refractory epithelial ovarian cancer. Cancer 2011;117:1661–9.

[165] Dizon DS, Damstrup L, Finkler NJ, et al. Phase II activity of belinostat (PXD-101), carboplatin, and paclitaxel in women with previously treated ovarian cancer. Int J Gynecol Cancer Off J Int Gynecol Cancer Soc 2012;22:979–86.

[166] Kummar S, Gutierrez M, Gardner ER, et al. Phase I trial of MS-275, a histone deacetylase inhibitor, administered weekly in refractory solid tumors and lymphoid malignancies. Clin Cancer Res Off J Am Assoc Cancer Res 2007;13:5411–17.

CHAPTER 4

Timing of Cytoreductive Surgery in the Treatment of Advanced Epithelial Ovarian Carcinoma

C.M. St. Clair[1], F. Khoury-Collado[1] and D.S. Chi[2]

[1]Maimonides Medical Center, Brooklyn, NY, United States [2]Memorial Sloan Kettering Cancer Center, New York, NY, United States

CONTENTS

INTRODUCTION

With an estimated 14,180 deaths in 2015, ovarian cancer is the leading cause of mortality from gynecologic cancers in the United States [1]. A clinically meaningful screening strategy is currently lacking. Therefore, patients tend to present with advanced-stage disease, at which point diagnosis is poor and disease is more difficult to treat. Standard practice in the treatment of ovarian cancer is cytoreductive surgery and debulking of gross disease, followed by cytotoxic chemotherapy [2–4]. Timing surgery to optimize the oncologic outcome is an area of active research. This chapter focuses on primary and interval debulking surgery PDS and IDS, respectively, where the current debate between neoadjuvant chemotherapy (NACT) and PDS stands, and how recent translational advances add to the debate.

HISTORY OF OVARIAN CANCER SURGERY: DEFINING "OPTIMAL"

In a majority of solid tumors, the presence of disseminated disease is a contraindication for surgical resection. In the ovarian cancer literature, however, studies dating back to 1975 have supported the clinical significance of maximal cytoreduction. Work by Griffiths et al. found an inversely proportional relationship between survival and tumor burden remaining after surgery [5]. They looked at 102 patients with stage II and III disease and showed that those with no residual disease had the best outcomes, while those with disease >1.5 cm did poorly and yielded no benefit from surgical resection.

Translational Advances in Gynecologic Cancers. DOI: http://dx.doi.org/10.1016/B978-0-12-803741-6.00004-5

The definition of "optimal" cytoreduction, however, has continuously shifted over the years. While Griffiths' data pointed to a cut-off of 1.5 cm, Hoskins et al. published a paper in 1994 demonstrating a survival advantage for those left with <2 cm of disease after surgery [6]. This secondary analysis of 297 patients from the Gynecologic Oncology Group (GOG) study 97 with sub-optimal debulked stage III disease showed a relative risk of death of 1.90 for tumors >2 cm, compared with those <2 cm following surgical cytoreduction ($P < 0.01$). They concluded that, when technically feasible, the tumor burden should be resected to <2 cm at the time of initial surgery.

Since the results of GOG 52 were published by Omura et al. in 1989, however, the GOG has defined optimal resection as residual disease ≤1 cm [2]. Numerous studies since have corroborated with a cut-off of 1 cm, including work by Chi et al. in 2001 that looked at prognostic factors in patients with advanced epithelial ovarian cancer undergoing upfront surgical cytoreduction [7]. After analyzing 282 women with stage III/IV disease who underwent primary surgery from 1987 to 1994, they found that only patient age, presence or absence of ascites, and size of residual tumor burden were statistically significant in terms of prognostication. Notably, a survival benefit to surgery was not achieved unless maximum residual tumor diameter was <1 cm.

As providers continued to strive for optimal surgical cytoreduction, it became evident via accumulating data that the less disease left behind the better the outcome [8]. Various groups began to publish data showing a distinct survival benefit in cases of complete gross resection (CGR). A meta-analysis by Bristow et al. published in the *Journal of Clinical Oncology* in 2002 included nearly 7000 patients from 81 studies treated with surgery and chemotherapy in the platinum era. They modeled that for every 10% increase in maximal cytoreduction, there was a resultant 5.5% increase in median survival time [9]. In 2006, Aletti and colleagues from the Mayo Clinic performed a retrospective study of 194 patients with International Federation of Gynecology and Obstetrics (FIGO) stage IIIC epithelial ovarian cancer, 144 of whom had peritoneal carcinomatosis. Optimal cytoreduction—defined in the study as residual disease ≤ 1 cm after surgery—was attained in 131 patients (67.5%). Residual disease was the only independent predictor of overall survival (OS), as depicted by their published survival curves [10]: CGR, >84 months; <1 cm, 34 months; 1–2 cm, 25 months; and >2 cm, 16 months. Chi et al. from Memorial Sloan Kettering Cancer Center published similar data the same year in an analysis of 465 patients with stage IIIC disease, excluding those deemed to have stage III disease based on lymph node involvement alone [11]. Median OS was as follows: CGR, 106 months; gross disease ≤0.5 cm, 66 months; 0.6–1.0 cm, 48 months; 1–2 cm, 33 months; and >2 cm, 34 months. Based on these findings, the authors concluded that CGR should be the goal of PDS in advanced ovarian cancer, and that when CGR is not feasible, the aim should be to reduce the tumor burden to the smallest diameter possible to optimize outcome.

THE QUEST FOR COMPLETE GROSS RESECTION: A RADICAL APPROACH

Despite evolving data demonstrating a clear survival advantage with optimal surgical cytoreduction (and CGR when feasible), in the year 2000, optimal debulking was attained in only 30–60% of advanced ovarian cancers [12]. In order to improve the rate of optimal debulking and to resect disease previously deemed unresectable, radical procedures in the chest, the upper abdomen, and the pelvis have been added to the surgical armamentarium. Incorporation of splenectomy, distal pancreatectomy, cholecystectomy, liver wedge resection, dissection of the portahepatis, and diaphragm stripping/resection has increased rates of optimal primary cytoreduction from 46% to 80%, and rates of CGR from 11% to 27% in one published institutional experience [13]; this translated to significantly improved progression-free survival (PFS) and OS in patients who would have been deemed unresectable or suboptimal in prior years [14].

A subset of patients with advanced epithelial ovarian cancer present at the time of diagnosis with a pleural effusion or mediastinal lymphadenopathy on imaging, constituting stage IV disease once biopsy-proven. Video-assisted thoracoscopic surgery (VATS) is used in many institutions to evaluate patients with moderate-to-large effusions as part of the initial workup. Roughly, two-thirds of these patients will have macroscopic disease in the chest at the time of VATS; in up to 73% of these cases, disease is larger than 1 cm [15,16]. The presence of a malignant pleural effusion alone portends a worse prognosis, even in cases of optimal cytoreduction, in comparison to patients with stage IIIC disease [17,18].

VATS can be used for diagnosis and also symptom management when combined with pleurodesis. However, perhaps the most critical role of VATS is in triaging patients for PDS versus NACT. When disease >1 cm is encountered in the chest, the patient is considered unresectable and can be referred for NACT. It would not be prudent to subject a patient to a prolonged intraabdominal debulking procedure while leaving thoracic disease >1 cm behind, resulting in a suboptimal debulking. If the disease encountered is subcentimeter and amenable to resection via VATS, a thoracic debulking can be performed with a plan to proceed with the abdominal surgery on the same day or a separate day, depending on the condition of the patient [16].

Finally, given the peritoneal spread of typical epithelial ovarian cancer, it is not uncommon for tumor to involve the small bowel, large bowel, or both. Small-volume disease on the serosa or within the mesentery can be ablated or excised without difficulty. Tumors infiltrating the mesentery or penetrating the intestinal wall, however, necessitate bowel resection if an optimal cytoreduction is to be undertaken. While intestinal procedures for ovarian cancer were considered radical at one time, bowel resections are performed in as many as 50% of patients in whom an optimal debulking is achieved [19]. A modified posterior exenteration entails resection of the recto-sigmoid colon en bloc with the uterus, adnexa, and peritoneum, and it may be necessary to resect a bulky pelvic mass, obliterating the posterior cul-de-sac. Ileocecectomy with or without extended right hemicolectomy is required at times to resect a right ovarian mass involving local structures. The transverse colon may be sacrificed if a plane cannot be created during total omentectomy. And finally, small bowel may be involved by a dominant mass or masses, or may be coated with miliary disease in the case of peritoneal carcinomatosis; in such a setting, one or more small bowel resections may be required. Available data suggest that intestinal surgery, as part of ovarian debulking, can be performed safely and with minimal complications when performed by an experienced surgeon [20–25]. The rate of anastomotic leak has been reported as 0.8–6.8%, depending on the series; this can be a lethal complication. However, placing a diverting loop ileostomy is only recommended in high-risk cases, for example, those involving very low anastomosis, high blood loss, previous radiation therapy, or multiple (large) bowel resections [26].

Undertaking more complex surgical procedures to achieve complete or optimal cytoreduction of ovarian carcinoma is associated with increased morbidity; yet the trade-off is a survival advantage compared with those patients who are suboptimally debulked, with decreased rates of relapse and death.

CHEMOTHERAPY THEN AND NOW: WHEN AND HOW?

To ultimately address the question of surgical timing, and to compare PDS with upfront NACT, one must consider the current chemotherapeutic regimens used. In the modern era of chemotherapy for ovarian carcinoma, platinum and taxane-based combination therapy is the standard of care. Omura et al. published two studies in 1986 and 1991 demonstrating the efficacy of cisplatin in advanced epithelial ovarian cancer [27,28]. In the first study—GOG 47—440 patients with stage III or IV suboptimal disease, as well as those with widespread recurrence, were randomized to cyclophosphamide and doxorubicin with or without cisplatin. For the cyclophosphamide and doxorubicin arm, the complete response rate was 26%; for the arm that included cisplatin, the complete response rate was 51%, with a complete plus partial response rate reaching 76%. Cisplatin-based therapy was again found to be a favorable prognostic factor in the long-term follow-up study published 5 years later [28]. While the results did not translate to an OS advantage, this was felt to be due to cross-over into the cisplatin arm in patients who progressed.

Taxanes were adopted into first-line treatment for advanced epithelial ovarian cancer as a result of two randomized phase III trials performed by the

GOG. In GOG 111, published in the *New England Journal of Medicine* in 1996 by McGuire et al., patients were randomly assigned to cisplatin and cyclophosphamide versus cisplatin and paclitaxel [29]. The study population also consisted of patients with stage III or IV suboptimally debulked disease, "suboptimal" defined as residual disease >1 cm. With a combined partial and complete response rate of 77%, the trial favored the taxane-containing arm, with a 28% reduced risk of progression and a 34% reduced risk of death compared with the nontaxane arm. Before these results were available, the GOG had embarked upon GOG 132, evaluating 648 patients in the same population randomized to cisplatin alone, paclitaxel alone, or the combination therapy regimens used in GOG 111 [30]. Cisplatin alone had a significantly improved response rate compared with paclitaxel alone (67% vs 42%). Notably, the combined regimens had an improved toxicity profile over the single agent drugs. Ultimately, with the publication of the findings of GOG 158 by Ozols et al. demonstrating noninferiority of carboplatin/paclitaxel to cisplatin/paclitaxel, carboplatin/paclitaxel in combination became the standard chemotherapeutic regimen for initial treatment of advanced epithelial ovarian cancer [3].

While intravenous (IV) chemotherapy was evolving to the standard of carboplatin and paclitaxel combination therapy, parallel trials were ongoing over the same timeframe evaluating the role of intraperitoneal (IP) chemotherapy. Given the typical spread pattern of advanced epithelial ovarian cancer, this was a logical step in the quest to optimize medical management of this disease; direct, concentrated delivery of drug to the peritoneal surfaces for patients with disseminated stage III carcinoma made sense. The first randomized phase III trial in the United States—GOG 104—looked at IV cisplatin and cyclophosphamide versus IP cisplatin and IV cyclophosphamide [31]. Median survival in the IP arm was 49 months, significantly improved over the 41 months of the IV arm, with a 24% reduction in the risk of death. However, the study results were published in 1996, at the same time GOG 111 results were published, extolling the benefits of paclitaxel; lacking a taxane component, the combination therapy in GOG 104 was deemed outdated. Furthermore, the study included patients with residual disease up to 2 cm. As IP chemotherapy penetrates a matter of millimeters, this was also a potential weakness in design.

Given the promising results of Alberts and colleagues, GOG 114 was undertaken to compare IV cisplatin and IV paclitaxel to IV carboplatin (area under the curve [AUC] 9) for two cycles followed by IP cisplatin and IV paclitaxel [32]. Patients in the study also had stage III disease, but residual disease ≤1 cm was considered optimal. The difference in OS was eye opening: 63 months versus 52 months in favor of the IP arm. However, with increased myelosuppression and a substantial number of patients unable to complete the treatment course, the authors concluded that this regimen could not be held as the new standard. Additionally, many questioned whether the IP cisplatin or the initial IV carboplatin with an AUC of 9 was responsible for the survival advantage observed in the experimental arm.

Finally, Armstrong et al. published findings from GOG 172 in 2006 comparing IV paclitaxel and IV cisplatin to IV paclitaxel followed by IP cisplatin and IP paclitaxel [4]. The median OS in the IP-containing arm was 66.9 months compared with 49.5 months in the IV-only arm, with a relative risk of death of 0.71. This was the third consecutive phase III trial to demonstrate superior oncologic outcomes with the IP administration of cisplatin in optimally resected advanced ovarian cancer. However, while the survival data were impressive, questions remained regarding the tolerability of IP therapy. Grade 3 and 4 toxicities were significantly more common in the IP arm, and less than half of the patients were able to complete 6 cycles of IP treatment. Though retrospective in design, modifications to the GOG 172 regimen published by Barlin et al. in 2012 suggested that survival outcomes could be maintained while improving the side effect profile and dosing schema of the original treatment algorithm [33]. Of 102 patients, 56 (55%) were able to complete 4 of 6 planned cycles, and OS was preserved at 67 months.

While various agents are used in the setting of recurrent, platinum-refractory disease, the therapeutic regimen for upfront therapy has remained constant since GOG 158 was published in 2003 [3]. Current trials are focused on dosing and delivery (standard carboplatin/paclitaxel vs dose-dense paclitaxel vs IP therapy) and on novel treatments, such as immune therapy, biologics including antiangiogenics, small molecule antagonists, and poly ADP ribose polymerase (PARP) inhibitors. While the role and incorporation of these targeted therapies continues to play out in the medical optimization of ovarian cancer treatment, a heated debate continues on the surgical side: is the historic standard of PDS still the standard, or is NACT a reasonable alternative with comparable oncologic outcomes? How does the concept of personalized medicine fit into the debate?

PRIMARY VERSUS INTERVAL CYTOREDUCTION: RATIONALE

Study after study has shown that aggressive surgical debulking to CGR, or as close to CGR as possible, drastically improves oncologic outcomes for ovarian cancer patients [7–14]. PDS carries intuitive benefits. Large, necrotic, hypoxic tumors with poor blood supply may not respond readily to chemotherapy given suboptimal drug delivery. In small, well-perfused tumors, cancer cells are likely to be dividing more rapidly, making them more sensitive to cytotoxic agents. Furthermore, smaller tumors may require fewer total cycles of a drug, decreasing the likelihood of developing chemotherapeutic resistance [34–36]. However, not all patients can proceed to cytoreductive surgery at the time of diagnosis, namely those who are medically inoperable due to comorbidities and those found to have such extensive disease burden on preoperative evaluation that the achievement of optimal cytoreduction is unlikely. The latter is left to the discretion of the treating surgeon and has historically been gestalt based on physical exam, CA-125 level, and imaging. An additional cohort of patients is made up of those who have undergone an initial surgical attempt of varying effort and who have been left with a significant (suboptimal) disease burden. NACT was proposed as a means to improve the rate of optimal debulking up from a meager 30–60% [12].

INTERVAL DEBULKING SURGERY: A SECOND CHANCE

Van der Burg and colleagues, under the auspices of the European Organisation for Research and Treatment of Cancer (EORTC), published the results of their study in the *New England Journal of Medicine* in 1995, which looked at outcomes in patients with advanced epithelial ovarian cancer who underwent IDS after three cycles of induction chemotherapy [36]. They studied patients with stage IIB to IV disease who underwent initial surgery and were left with lesions measuring >1 cm. They all received three cycles of cisplatin and cyclophosphamide; those with clinical response and/or stable disease based on CA-125 and imaging were randomized to IDS versus completion of chemotherapy alone. All randomized patients received a total of [at least] six cycles of cisplatin and cyclophosphamide. The authors demonstrated a survival advantage in both PFS and OS for the patients who underwent surgery versus those who did not: 18 months versus 13 months, respectively, for PFS, and 26 months versus 20 months, respectively, for OS.

The GOG went on to publish the results of a similar study in 2004, also in the *New England Journal of Medicine*, addressing the same question in the era of platinum- and taxane-based chemotherapy [37]. They evaluated patients who had undergone suboptimal initial surgical attempts and went on to receive three cycles of carboplatin and paclitaxel followed by randomization to a second attempt at surgical debulking versus no further surgery. Following the randomization, both groups were to complete six total cycles of chemotherapy. The differences in PFS and OS were not statistically significant between those who underwent IDS and those who received chemotherapy alone; PFS was 10.5 versus 10.7 months, respectively, and median OS was 33.9 and 33.7 months, respectively.

The role of a second surgical attempt after chemotherapy remained uncertain given opposite conclusions from these two large randomized trials in the United States and Europe [36,37]. Notably, the GOG protocol incorporated paclitaxel, which had not yet been adopted at the time of the EORTC trial. Additionally, the US study dictated that the primary surgical attempt constitutes a "maximal surgical effort"; the surgery was performed via laparotomy by a trained gynecologic oncologist, and the remaining tumor burden was <5 cm in a majority of cases [37]. The difference in oncologic outcome may have been a reflection of the difference in effort put forth in the initial debulking. Rose et al. concluded that if the primary procedure was performed by a qualified surgeon attempting to achieve an optimal cytoreduction, that reoperation after three cycles of chemotherapy was unlikely to improve survival. However, if the initial surgical attempt was by a surgeon ill-equipped to proceed with PDS, then proceeding with chemotherapy followed by interval cytoreduction may be indicated, as supported by the European data.

NEOADJUVANT CHEMOTHERAPY: PLANNED FROM THE START

The question remained, however, as to whether there was a role for initiating treatment with upfront chemotherapy in the setting of bulky stage IIIC or IV disease. The aforementioned studies looked at patients who had already undergone a failed attempt at optimal PDS. But many theorized that NACT would serve to decrease the tumor burden and, with it, the likelihood of a suboptimal cytoreduction once surgical resection was undertaken, ultimately improving survival outcomes. This theory was tested in a meta-analysis by Bristow and colleagues published in 2006 in which 22 cohorts totaling over 800 patients treated with upfront chemotherapy followed by IDS were evaluated [38]. The authors demonstrated a median OS of 24.5 months for all cohorts and showed that as the number of chemotherapy cycles increased, the median survival decreased. They concluded that NACT was associated with inferior survival compared with PDS, and that when NACT is initiated, definitive surgical cytoreduction should be completed as early in the course as possible. More recent studies have corroborated the conclusion that increasing cycles of NACT are associated with worse outcomes in bulky, late-stage disease [39], potentially due to earlier development of platinum resistance [40,41].

A landmark paper by Vergote et al. of the EORTC published in the *New England Journal of Medicine* in 2010, however, contradicted these conclusions and added to the controversy surrounding the role of NACT versus PDS [42]. In the study, 670 patients with biopsy-proven advanced ovarian cancer were randomly assigned to either PDS followed by six cycles of platinum-based chemotherapy or three cycles of NACT followed by IDS (NACT-IDS) and at least three additional cycles of adjuvant chemotherapy. Median PFS was

12 months for both groups. Overall survival for the PDS arm was 29 months compared with 30 months for the NACT-IDS arm, again showing no statistically significant difference. Per-protocol analysis of OS according to the treatment received was the same. The authors concluded that NACT-IDS is not inferior to PDS in the treatment of advanced, stage IIIC or IV, ovarian carcinoma. While they posit dismal survival outcomes were likely secondary to a study population composed of entirely advanced-stage patients, in comparison to other studies that have included stage II, IIIA, and IIIB disease, the poor median OS of 29 months in the PDS arm has remained a source of criticism in the ongoing discussion of PDS versus NACT. Critics contend that the EORTC trial was a negative trial, because the outcomes of the surgical arm were inferior to those of other published trials and not because the NACT arm was truly noninferior to PDS.

Following the EORTC trial, Dewdney et al. surveyed members of the Society of Gynecologic Oncology (SGO) to assess patterns of care for patients with advanced-stage ovarian cancer and, specifically, their clinical practice as it pertained to NACT [43]. Of the 339 respondents (30% of surveyed members), 60% reported using NACT in <10% of their advanced-stage cases. The majority (82%) did not feel that available evidence was sufficient to justify NACT-IDS. Sixty-two percent reported that it is impossible to predict optimal debulking preoperatively, and 86% felt that both biological factors and surgical skill play a part in resectability and patient outcomes. Taken together, SGO responders reported that the evidence for NACT-IDS was insufficient to support a change in practice.

Chi and colleagues went on to publish their results of a comparison to the EORTC trial, which looked at an identical patient population treated during the same time period but with PDS [44]. Published in 2012, their study evaluated 316 patients, of whom 285 (90%) underwent PDS and 31 (10%) received NACT. The majority had high-grade serous tumors, and optimal cytoreduction was achieved in 203 patients (71%). Median PFS was 17 months, while median OS was 50 months. The authors concluded that even in patients with bulky, advanced-stage ovarian, tubal or peritoneal carcinoma, PDS confers a survival benefit and should continue to be the preferred initial management.

Additional retrospective analyses have been published since the Vergote study looking at NACT-IDS versus PDS [45,46]. Most show an improved rate of optimal cytoreduction at the time of IDS when NACT is employed, but no difference in PFS or OS. In a recent publication by Bian et al., of 339 patients with advanced disease, survival in those who underwent PDS versus those who underwent NACT-IDS was similar, with a median PFS of 10 versus 11 months, and a disappointing 25 months for OS in both groups [47]. They did report, notably, that there was a survival advantage to resecting disease down to small residual tumors under 1 cm in the PDS group, but not in the NACT-IDS group; they concluded that the goal of IDS after NACT should be CGR. Conclusions like these frame the discussion around what constitutes "optimal debulking" in the current management of advanced ovarian carcinoma. While data support the goal of CGR, most authors feel that visible disease of millimeters is acceptable, optimal, and confers a benefit to survival.

A recent publication in *The Lancet* of the Chemotherapy or Upfront Surgery (CHORUS) trial, a randomized, controlled, noninferiority trial of primary chemotherapy versus primary surgery for newly diagnosed advanced ovarian carcinoma, attempted to readdress the question of NACT as a reasonable alternative to PDS, with reduced morbidity [48]. More than 550 women were enrolled and randomized, 276 to primary surgery and 274 to primary chemotherapy, and both intention-to-treat and per-protocol analyses were performed. Median OS was 22.6 months in the primary surgery group versus 24.1 months in the primary chemotherapy group. The authors reported significantly more grade 3 and 4 postoperative adverse events, as well as a great number of grade 3 and 4 chemotherapy-related toxicities in the primary surgical arm compared with the primary chemotherapy arm. The authors concluded that when treating newly diagnosed, advanced ovarian cancer, outcomes with primary chemotherapy are noninferior to those with primary surgical resection, and that NACT is an acceptable standard of care in this patient population. These conclusions are in line with previously reported data finding the two strategies to be comparable, and yet again, OS was 22–24 months, on par with suboptimal debulking in other studies.

Prior to the CHORUS trial, Vergote and du Bois and colleagues from their institutions collaborated and published a review article entitled "Neoadjuvant chemotherapy for advanced ovarian cancer: on what do we agree and disagree?" [49] The authors discussed criteria they deemed appropriate for selecting either PDS or NACT for upfront therapy (Table 4.1). Patients they considered appropriate candidates for PDS were those with disease confined to the pelvis or with spread beyond the pelvis <2 cm (stage IIIB or less); those with stage IIIC disease based on lymph node disease alone; and those with abdominal disease up to 5 cm in maximum diameter. Furthermore, the patient had to be evaluated by an experienced gynecologic oncology surgeon, and there had to be a high likelihood of achieving CGR or leaving behind as little disease as possible (<1 cm). The authors agreed that issues of surgical expertise and convenience of treatment planning should never impact the decision to proceed with PDS versus NACT-IDS.

The authors were unable to agree upon the best management approach for approximately 15–20% of the patient population, essentially women with very advanced ovarian cancer and extrapelvic disease >5 cm. The Leuven group

Table 4.1 Criteria for Primary Chemotherapy and for Interval Debulking Surgery in FIGO Stages IIIC and IV Ovarian Carcinoma

Criteria	Essen Criteria	Leuven Criteria
Diagnosis	Biopsy with histologically proven epithelial ovarian (or tubal or peritoneal cancer FIGO stage IIICIV	
	–	Or fine needle aspiration proving the presence of carcinoma cells in patients with suspicious pelvic mass if CA-125 (KU/L)/CEA (ng/mL) ratio is >25. If the serum CA-125/CEA ratio is ≤25, imaging or endoscopy is obligatory to exclude a primary gastric, colon, or breast carcinoma
Abdominal metastases	Involvement of the superior mesenteric artery	
	Diffuse deep infiltration of the radix mesenterii of the small bowel	
	Diffuse and confluent carcinomatosis of the stomach and/or small bowel involving such large parts that resection would lead to a short bowel syndrome or a total gastrectomy	
	Multiple parenchymatous liver metastases in both lobes	Intrahepatic metastases
	Tumor involving large parts of the pancreas (not only tail) and/or the duodenum	Infiltration of the duodenum and/or pancreas and/or the large vessels of the ligamentum hepatoduodenale, truncus coaeliacus or behind the portahepatis
	Tumor infiltrating the vessels of the lig. Hepatoduodenale or truncus coeliacus	
Extraabdominal metastases	Not completely resectable metastases, as e.g., ■ Multiple parenchymal lung metastases (preferably histologically proven) ■ Nonresectable lymph node metastases ■ Brain metastases	All excluding: ■ Resectable inguinal lymph nodes ■ Solitary resectable retrocrual or paracardial nodes ■ Pleural fluid containing cytologically malignant cells without proof of the presence of pleural tumors
Patients characteristics/ others	Impaired performance status and comorbidity not allowing a "maximal surgical effort" to achieve a complete resection	
	Patients' nonacceptance of potential supportive measures as blood transfusions or temporary stoma	
Criteria for interval debulking	■ Upfront surgical effort in an institution without expert surgical skills/infrastructure ■ Barrier for initial surgery has disappeared (e.g., improved medical condition) ■ Not, if reason for primary chemotherapy was tumor growth pattern diagnosed during open surgery by an experienced gynecologic oncologist under optimal circumstances (as in GOG study 152)	■ No progressive disease ■ In case of extraabdominal disease at diagnosis, the extraabdominal disease should be in complete response or resectable ■ Performance status and comorbidity allowing a maximal surgical effort to no residual diseases

With permission from Elsevier: Vergote I, du Bois A, Amant F, et al. Neoadjuvant chemotherapy in advanced ovarian cancer: on what do we agree and disagree? Gynecol Oncol 2013;128(1):6–11.

endorses a strategy incorporating diagnostic laparoscopy, acknowledging that some patients who could be optimally cytoreduced may be aborted and given NACT based on findings at laparoscopy. The Essen group favors giving these patients the benefit of a surgical attempt, knowing that with it, there will be a certain percentage of patients who undergo laparotomy that will be aborted due to the extent of disease. To limit these occurrences, they have developed a standardized approach for exploration, beginning with a two-point evaluation of the upper abdomen and small bowel. If the patient is deemed inoperable at either of these junctures, the surgery is terminated within the first 60–90 minutes and harm is minimized. The criteria used by both institutions to select candidates better suited to NACT are listed in Table 4.1.

PREDICTING SURGICAL SUCCESS: TRANSLATIONAL ADVANCES

Both prospective and retrospective data continue to build on the paradigm that once surgical debulking is undertaken, whether in the primary or interval setting, optimal debulking down to as little disease as possible is the goal. Predicting which patients will be optimally versus suboptimally cytoreduced at the completion of debulking surgery has become an area of active research, as patients who are likely to undergo a suboptimal debulking may be triaged accordingly to NACT or a clinical trial. CA-125 levels, computed tomography, MRI and PET scans, and various combinations of patient factors, tumor markers and imaging findings have all been investigated in an attempt to preoperatively predict the likelihood of achieving an optimal versus suboptimal debulking. One recent multicenter prospective trial identified nine factors—three clinical and six radiographic—that when taken together yielded a predictive accuracy of 0.758 [50]. Those with a total score ≥9 had a 74% rate of suboptimal debulking, while patients with a score of 0 had only a 5% rate of suboptimal debulking (Tables 4.2 and 4.3).

While clinical models such as these have given surgeons a way in which to evaluate patients preoperatively, their utility remains limited. Issues with predictive accuracy, reproducibility, and ease of use in a busy practice hamper widespread adoption, and many clinicians continue to use patient symptoms, physical exam, tumor markers, and imaging together to formulate a decision for or against PDS, without the use of a predictive model. Researchers have turned their attention instead to the make-up of the tumor and its surrounding cellular environment. Since the genomic revolution began, it has played out in diseases like breast cancer, in which gene expression profiles are used regularly to predict chemosensitivity and recurrence, and to inform treatment decisions [51,52]. While we are not there yet with ovarian cancer, various gene signatures have been identified and associated with response to

Table 4.2 Multivariate Model of Significant Clinical and Radiologic Criteria Predictive of Suboptimal Cytoreduction

Criteria	OR	95% CI	p	Predictive Value Score[a]
Age ≥60 years	1.32	1.06–1.63	0.01	1
CA-125 ≥500 U/mL	1.47	1.28–1.69	<0.001	1
ASA 3–4	3.23	1.76–5.91	<0.001	3
Retroperitoneal lymph nodes above the renal hilum (including supradiaphragmatic) >1 cm	1.59	1.58–1.6	<0.001	1
Diffuse small bowel adhesions/thickening	1.87	1.86–1.87	<0.001	1
Perisplenic lesion >1 cm	2.27	1.7–3.03	<0.001	2
Small bowel mesentery lesion >1 cm	2.28	1.08–4.8	0.03	2
Root of the superior mesenteric artery lesion >1 cm	2.4	1.34–4.32	0.003	2
Lesser sac lesion >1 cm	4.61	4.39–4.84	<0.001	4

With permission from Elsevier: Suidan RS, Ramirez PT, Sarasohn DM, et al. A multicenter prospective trial evaluating the ability of preoperative computed tomography scan and serum CA-125 to predict suboptimal cytoreduction at primary debulking surgery for advanced ovarian, fallopian tube, and peritoneal cancer. Gynecol Oncol 2014;134(3):455–61.
ASA, American Society of Anesthesiologists.

Table 4.3 Predictive Value Score and Suboptimal Cytoreduction (N = 349)[a]

Total Predictive Value Score	Total Patients n (%)	Optimal (n)	Suboptimal (n)	Suboptimal Rate
0	22/349 (6%)	21	1	5%
1–2	79/349 (23%)	71	8	10%
3–4	109/349 (31%)	91	18	17%
5–6	85/349 (24%)	56	29	34%
7–8	31/349 (9%)	15	16	52%
≥9	23/349 (7%)	6	17	74%

With permission from Elsevier: Suidan RS, Ramirez PT, Sarasohn DM, et al. A multicenter prospective trial evaluating the ability of preoperative computed tomography scan and serum CA-125 to predict suboptimal cytoreduction at primary debulking surgery for advanced ovarian, fallopian tube, and peritoneal cancer. Gynecol Oncol 2014;134(3):455–61.
[a]*One patient excluded for a missing American Society of Anesthesiologists class.*

platinum chemotherapy, time to recurrence, OS, and even debulking status, although they are not yet used to navigate clinical management [53–57].

In a recent publication, Liu et al. out of Cedars-Sinai Medical Center evaluated gene expression levels in late-stage serous ovarian cancer patients who had undergone primary optimal surgical cytoreduction versus suboptimal surgical cytoreduction [57]. To do so, they analyzed the three largest datasets for ovarian carcinoma: The Cancer Genome Atlas (TCGA), GSE26712 and GSE9891, in which the status of cytoreduction was known. Molecular signatures were identified using the first two datasets and then validated using the third and final datasets. In the analysis, the group looked not only at specific genes that were differentially expressed but also at gene–gene interactions and molecular networks. Comparing the profiles of patients who were optimally debulked versus suboptimally debulked, there were just over 1200 differentially expressed genes within the TCGA dataset and roughly 980 from the GSE26712 dataset; 136 of these differentially expressed genes overlapped between the two datasets. Using this information, along with data regarding overlapping networks, the authors were able to create the suboptimal cytoreduction associated network (SCAN), which is composed of 11 differentially expressed genes (Fig. 4.1). Serving as a candidate gene signature, the 11 SCAN genes were externally validated using the GSE9891 dataset; the four with the lowest P values—POSTN, FAP, TIMP3, and COL11A1—went on to further testing. Transcript levels of each of the four increased proportionally with increasing levels of residual disease. Furthermore, within the TCGA data, these SCAN genes were found to be highly expressed in the mesenchymal molecular subtype, the subtype associated with the worse survival and enriched within invasive and metastatic ovarian cancer. While this suboptimal signature is not yet used clinically to predict surgical outcome, further evaluation of the top candidate genes will provide additional insight into the biology of unresectable disease.

Accurately predicting which patients are unlikely to be optimally debulked may help direct certain patients toward NACT or clinical trials; however, taken another way, accurately predicting which patients are *likely* to undergo optimal debulking may give clinicians the information they need to pursue PDS aggressively, with a goal of CGR.

IN CONCLUSION: WHAT NOW?

While researchers and clinicians continue to debate the relative contributions of tumor biology and surgical expertise in the management of patients with advanced-stage ovarian carcinoma, the reality is that both factors are likely to play a major role in "debulkability" and survival outcomes. Translational

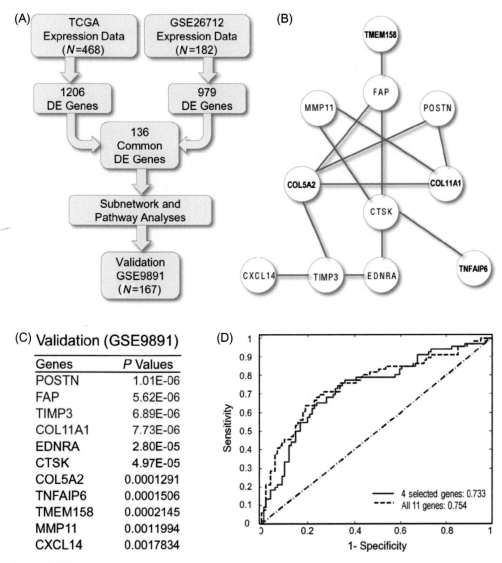

FIGURE 4.1

Identification and validation of the suboptimal cytoreduction associated network (SCAN). (A) Statistical analysis workflow chart. (B) Selected biomarkers with both differentially expressed genes and differential networks. (C) External validation of the network genes in the validation dataset (GSE9891). The top four genes with the lowest *P* values are highlighted in red. (D) Predicted Area Under the ROC Curve (AUC) for the SCAN genes in the validation dataset (GSE9891). *With permission from Elsevier: Liu Z, Beach JA, Agadjanian H, et al. Suboptimal cytoreduction in ovarian carcinoma is associated with molecular pathways characteristic of increased stromal activation. Gynecol Oncol 2015, doi:10.1016/j.ygyno.2015.08.026.*

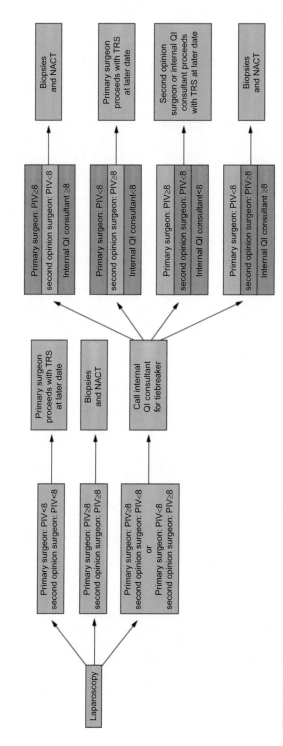

FIGURE 4.2

MD Anderson Cancer Center's triage algorithm, in which patients deemed to be appropriate surgical candidates for primary debulking surgery first undergo diagnostic laparoscopy for peritoneal disease assessment. The decision to proceed with primary debulking versus biopsies and primary chemotherapy is based upon a predictive score from two independent surgeons. *With permission from the Nature Publishing Group: Nick AM, Coleman RL, Ramirez PT, et al. A framework for a personalized surgical approach to ovarian cancer. Nat Rev Clin Oncol 2015;12(4):239—45.*

research continues to evolve, and in the months and years to come, we will benefit from prognostically accurate -omics signatures in order to tailor our patients' treatment plans, including the timing of surgical cytoreduction. Until that time, treatment for patients with advanced epithelial ovarian cancer should include PDS whenever possible, deferred only in cases of medical or surgical contraindication and when the preoperative assessment is highly predictive of a suboptimal debulking. In those circumstances, the patient should be triaged to NACT, with a plan for interval cytoreductive surgery as soon as clinically feasible. Groups from two US cancer centers and others internationally have implemented treatment algorithms incorporating laparoscopy to triage patients with clinically advanced ovarian carcinoma to PDS versus NACT-IDS [50,58,59] (Fig. 4.2). While some institutions have recommended NACT for any patient in whom a CGR is felt to be unlikely, these authors feel that there is sufficient evidence to support upfront surgical cytoreduction when an *optimal* PDS is likely, even when the degree of optimization is yet unknown. Efforts are ongoing to bring to fruition a definitive phase III randomized controlled trial in the United States to study PDS versus NACT in advanced-stage ovarian cancer. Until then, we must continue to usher these patients to centers that will provide them with maximum surgical effort whenever the decision is made to proceed to the operating room [60,61].

References

[1] Siegel RL, Miller KD, Jemal A. Cancer statistics, 2015. CA Cancer J Clin 2015;65(1):5–29.

[2] Omura GA, Bundy BN, Berek JS, et al. Randomized trial of cyclophosphamide plus cisplatin with or without doxorubicin in ovarian carcinoma: a GOG study. J Clin Oncol 1989;7:457–65.

[3] Ozols RF, Bundy BN, Greer BE, et al. Phase III trial of carboplatin and paclitaxel compared with cisplatin and paclitaxel in patients with optimally resected stage III ovarian cancer: a GOG study. J Clin Oncol 2003;21(17):3194–200.

[4] Armstrong DK, Bundy B, Wenzel L, et al. Intraperitoneal cisplatin and paclitaxel in ovarian cancer: a GOG study. N Engl J Med 2006;354:34–43.

[5] Griffiths CT. Surgical resection of tumor bulk in the primary treatment of ovarian carcinoma. Natl Cancer Inst Monogr 1975;42:101–4.

[6] Hoskins WJ, McGuire WP, Brady MF, et al. The effect of diameter of largest residual disease on survival after primary cytoreductive surgery in patients with suboptimal residual epithelial ovarian carcinoma. Am J Obstet Gynecol 1994;170:974–9.

[7] Chi DS, Liao JB, Leon LF, et al. Identification of prognostic factors in advanced epithelial ovarian carcinoma. Gynecol Oncol 2001;82:532–7.

[8] Chang SJ, Bristow RE. Evolution of surgical treatment paradigms for advanced-stage ovarian cancer: redefining "optimal" residual disease. Gynecol Oncol 2012;125(2):483–92.

[9] Bristow RE, Tomacruz RS, Armstrong DK, et al. Survival effect of maximal cytoreductive surgery for advanced ovarian carcinoma during the platinum era: a meta-analysis. J Clin Oncol 2002;20:1248–59.

[10] Aletti GD, Dowdy SC, Gostout BS, et al. Aggressive surgical effort and improved survival in advanced-stage ovarian cancer. Obstet Gynecol 2006;107(1):77–85.

[11] Chi DS, Eisenhauer EL, Lang J, et al. What is the optimal goal of primary cytoreductive surgery for bulky stage IIIC epithelial ovarian carcinoma (EOC)? Gynecol Oncol 2006;103:559–64.

[12] Dauplat J, Le Bouedec G, Pomel C, et al. Cytoreductive surgery for advanced stages of ovarian cancer. Semin Surg Oncol 2000;19(1):42–8.

[13] Chi DS, Eisenhauer EL, Zivanovic O, et al. Improved progression-free and overall survival in advanced ovarian cancer as a result of a change in surgical paradigm. Gynecol Oncol 2009;114:26–31.

[14] Eisenhauer EL, Abu-Rustum NR, Sonoda Y, et al. The addition of extensive upper abdominal surgery to achieve optimal cytoreduction improves survival in patients with stages IIIC–IV epithelial ovarian cancer. Gynecol Oncol 2006;103:1083–90.

[15] Juretzka MM, Abu-Rustum NR, Sonoda Y, et al. The impact of video-assisted thoracic surgery (VATS) in patients with suspected advanced ovarian malignancies and pleural effusions. Gynecol Oncol 2007;104:670–4.

[16] Diaz JP, Abu-Rustum NR, Sonoda Y, et al. Video-assisted thoracic surgery (VATS) evaluation of pleural effusions in patients with newly diagnosed advanced ovarian carcinoma can influence the primary management choice for these patients. Gynecol Oncol 2010;116:483–8.

[17] Eitan R, Levine DA, Abu-Rustum N, et al. The clinical significance of malignant pleural effusions in patients with optimally debulked ovarian carcinoma. Cancer 2005;103(7):1397–401.

[18] Mironov O, Ishill NM, Mironov S, et al. Pleural effusion detected at CT prior to primary cytoreduction for stage III or IV ovarian carcinoma: effect on survival. Radiology 2011;258(3):776–84.

[19] Hoffman MS, Zervose E. Colon resection for ovarian cancer: intraoperative decisions. Gynecol Oncol 2008;111(2 Suppl.):S56–65.

[20] Shimada M, Kigawa J, Minagawa Y, et al. Significance of cytoreductive surgery including bowel resection for patients with advanced ovarian cancer. Am J Clin Oncol 1999;22(5):481–4.

[21] Mourton SM, Temple LK, Abu-Rustum NR, et al. Morbidity of rectosigmoid resection and primary anastomosis in patients undergoing primary cytoreductive surgery for advanced epithelial ovarian cancer. Gynecol Oncol 2005;99(3):608–14.

[22] Tebes SJ, Cardosi R, Hoffman MS. Colorectal resection in patients with ovarian and primary peritoneal carcinoma. Am J Obstet Gynecol 2006;195(2):585–9.

[23] Aletti GD, Podratz KC, Jones MB, et al. Role of rectosigmoidectomy and stripping of pelvic peritoneum in outcomes of patients with advanced ovarian cancer. J Am Coll Surg 2006;203(4):521–6.

[24] Park JY, Seo SS, Kang S, et al. The benefits of low anterior en bloc resection as part of cytoreductive surgery for advanced primary and recurrent epithelial ovarian cancer patients outweigh morbidity concerns. Gynecol Oncol 2006;103(3):977–84.

[25] Estes JM, Leath CA, Straughn Jr JM, et al. Bowel resection at the time of primary debulking for epithelial ovarian carcinoma: outcomes in patients treated with platinum and taxane-based chemotherapy. J Am Coll Surg 2006;203(4):527–32.

[26] Kalogera E, Dowdy SC, Mariani A, et al. Multiple large bowel resections: potential risk factor for anastomotic leak. Gynecol Oncol 2013;130(1):213–18.

[27] Omura GA, Blessing JA, Ehrlich CE, et al. A randomized trial of cyclophosphamide and doxorubicin with or without cisplatin in advanced ovarian carcinoma: a GOG study. Cancer 1986;57:1725–30.

[28] Omura GA, Brady MF, Homesley HD, et al. Long-term follow-up and prognostic factor analysis in advanced ovarian carcinoma: the GOG experience. J Clin Oncol 1991;9:1138–50.

[29] McGuire WP, Hoskins WJ, Brady MF, et al. Cyclophosphamide and cisplatin compared with paclitaxel and cisplatin in patients with stage III and stage IV ovarian cancer. N Engl J Med 1996;334:1–6.

[30] Muggia FM, Braly PS, Brady MF, et al. Phase III randomized study of cisplatin versus paclitaxel versus cisplatin and paclitaxel in patients with suboptimal stage III or IV ovarian cancer: a GOG study. J Clin Oncol 2000;18(1):106–15.

[31] Alberts DS, Liu PY, Hannigan EV, et al. Intraperitoneal cisplatin plus intravenous cyclophosphamide versus intravenous cisplatin plus intravenous cyclophosphamide for stage III ovarian cancer. N Engl J Med 1996;335:1950–5.

[32] Markman M, Bundy BN, Alberts DS, et al. Phase III trial of standard-dose intravenous cisplatin plus paclitaxel versus moderately high-dose carboplatin followed by intravenous paclitaxel and intraperitoneal cisplatin in small-volume stage III ovarian carcinoma: an intergroup study of the Gynecologic Oncology Group, Southwestern Oncology Group, and Eastern Cooperative Oncology Group. J Clin Oncol 2001;19(4):1001–7.

[33] Barlin JN, Dao F, Bou Zgheib N, et al. Progression-free and overall survival of a modified outpatient regimen of primary intravenous/intraperitoneal paclitaxel and intraperitoneal cisplatin in ovarian, fallopian tube and primary peritoneal carcinoma. Gynecol Oncol 2012;125(3):621–4.

[34] Skipper HE. Thoughts on cancer chemotherapy and combination modality therapy. JAMA 1974;230:1033–5.

[35] Goldie JH, Coldman AJ. A mathematic model for relating the drug sensitivity of tumors to their spontaneous mutation rate. Cancer Treat Rep 1979;63:1727–33.

[36] Van der Burg ME, van Lent M, Buyse M, et al. The effect of debulking surgery after induction chemotherapy on the prognosis in advanced epithelial ovarian cancer. N Engl J Med 1995;332(10):629–34.

[37] Rose PG, Nerenstone S, Brady M, et al. Secondary surgical cytoreduction in advanced ovarian carcinoma: a GOG study. N Engl J Med 2004;351(24):2489–97.

[38] Bristow R, Chi DS. Platinum-based neoadjuvant chemotherapy and interval surgical cytoreduction for advanced ovarian cancer: a meta-analysis. Gynecol Oncol 2006;103(3):1070–6.

[39] Ren Y, Shi T, Jiang R, et al. Multiple cycles of neoadjuvant chemotherapy associated with poor survival in bulky stage IIIC and IV ovarian cancer. Int J Gynecol Cancer 2015;25(8):1398–404.

[40] Rauh-Hain JA, Nitschmann CC, Worley Jr MJ, et al. Platinum resistance after neoadjuvant chemotherapy compared to primary surgery in patients with advanced epithelial ovarian carcinoma. Gynecol Oncol 2013;129(1):63–8.

[41] Lim MC, Song YJ, Seo SS, et al. Residual cancer stem cells after interval cytoreductive surgery following neoadjuvant chemotherapy could result in poor treatment outcomes for ovarian cancer. Onkologie 2010;33(6):324–30.

[42] Vergote I, Trope CG, Amant F, et al. Neoadjuvant chemotherapy or primary surgery in stage IIIC or IV ovarian cancer. N Engl J Med 2010;363(10):943–53.

[43] Dewdney SB, Rimel BJ, Reinhart AJ, et al. The role of neoadjuvant chemotherapy in the management of patients with advanced stage ovarian cancer: survey results from members of the Society of Gynecologic Oncologists. Gynecol Oncol 2010;119:18–21.

[44] Chi DS, Musa F, Dao F, et al. An analysis of patients with bulky advanced stage ovarian, tubal and peritoneal carcinoma treated with primary debulking surgery (PDS) during an identical time period as the randomized EORTC-NCIC trial of PDS vs neoadjuvant chemotherapy (NACT). Gynecol Oncol 2012;124:10–14.

[45] Rauh-Hain JA, Rodriguez N, Growdon WB, et al. Primary debulking surgery versus neoadjuvant chemotherapy in stage IV ovarian cancer. Ann Surg Oncol 2012;19(3):959–65.

[46] Taskin S, Gungor M, Ortac F, et al. Neoadjuvant chemotherapy equalizes the optimal cytore-duction rate to primary surgery without improving survival in advanced ovarian cancer. Arch Gynecol Obstet 2013;288(6):1399–403.

[47] Bian C, Yao K, Li L, et al. Primary debulking surgery vs. neoadjuvant chemotherapy followed by interval debulking surgery for patients with advanced ovarian cancer. Arch Gynecol Obstet 2015 Epub ahead of print.

[48] Kehoe S, Hook J, Nankivell M, et al. Primary chemotherapy versus primary surgery for newly diagnosed advanced ovarian cancer (CHORUS): an open-label, randomised, con-trolled, non-inferiority trial. Lancet 2015;386(9990):249–57.

[49] Vergote I, du Bois A, Amant F, et al. Neoadjuvant chemotherapy in advanced ovarian cancer: on what do we agree and disagree? Gynecol Oncol 2013;128(1):6–11.

[50] Suidan RS, Ramirez PT, Sarasohn DM, et al. A multicenter prospective trial evaluating the ability of preoperative computed tomography scan and serum CA-125 to predict subopti-mal cytoreduction at primary debulking surgery for advanced ovarian, fallopian tube, and peritoneal cancer. Gynecol Oncol 2014;134(3):455–61.

[51] Paik S, Shak S, Tang G, et al. A multigene assay to predict recurrence of tamoxifen-treated, node-negative breast cancer. N Engl J Med 2004;351(27):2817–26.

[52] van de Vijver MJ, He YD, van't Veer LJ, et al. A gene-expression signature as a predictor of survival in breast cancer. N Engl J Med 2002;347(25):1999–2009.

[53] Network CGAR Integrated genomic analyses of ovarian carcinoma. Nature 2011; 474(7353):609–15.

[54] Mankoo PK, Shen R, Schultz N, et al. Time to recurrence and survival in serous ovarian tumors predicted from integrated genomic profiles. PLoS ONE 2011;6(11):e24709.

[55] Sung CO, Song IH, Sohn I. A distinctive ovarian cancer molecular subgroup characterized by poor prognosis and somatic focal copy number alterations at chromosome 19. Gynecol Oncol 2014;132(2):343–50.

[56] Verhaak RG, Tamayo P, Yang JY, et al. Prognostically relevant gene signatures of high-grade serous ovarian carcinoma. J Clin Invest 2013;123(1):517–25.

[57] Liu Z, Beach JA, Agadjanian H, et al. Suboptimal cytoreduction in ovarian carcinoma is associated with molecular pathways characteristic of increased stromal activation. Gynecol Oncol 2015. http://dx.doi.org/10.1016/j.ygyno.2015.08.026.

[58] Petrillo M, Vizzielli G, Fanfani F, et al. Definition of a dynamic laparoscopic model for the prediction of incomplete cytoreduction in advanced epithelial ovarian cancer: proof of a concept. Gynecol Oncol 2015;139(1):5–9.

[59] Nick AM, Coleman RL, Ramirez PT, et al. A framework for a personalized surgical approach to ovarian cancer. Nat Rev Clin Oncol 2015;12(4):239–45.

[60] Bristow RE, Palis BE, Chi DS, et al. The National Cancer Database report on advanced-stage epithelial ovarian cancer: impact of hospital surgical case volume on overall survival and surgical treatment paradigm. Gynecol Oncol 2010;118:262–7.

[61] Bristow RE, Chang J, Ziogas A, et al. High volume ovarian cancer care: survival impact and disparities in access for advanced-stage disease. Gynecol Oncol 2014;132(2):403–10.

CHAPTER 5

Angiogenesis

A.A. Secord and S. Siamakpour-Reihani
Duke University Medical Center, Durham, NC, United States

CONTENTS

INTRODUCTION

Angiogenesis, the development of new blood vessels from preexisting vessels, is a complex multifaceted process that is essential for continued primary tumor growth, facilitation of metastasis, subsequent support for metastatic tumor growth, and cancer progression. Angiogenesis is regulated by both pro- and antiangiogenic factors and is promoted during tumor development when an imbalance in these factors confers an "angiogenic switch" favoring a proangiogenic milieu [1]. In addition to angiogenesis, several different types of vascularization exist such as vessel cooption, vascular mimicry, intussusception or vessel splitting, endothelial cell differentiation, and vasculogenesis [2]. This chapter will focus on tumor angiogenesis via vascular sprouting and current research that has uncovered the incredible complexity of this process which involves numerous signaling pathways and tumor microenvironment interactions. The contemporary findings have led to new molecular insights understanding the development of resistance to antiangiogenic therapy, and identified new potential biomarkers and molecular targets. The translational advances with antiangiogenic therapies in clinical trials as well as future directions in gynecologic cancer will be explored.

ANGIOGENESIS

In 1971, Folkman proposed that tumor angiogenesis was essential to provide essential nutrients beyond the limit of simple diffusion and to allow for growth >2 mm [3]. Angiogenesis via vessel sprouting is the most studied model of neovasculature. Essentially, blood vessels are composed of endothelial cells (ECs) that form the inner lining of the vessel wall and the pericytes, or perivascular cells that interact with the ECs and envelop the surfaces of the

Translational Advances in Gynecologic Cancers. DOI: http://dx.doi.org/10.1016/B978-0-12-803741-6.00005-7

vascular tubes [2]. ECs are activated by a variety of proangiogenic cytokines in concert with simultaneous remodeling of cell–cell junctions and the extracellular matrix (ECM), and pericyte detachment. There are many pathways involved including the VEGF (vascular endothelial growth factor), platelet-derived growth factor (PDGF), and fibroblast growth factor (FGF) signaling pathways, however, the VEGF pathway plays the most prominent role is the multistep process.

The VEGF family includes seven different growth factors, (VEGF A–E) and placental growth factor (PlGF 1–2). VEGFA is the predominant isoform and will be referred to as VEGF [4]. The VEGF ligands can bind to three different transmembrane receptor tyrosine kinases, the VEGF receptors (VEGFR 1–3). VEGFA binds to both VEGFR1 and VEGFR2, while VEGFB binds to VEGFR1, and VEGFE to VEGFR2. VEGFC binds to VEGFR3, and weakly to VEGFR2, and plays a role in tip cell formation as well as lymphangiogenesis. VEGFD also stimulates lymphangiogenesis via binding to VEGFR3. PlGF promotes M2-polarized macrophages, which release proangiogenic factors [2]. VEGF can be induced by numerous growth factor signaling pathways, which are activated by tumor and stromal-induced growth factors, hypoxic conditions in the tumor microenvironment, and may be regulated the oncogenic and tumor suppressor pathways [5].

The VEGF pathway is also joined by other key modulators of angiogenesis such as the NOTCH/DLL4 pathway, the Angiopoietin (Ang)/Tie2 pathway, and the axonal guidance signaling pathways as well as multiple secondary pathways [2]. The ECM also plays a critical role in tumor angiogenesis, metastasis, invasion, and growth.

VESSEL SPROUTING MODEL

The vessel sprouting model involves the formation and migration of specific ECs known as tip and stalk cells. The model described is based on physiologic angiogenesis. Tip cells are present at the leading edge of the sprout and express high levels of VEGFR2, VEGFR3, and PDGF-B. The tip cells are characterized by filopodia, cytoplasmic projections that extend beyond the leading edge of the cells, and are a feature of migrating cells. Stalk cells proliferate and line the developing lumen [2,6]. Essential initial steps also include ECM remodeling, loosing of EC junctions, and pericyte detachment (Fig. 5.1) [2]. These initial steps are followed by a cascade of events including collective EC migration, tip cell guidance/adherence, stalk elongation, lumen formation, and pericyte recruitment mediated by release of angiogenic factors from the tumor microenvironment including the extracellular matrix, macrophages, and myeloid cells (Fig. 5.1). The pericytes are critical players in angiogenesis and are associated with the regulation of blood vessel diameter, vessel

permeability, and endothelial cell proliferation. PDGF-B/PDGFR-β, Ang1/
Tie2, and transforming growth factor-β (TGF-β) are the prominent signaling
pathways for pericyte recruitment [2].

VEGF promotes the migration of filopodial extensions from ECs at the edge
of the vascular sprouts and stimulates tip cell migration via VEGFR2. Stalk
cells, adjacent to the tip, express high levels of VEGFR1, proliferate, and then
elongate from the sprout to form a lumen that is stabilized by pericytes. The
tip cells from adjacent sprouts eventually join and form a vascular branch.
With perfusion the ECs become quiescent phalanx cells, the basement membrane is reformed, and ECs are covered by mature pericytes [2,6].

VEGF in coordination with Notch-DLL4 synchronizes the differentiation
of ECs into tip and stalk cells via a VEGF-feedback loop. VEGFR2 activated
ECs compete for the tip position by increasing their expression of DLL4 in
response to VEGF. DLL4 is produced by tumor ECs, regulated by cellular
and matrix components, and is required for normal vascularization. Upon
release DLL4 binds to and activates the Notch receptors on adjacent stalk
ECs resulting in a reduction of VEGFR2 and -3 expression and induction of
VEGFR1 on the stalk cells. The lack of Notch signaling on the tip cells maintains high VEGFR2 expression and sensitivity to VEGF. The ECs with the highest VEGFR2/VEGFR1 ratio migrate to the tip position. The feedback loop
ensures coordinated vascular sprouting. Notch and DLL4 blockade leads to
vessel hyperbranching and hypoperfused tumor vasculature [2,6]. VEGFC
and VEGFR3, once thought to be involved only in lymphangiogenesis, are
now recognized to have a role in tumor angiogenesis. The interplay among
VEGFR2, VEGFR3, and Notch signaling is very complex and there are conflicting findings regarding the interaction. It appears that VEGFR3 has both
active VEGFC ligand-dependent and passive ligand-independent signaling functions. Matrix-dependent kinases activate VEGFR3, in turn activating
Notch. Activated Notch can downregulate expression of VEGFR3, while low
Notch signaling can stimulate VEGFR3-induced angiogenesis independent of
VEGFR2 and VEGFR3 ligand-independent activities. VEGFC can also activate
Notch target genes and decreases VEGF sensitivity, allowing for vascular loop
assembly. At points of sprout fusion, VEGFR3 signaling converts tip cells to
stalk cells allowing for opposing branch anastomoses [2].

The Ang1/Ang2/Tie2 signaling pathway is integral to angiogenesis. Tie2 is a
tyrosine kinase receptor (TKR) that is expressed on stalk and phalanx ECs,
and serves as the receptor for angiopoietins as well as an endothelial receptor protein tyrosine phosphatase, vascular endothelial protein tyrosine
phosphatase (VE-PTP). Tie2 is constitutively expressed and activated in the
healthy, quiescent vasculature, where it maintains endothelial barrier function [7]. Tie2 function varies based on ligand activation. Expression of Ang1,

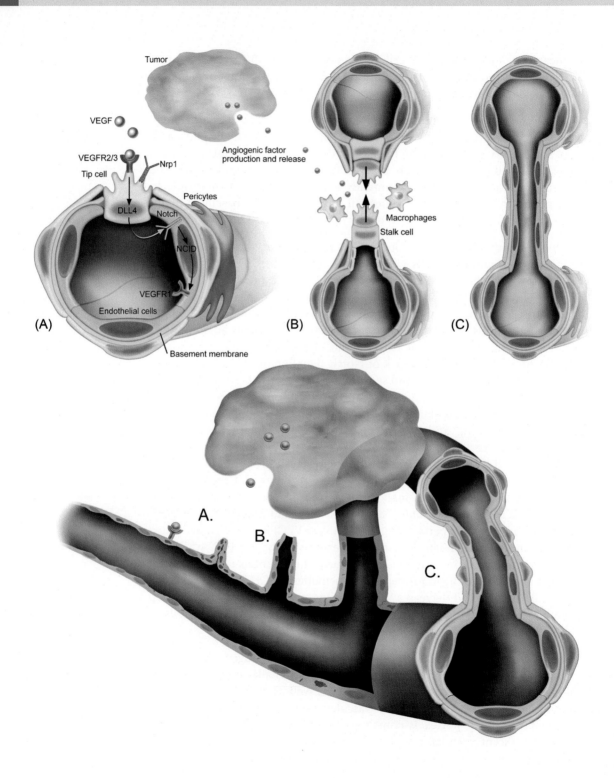

➤ **FIGURE 5.1**

Vascular sprouting model of angiogenesis. (A) Tumor and the surrounding microenvironment release proangiogenic factors that initiate the angiogenic process. ECs at the leading edge of the vascular sprout migrate toward angiogenic signals. Tip cell migration is stimulated by VEGF activation of VEGFR2 signaling which is enhanced by the Nrp1 coreceptor. ECs differentiate into tip cells or the proliferating stalk cell via Notch regulation. ECs with activated VEGFR2 signaling compete for the tip cell position by increasing their expression of DLL4. DLL4 binds to Notch receptors on neighboring ECs, releasing the transcription regulator, Notch intracellular domain (NICD). NCID causes stalk cells to be less responsive to VEGF by increasing stalk cell VEGFR1 expression and decreasing VEGFR2 and Nrp1 expression. (B) Tip cells extend filopodia and migrate forward dependent on the VEGFR1/VEGFR2 ratio. The tip cells need high levels of VEGFR2 expression and migrate toward a VEGF gradient. Basement membrane degradation, EC junction loosening, and pericyte recruitment are also essential in the process. Macrophages act as bridge cells when two tip cells meet and assist with anastomosis by connecting the filopodia of adjacent tip cells, and release angiogenic factors that further stimulate sprouting. (C) Once a connected lumen is formed, blood flow perfuses hypoxic tissue, increases oxygen and nutrient concentrations leading to decreased proangiogenic signals, and established phalanx EC quiescence. Basement membrane deposition, cell–cell junction establishment, and pericytes stabilize the forming vessels. Autocrine and paracrine signaling from ECs and the extracellular matrix by VEGF, angiopoietins, FGF, and Notch maintain the quiescent EC phenotype. *Adapted from Welti J, Loges S, Dimmeler S, Carmeliet P. Recent molecular discoveries in angiogenesis and antiangiogenic therapies in cancer. J Clin Invest 2013;123(8):3190–200.*

a Tie2 ligand and agonist, results in increased vascular maturity, microvessel density, vessel enlargement, and protection against VEGF-induced vascular leak [7]. Ang1 expression on perivascular cells stabilizes and tightens the EC barrier by recruiting Tie2–VE-PTP complexes and preventing VEGFR2-induced internalization of the junctional molecule VE-cadherin [2]. In contrast, Ang2 which is primarily a Tie2 antagonist blocks Ang1-induced vascular stability thus promoting vascular leak and facilitates VEGF-mediated angiogenesis. In addition to stimulating vessel destabilization, Ang2 sensitizes ECs to proangiogenic signals. However, Ang2 can also function as a partial agonist and stimulate Tie2 activation when Ang1 expression is low or absent (in immature pericyte-deprived tumor vessels). Ang2 also can stimulate tip cell migration by activating integrins and promote macrophage association with tumor blood vessels. The interaction between Ang1 and Ang2 is complex and includes complementary yet counterregulatory roles in angiogenesis regulation. Furthermore, VE-PTP further negatively regulates Tie2 signal transduction via dephosphorylation and VE-PTP inhibition enhances TIE2 activity in the presence of either Ang1 or Ang2, thereby effectively turning Ang2 into a potent TIE2 agonist [8].

Guidance molecules have recently been identified to have a role in tumor angiogenesis. Examples include molecules initially identified to be involved in neural development and axon guidance such as semaphorins (SEMAs), ephrins (Eph), netrins, Slit proteins, and their respective receptors neuropilin (NRP1), EphB, UNC5, and Robo1-4 [2]. There are numerous feedback loops and cross-talk interaction. Semaphorins interact with NRP1, a transmembrane receptor for VEGF, and a coreceptor which improves VEGF signaling through VEGFR2 and VEGFR3. The semaphorin family is comprised of several growth factors. Sema3A, Sema6A, and Sema3E have important roles in both neuronal and vascular development. Sema6A regulates VEGFR2 expression and its downstream signaling, while Sema3E maintains the balance of tip/stalk cells by controlling VEGF activity and DLL4 expression [2].

Ephrin and the ephrin receptors are a family of membrane-bound tyrosine kinases and TKRs that are capable of binding to numerous ephrin ligands and critical role in embryologic vascular development and tumor angiogenesis [2,9,10]. Since both receptors and ligands are membrane bound, the receptor ligand interactions are capable of bidirectional signaling [10]. Ephrin-B2 activates Eph receptors in a positive feedback loop that is important for EC morphology (arterial vs venous vessels) and motility. Slit proteins are ligands for the roundabout receptors (Robo). Robo4 can inhibit VEGF signaling and reduce angiogenesis when bound to a netrin receptor, UNC5b. VEGF can also promote the expression of the transcription factor Hlx1, which is expressed on sprouting ECs and maintains the stalk cell phenotype, and results in increased expression of Sema3G [2].

In addition to the importance of tip and stalk EC migration and maturation, the role of the ECM in angiogenesis is crucial. Degradation of the ECM also facilitates tumor cells migration, invasion, and successful metastasis. The ECM is a dynamic complex substance composed of over thousands of proteins including proteoglycans (collagen and elastin) and ECM-associated proteins produced and surrounding by the embedded cells [11]. The ensemble of the ECM and its ECM-associated proteins is referred to as the matrisome [11]. The ECM functions as a cellular structural scaffold, binds to growth factors, modulates intercellular communication, and is continually undergoing remodeling. Alteration in the ECM is essential for physiologic and pathologic angiogenesis. Remodeling of the ECM is influenced by vascular ECs that produce various matrix metalloproteinases (MMPs) and proteases, enzymes which are crucial for ECM degradation. Metalloproteinases include five categories: interstitial collagenases, gelatinases, stromelysins, membrane-type, and elastases which have both activating and inhibitory effects on angiogenesis, regulating the cell environment and signaling [12]. Reversion-inducing cysteine-rich protein with Kazal motifs (RECK) is a glycoprotein that controls the activity of MMPs, regulates ECM remodeling and angiogenesis [13]. ECM remodeling is also very complex and further research detailing the regulation of MMPs and other proteases during vascular development is needed.

TUMOR ANGIOGENESIS

The degree of tumor angiogenesis based on the surrogate microvessel density has been shown to correlate with disease progression and/or survival in epithelial ovarian cancer (EOC) [14], cervical cancer [15], and endometrial cancer [16]. The underlying mechanisms that regulate tumor angiogenesis have yet to be completely understood but most likely involve interactions controlled by convergent and divergent pathways in the tumors and their microenvironment. The process of pathologic angiogenesis, while similar to normal physiologic angiogenesis, varies beyond heighted expression of proangiogenic factors and their ligands. Phenotypically, tumor-related angiogenic vessels are different from normal vascular, and are typically thin-walled and leaky which facilitate entry of tumor cells into the vascular system and metastasis to distant sites. Growth factors and their respective receptors not essential for physiologic vascular development may regulate pathologic angiogenesis and tumor growth. For example, tumor-induced PlGF production by stromal increases pathogenic angiogenesis in leukemic bone marrow and promotes tumor cell proliferation via Nrp1 signaling [2]. Angiogenic growth factors may be dysregulated by tumor-related genetic alterations (inactivated tumor suppressor and activated oncogene pathways) such as *TP53* and *ras*, respectively [5]. Furthermore, the tumor microenvironment, the complex milieu of ECM components and stromal cells including endothelial cells, cancer-associated fibroblasts, and inflammatory cells, plays a key role in angiogenesis [17]. Tumoral microenvironment hypoxic conditions stimulate angiogenesis in a variety of cancers including EOC [18], endometrial [19], and cervical cancers [18]. In summary, tumor angiogenesis is exceedingly complex and driven by molecular and environmentally responsive pathways.

GROWTH FACTOR PATHWAYS IN GYNECOLOGIC CANCERS

The importance of growth factors or chemokines in the process of angiogenesis was described earlier. Growth factors are released by cancer cells as well as noncancer cells in the tumor microenvironment and can act in an autocrine or paracrine manner with multiple feedback loops [2]. Growth factors and their receptors can also mediate tumor cell infiltration of the ECM, and passage into the vasculature, and/or lymphatics and have a role in tumor invasion and migration and ascites formation [20]. The prognostic significance of angiogenic growth factors in gynecologic cancers is reviewed in Table 5.1. Several clinical trials are exploring the antitumor activity of Notch and Ang1/Ang2/Tie2 pathway inhibitors.

Table 5.1 Expression of Angiogenic Growth Factors in Ovarian, Endometrial, and Cervical Cancers

Pathways	Growth Factors	Diseases	Platforms	Findings
VEGF pathway	VEGF	EOC	Serum	High VEGF associated with advanced stage disease, poorly differentiated histology, ascites, and decreased survival [21–24]
			Tumor protein expression	Higher VEGF expression in ovarian cancers compared to benign lesions; increased with advanced stage and associated with worse survival [25]
		Endometrial	Serum	Higher VEGF levels in endometrial cancer patients compared to normal healthy subjects and elevated VEGF significantly associated with advanced stage [26]
		Cervical	Tumor protein expression	Increased incremental VEGF expression from normal cervical epithelium through mild-to-severe CIN to invasive squamous cell cervical carcinoma [27]. VEGF expression correlated with advanced stage, lymphvascular space invasion, parametrial involvement, and lymph node metastasis [28]
	VEGFR2	EOC	Serum	High Ang2/sVEGFR2 ratio was associated with ascites, higher stage, and disease recurrence [29]
	VEGFD, and VEGFR3	Endometrial	Tumor protein expression	Higher VEGFD expression and VEGFR3 expression in cancer compared to normal endometrium. High levels of VEGFD and VEGFR3 significantly associated with myometrial invasion, lymph node metastasis, and worse survival [30]
PDGF pathway	PDGF network	EOC	Tumor gene expression	Higher PDGF pathway activation associated with worse survival [31]
	PDGFBB	EOC	Serum	High PDGFBB associated with worse overall survival [32]
	PDGFD	Endometrial	Tumor protein expression	PDGFD overexpressed in endometrial cancer compared to normal endometrial tissue and associated with nonendometrioid histology, advanced stage, pelvic node metastasis, VEGFA expression, and worse survival [33]
	PDGFRα, PDGFRβ, and PDGFA	Cervical	Tumor protein expression	Expression of the PDGFRα, PDGFRβ, and PDGFA noted in 41.6%, 52.7%, and 60% cancer specimens, respectively. Statistically significant reduction in vitro cellular viability noted with pharmacologic inhibition of PGDF ligands [34]

Pathway	Gene/Marker	Assay	Cancer	Finding
FGF pathway	FGFR1	Tumor gene amplification	EOC	5–7% *FGFR1* amplification [35]
	FGF2	Serum	EOC	Levels greater than 75th quartile associated with worse survival [32]
	FGF2	Serum	Endometrial	Higher FGF2 levels in cancers than in controls. FGF2 levels associated with higher stage, worse disease-free, and overall survival [26]
	FGFR2	Tumor gene mutations	Endometrial	FGFR2 mutations in 10–16% of endometrial cancer cases [35,36]
	FGF1, FGF2, FGFR2	Tumor protein and mRNA expression	Cervical	Protein expression of pathway isoforms reported and strong *FGFR2-IIIc* mRNA expression noted in areas of cancer cell infiltration. 86% of cervical carcinomas express both FGFR2-IIIb and FGFR2-IIIc which are correlated with CIN progression [36]
	FGFR3	Tumor gene mutations	Cervical	FGFR3 mutations in 5% of cervical cancers [35]
Notch pathway	DLL4, Notch1, and Notch3	Tumor protein expression	EOC	Significantly higher expression in EOC compared to normal ovary. DLL4 expression associated with VEGFR1 expression, while Notch1 expression was positively associated with VEGFR2, as well as with the degree of tumor angiogenesis [37]
	DLL4, Notch1, Notch3, and JAG1	Tumor protein expression	Endometrial	Significantly higher expression in endometrial cancer specimens compared to normal endometrial tissue. Tumors highly expressed both Notch1 and JAG1 had a significantly worse prognosis [38]
	DLL4	Tumor protein and mRNA expression	Cervical	*DLL4* mRNA and protein expression was significantly higher in cervical carcinoma specimens compared to normal cervical tissue. High DLL4 protein expression was associated with higher stage, positive lymphvascular space involvement, pelvic node metastasis, increased disease recurrence, and worse overall survival [39]

(Continued)

Table 5.1 Expression of Angiogenic Growth Factors in Ovarian, Endometrial, and Cervical Cancers (Continued)

Pathways	Growth Factors	Diseases	Platforms	Findings
Ang1/2/Tie2 pathway	Ang1/2 and Tie2	EOC	Serum	Levels of Ang1 and 2 were elevated in EOC patients compared to in women with normal ovaries, benign and/or borderline ovarian neoplasms. High Ang2/sVEGFR2 ratio was associated ascites, higher stage, and disease recurrence. Elevated Ang2, Ang2/VEGF, and Ang2/sVEGFR2 ratios were associated with worse recurrence free survival and overall survival [29]
			Tumor gene expression	Low Ang1/Ang2 gene expression ratio associated with higher MVD. Ang1/Ang2 gene expression ratio associated with a poor prognosis in univariate analysis and retained marginal association in multivariate Cox regression analysis [40]
	Ang2	Endometrial	Tumor mRNA expression	Higher *Ang2* expression in poorly differentiated endometrial cancers [41]
	Ang1/2 and Tie2	Cervical	Plasma	Ang1/2, Tie2, and Ang1/Ang2 ratios were significantly higher in cervical cancer patients than in controls [42]
Ephrin pathway	EPHB2 and EPHB4	EOC	Tumor protein, mRNA, and gene expression	Overexpression of *EPHB2* and *EPHB4* receptor associated with worse survival. Strong EPHB4 expression associated with decreased response to chemotherapy [9,43]
		Endometrial	Tumor protein, mRNA, and gene expression	High *EPHB2* and *EPHB4* associated with increased stages, dedifferentiation, myometrial invasion, and worse survival [44]
		Cervical	Tumor protein, mRNA, and gene expression	High *EPHB4* and *EPHB2* expression were associated with worse survival [45]
	EPHB1	EOC	Tumor protein expression	*EPHB1* has been associated with increased tumor angiogenesis, higher disease recurrence, and worse overall survival [46]
	EPHA2	Endometrial	Tumor protein expression	Higher EPHA2 expression in endometrial cancers compared to benign tissue. High EPHA2 expression was associated with increased stage, higher grade, increased depth of myometrial invasion, and worse disease-specific survival [47]
	EPHA2 and EPHA1	Cervical	Tumor protein and mRNA expression	High levels of EPHA2 and moderate-to-high EPHA1 levels are associated with worse overall survival [48]

Ang, angiopoietin; DLL, Delta like ligand; EOC, epithelial ovarian cancer; Eph, Ephrin; FGF, fibroblast growth factor; FGFR, fibroblast growth factor receptor; CIN, cervical intraepithelial neoplasia; Jag, Jagged; MVD, microvessel density; mRNA, messenger RNA.PDGF, platelet-derived growth factor; PDGFR, platelet-derived growth factor receptor; VEGF, vascular endothelial growth factor; VEGFR, vascular endothelial growth factor receptor.

ANTIANGIOGENIC THERAPIES IN CLINICAL TRIALS

There is strong rational for targeting angiogenesis in combination with chemotherapy including the potential to enhance delivery of chemotherapy to tumor, sensitize tumor endothelial cells to chemotherapy, decrease production of endothelial cell growth factors, decrease secretion of soluble tumor growth factors, and independent mechanisms of action. Several randomized clinical trials have demonstrated that combination treatment with chemotherapy and different angiogenic agents significantly improved progression-free survival (PFS), and in some cases improved overall survival (OS), compared with chemotherapy alone in women with EOC [49–51], PFS in endometrial cancer [52–54], and improved OS in women with cervical cancer [55]. A meta-analysis of 12 randomized trials in EOC (bevacizumab, $n = 4$; VEGFR tyrosine kinase inhibitors (TKIs), $n = 6$; trebananib, $n = 2$) demonstrated improved PFS (HR = 0.61; CI = 0.48–0.79; $P < 0.001$) for bevacizumab, VEGFR TKIs (HR = 0.71; CI = 0.59–0.87; $P = 0.001$), and trebananib (HR = 0.67; CI = 0.62–0.72; $P < 0.001$). There was no significant improvement for OS with either bevacizumab or VEGFR TKIs [56].

At this time, there are over 260 interventional clinical trials of antiangiogenic agents in ovarian, endometrial, and cervical cancers listed in the NCI clinical trials.gov (accessed November 25, 2015). Assessment of Notch and Ephrin pathway inhibitors is ongoing in early clinical trial development. An overview of select randomized phase II and III clinical trials of antiangiogenic agents in gynecologic cancer will be reviewed (Tables 5.2 and 5.3).

MONOCLONAL ANTIBODIES

Bevacizumab

Bevacizumab is a recombinant humanized version of the murine antihuman VEGF monoclonal antibody. Promising activity of single agent bevacizumab and 6-month PFS rates in ovarian [73], cervical [74], and endometrial cancers [75] prompted randomized clinical trials of bevacizumab with concurrent chemotherapy. Bevacizumab received Food and Drug Administration (FDA) approval in combination with paclitaxel, pegylated liposomal doxorubicin, or topotecan for the treatment of patients with platinum-resistant recurrent EOC, fallopian tube cancer (FTC), or primary peritoneal cancer (PPC) and in combination with paclitaxel and either cisplatin or topotecan for the treatment of persistent, recurrent, or metastatic cervical cancer. Bevacizumab has received European Medicines Agency (EMEA) approval for these indications as well as for advanced EOC and platinum-sensitive recurrent EOC in combination with chemotherapy.

Table 5.2 Select Phase II/III Clinical Trials of Targeted Antiangiogenic Agents in Epithelial Ovarian Cancer

Agents Trial	Targets	Diseases	Phases (N)	Description of Study	PFS	OS	Adverse Events
Bevacizumab GOG 218 [50] NCT00262847	VEGF	First-line stage III/IV EOC	III (1873)	A. Paclitaxel (175 mg/m^2) + carboplatin (AUC 6) every 3 weeks for 6 cycles—placebo in cycles 2–22 B. 3-weekly chemotherapy for 6 cycles + BEV 15 mg/kg in cycles 2–6–3-weekly placebo in cycles 7–22 C. 3-weekly chemotherapy for 6 cycles—3-weekly BEV 15 mg/kg in cycles 2–22	10.3 vs 11.2 vs 14.1 months C vs A: HR = 0.717; CI = 0.625 to 0.824; $P<0.0001$	39.3 vs 38.7 vs 39.7 months C vs A: HR = 0.915; CI = 0.727–1.152; $P = 0.45$	Grade ≥ 2 HTN 22.9% (C) vs 16.5% (B) vs 7.2% (A)
Bevacizumab ICON 7 [49] NCT00483782		First-line stage IA–IIA clear-cell or grade 3 or with stage IIB–IV EOC	III (1528)	Paclitaxel (175 mg/m^2) + carboplatin (AUC 5–6) every 3 weeks for 6 cycles or the same regimen plus BEV 7.5 mg/kg given concurrently and continued for 12 additional cycles or until progression	19 vs 17.3 months HR = 0.81; CI = 0.70–0.94; $P = 0.004$ Subgroup analysis[a] 16.0 vs 10.5 months HR = 0.7; 95% CI, 0.60–0.93; $P = 0.002$	HR = 0.81; CI = 0.63–1.04; $P = 0.098$ 36.6 vs 28.8 months HR = 0.64; CI = 0.48–0.85; $P = 0.002$	Grade ≥ 2 HTN 18% vs 2% –
Bevacizumab OCEANS [51] NCT00434642		Platinum-sensitive recurrent EOC, FTC, PPC	III (484)	Gemcitabine (1000 mg/ m^2 day 1 and 8) + carboplatin (AUC 4 day 1) with either BEV (15 mg/kg) or placebo every 3 weeks for 6–10 cycles	12.4 vs 8.4 months HR, 0.484; CI, 0.388–0.605; $P<0.0001$	33.3 vs 35.2 months	Grade ≥ 3 HTN (17.4% vs 0.4%) and proteinuria (8.5% vs 0.9%)

Drug/Trial	Population	Regimen	PFS	OS	Safety
Bevacizumab GOG 213 [57] NCT00565851	Platinum-sensitive recurrent EOC, FTC, PPC	Paclitaxel (175 mg/m^2) or gemcitabine (Days 1 and 8) plus carboplatin (AUC 4–5) with or without BEV (15 mg/kg) followed by maintenance BEV until disease progression	13.8 vs 10.4 months HR = 0.61; P<0.0001	42.2 vs 37.3 months HR = 0.83; P = 0.056	No new safety concerns
Bevacizumab AURELIA [54] NCT00976911	Platinum-resistant EOC, FTC, PPC ≤2 prior regimens	The addition of BEV to standard of chemotherapy (PLD [40 mg/m^2 every 4 weeks] or paclitaxel (80 mg/m^2 weekly) or topotecan (4 mg/m^2 on days 1, 8, and 15 every 4 weeks or 1.25 mg/ m^2 on days 1–5 every 3 weeks)	6.7 vs 3.4 months HR = 0.48; CI = 0.38–0.60; P<0.001	16.6 vs 13.3 months HR = 0.85; CI = 0.66–1.08; P<0.17	Grade ≥ 2 HTN (20% vs 7%) and proteinuria (2.0% vs 0%)
Bevacizumab Everolimus GOG186G [58] NCT00886691	Recurrent or persistent EOC, FTC, PPC	BEV 10 mg/kg IV every 2 weeks in 4 week cycles + oral Everolimus (10 mg daily) or placebo	5.9 vs 4.5 months HR = 0.95; CI = 0.66–1.37; P =0.39	16.6 vs 17.3 months HR = 1.16; CI = 0.72–1.87; P = NS	Grade ≥ 2 with Everolimus increased: anemia, neutropenia, oral mucositis, nausea, fatigue
Bevacizumab Erlotinib STAC [59] NCT00520013	Advanced EOC, FTC, PPC and Papillary Serous or Clear Cell Mullerian Tumors	BEV ± erlotinib consolidation chemotherapy after carboplatin, paclitaxel, and BEV	–	–	–
Bevacizumab Fosbretabulin GOG186I [60] NCT01305213	Recurrent or persistent EOC, FTC, PPC	BEV 15 mg/kg + fosbretabulin 60 mg/m^2 IV every 3 weeks	7.3 vs 4.8 months HR = 0.95; CI = 0.47–1.0; P = 0.049	NR vs 21.2 months HR = 1.03; CI = 0.56–1.89; P = 0.94	Grade ≥ 3 HTN 19.6% vs 32.7%

(Continued)

Table 5.2 Select Phase II/III Clinical Trials of Targeted Antiangiogenic Agents in Epithelial Ovarian Cancer (Continued)

Agents Trial	Targets	Diseases	Phases (N)	Description of Study	PFS	OS	Adverse Events
Pazopanib AGO-OVAR 16 [61] NCT00866697	VEGFR-1/2/3, PDGFR-α/β, FGFR-1/3, and c-Kit	Stage II–IV EOC, FTC, and PPC with no evidence of progression after first-line platinum- and taxane-based chemotherapy	III (n = 940)	Pazopanib 800 mg/day versus placebo as maintenance for up to 24 months after first-line chemotherapy	17.9 vs 12.3 HR = 0.77; CI = 0.64–0.91; P = 0.0021	HR = 1.08; CI = 0.87–1.33; P = 0.499	Grade ≥ 3 HTN, liver toxicity, neutropenia, diarrhea
Pazopanib NCT01227928 [62]		Stage II–IV disease after surgery and first-line platinum- and taxane-based chemotherapy	III (n = 145) Limited to Asian patients	Pazopanib 800 mg/day versus placebo as maintenance for up to 24 months after first-line chemotherapy	18.1 vs 18.1 months	–	Grade > 1 HTN (76%), neutropenia (64%), diarrhea (47%), PPE (29%) thrombocytopenia (24%), elevated liver enzymes (22–28%)
Pazopanib MITO-11 [63] NCT01644825		Platinum-resistant or platinum-refractory EOC	II (74)	Weekly paclitaxel 80 mg/m² with or without pazopanib 800 mg daily	6.4 vs 3.5 months HR = 0.42; CI = 0.25–0.69; P = 0.0002	19.1 vs 13.7 months HR = 0.60; CI = 0.32–1.13; P = 0.056	Grade > 3 neutropenia (30% vs 3%), fatigue (11% vs 6%), HTN (8% vs 0%), increased LFTs (8% vs 0%)
Nintedanib NCT00710762 [64]	VEGFR-1/2/3, PDGFR-α/β, FGFR-1/2/3, and SRC family	Recurrent EOC with response to second/ further regimen	II (83)	Nintedanib 250 mg bid vs placebo up to 36 weeks	16.3% vs 5%[b] HR = 0.6; CI = 0.42–1.02; P = 0.06	HR = 0.84; CI = 0.51–1.39; P = 0.51	Grade > 3 hepatotoxicity (51.2% vs 7.5%)
Nintedanib LUME-Ovar 1; AGO-OVAR 12 [65] NCT01015118		First-line stage IIB–IV EOC, FTC, PPC	III (1366)	Paclitaxel (175 mg/m²) + carboplatin (AUC 5 or 6) plus either nintedanib or placebo for six cycles, followed by maintenance daily nintedanib or placebo for a maximum of 120 weeks	17.3 vs 16.6 months HR = 0.84; CI = 0.72–0.98; P = 0.0239	No significant difference	Grade ≥ 3 diarrhea (21% vs 2%), hepatotoxicity, hypertension, and fatigue

Agent/Trial	Target	Phase (n)	Setting	Intervention	PFS	OS	Toxicity
Cediranib ICON 6 [66] NCT00532194	VEGFR-1/2/3	III (456)	Platinum-sensitive recurrent EOC	3-weekly platinum-based chemotherapy + Cediranib 20mg daily vs placebo for 6 cycles followed by daily cediranib vs placebo up to 18 months	9.4 vs 12.6 months HR = 0.57; CI = 0.45–0.74; $P<0.01$	17.6 vs 20.3 months HR, 0.70; $P = 0.04$	HTN, diarrhea, hypothyroidism, hoarseness, bleeding, proteinuria, and fatigue
Cediranib Olaparib NCT01116648 [67]		II (90)	Platinum-sensitive recurrent high-grade serous or endometrioid EOC, or those with germline BRCA1/2 mutations	Olaparib (400mg twice daily) or the combination of cediranib (30mg daily) and olaparib (200mg twice daily)	17.7 vs 9.0 months HR = 0·42; CI = 0.23–0.76; $P = 0·005$ Subgroup analysis: BRCA wild-type or unknown 16.5 vs 5.7 months HR = 0.32; CI = 0.14–0.74; $P = 0.008$	2-year OS: 81% vs 65% Subgroup analysis: BRCA wild-type or unknown –	Grade ≥ 3 more frequent with combination: fatigue (27.3% vs 10.9%), diarrhea (22.7% vs 0%), and HTN (40.9% vs 0%) –
Trebananib Trinova1 [68] NCT01204749	Ang1 and Ang2	III (919)	Recurrent EOC	Weekly paclitaxel (80mg/m²) plus either weekly trebananib (15mg/kg) or placebo	7.2 vs 5.4 months HR = 0.66; CI = 0.57–0.77; $P<0.0001$	19.0 vs 17.3 months HR = =0.86; CI = 0.69–1.08; $P = 0.19$	Any grade edema (64% vs 28%) and AE-related treatment discontinuations (17% vs 6%)

AGO, Arbeitsgemeinschaft Gynaekologische Onkologie; Ang, angiopoietin; AUC, area under the curve; BEV, bevacizumab; BRCA, breast cancer; CI, confidence interval; EOC, epithelial ovarian cancer; FGFR, fibroblast growth factor receptor; FTC, fallopian tube; GOG, Gynecologic Oncology Group; HR, hazard ratio; HTN, hypertension; ICON, International Collaborative Ovarian Neoplasm; KIT, v-kit Hardy-Zuckerman 4 feline sarcoma viral oncogene homolog; MITO, Multicenter Italian Trials in Ovarian cancer and gynecologic malignancies; PPE, palm–plantar erythrodysaethesia syndrome; PDGFR, platelet-derived growth factor receptor; PLD, pegylated liposomal doxorubicin; PPC, primary peritoneal cancer; VEGF, vascular endothelial growth factor; VEGFR, vascular endothelial growth factor receptor.
[a]Subgroup analysis (n=465) high-risk women (stage IV or stage III with residual disease > 1 cm).
[b]36-week PFS.

Table 5.3 Select Phase II/III Clinical Trials of Targeted Antiangiogenic Agents in Uterine and Cervical Malignancies

Agents	Targets	Disease types	Phases (N)	Description of Study	PFS	OS	Adverse Events
Bevacizumab GOG86P [53] NCT00977574	VEGF	Advanced (stage III–IV) or recurrent endometrial cancer	II (349)	A. Carboplatin AUC 6 + Paclitaxel 175 mg/m²+BEV 15 mg/kg every 3 weeks × 6 cycles + BEV 15 mg/kg B. Carboplatin AUC 5 + Paclitaxel 175 mg/m²+TEM 25 mg days 1 and 8 every 3 weeks × 6 cycles + TEM 25 mg IV weekly days 1, 8, and 15 every 3 weeks C. Carboplatin AUC 6 + Ixabepilone 30 mg/m²+BEV 15 mg/kg every 3 weeks × 6 cycles + BEV 15 mg/kg	A. HR = 0.81; CI = 0.63–1.02 B. HR = 1.22; CI = 0.96–1.55 C. HR = 0.87; CI = 0.68–1.11	A. HR = 0.71; CI = 0.55–0.91 B. HR = 0.99; CI = 0.78–1.26 C. HR = 0.97; CI = 0.77–1.23	Grade > 3 HTN (16% BEV vs 3% TEM), pneumonitis and oral mucositis were more common with TEM
Bevacizumab MITO END-2 [52] NCT01770171	VEGF	Advanced (stage III–IV) or recurrent endometrial cancer	II (108)	Carboplatin AUC 5 + Paclitaxel 175 mg/m²+BEV 15 mg/kg every 3 weeks for 6–8 cycles + BEV 15 mg/kg	13.0 vs 8.7 months HR = 0.57; CI = 0.34–0.96; P = 0.036		Grade 3 cardiac toxicity was documented in 4 cases in BEV arm vs no cases in control arm
Bevacizumab GOG240 [55] NCT00803062	VEGF	Recurrent, persistent or metastatic cervical cancer	III (452)	Cisplatin 50 mg/m² + paclitaxel 135 or 175 mg/m² every 3 weeks or topotecan 0.75 mg/m² on days 1–3 + paclitaxel 175 mg/m² on day 1 every 3 weeks + BEV 15 mg/kg every 3 weeks	8.2 vs 5.9 months HR = 0.67; CI = 0.54–0.82; P = 0.002	17.0 vs 13.3 months HR = 0.71; CI = 0.54–0.95; P = 0.004	Grade ≥ 2 HTN (25% vs 2%), GIP or genitourinary fistulas (6% vs 0%), and grade ≥ 3 thromboembolic events (8% vs 1%)

Drug/Trial	Target	Cancer type	Phase (n)	Treatment	Efficacy (PFS)	Efficacy (OS)	AEs/Toxicity
Bevacizumab RTOG0417 [69] NCT00369122	VEGF	Locally advanced cervical cancer	II (49)	BEV (10mg/kg every 2 weeks) for 3 cycles during chemoradiation (WPRT + brachytherapy + weekly cisplatin 40mg/m²)	3-year DFS: 68.7% LRF: 23.2%	3-year OS: 81.3%	Grade 3 and 4 toxicity occurred in 26.5% and 10.2%, respectively. Mostly hematologic toxicity
Pazopanib NCT00430781 [70]	VEGFR-1/2/3, PDGFR-α/β, FGFR-1/3, and c-Kit	Stage IVb, persistent or recurrent cervical cancer	II (152)	A. Lapatinib 1500mg daily B. Pazopanib 800mg daily C. Lapatinib + pazopanib (Arm C closed for futility)	17.1 vs 18.1 weeks HR = 0.66; CI = 0.48–0.91; $P = 0.013$	39.1 vs 50.7 weeks HR = 0.67; CI = 0.46–0.99; $P = 0.045$	Most common Grade 3 AEs: diarrhea (11% for pazopanib and 13% for lapatinib)
Pazopanib PALETTE [71] NCT00753688		Uterine LMS	III (369)	Pazopanib 800mg daily	4.6 vs 1.6 months HR = 0.35; CI = 0.26–0.48; $P<0.001$ LMS Subgroup Analysis HR = 0.37; CI = 0.23–0.60)	12.6 vs 10.7 months HR = 0.87; CI = 0.67–1.12; $P = 0.25$ —	Most common AEs: fatigue, diarrhea, nausea, weight loss, hypertension, decreased appetite, vomiting, tumor pain, hair color changes, musculoskeletal pain, headache, dysgeusia, dyspnea, and skin hypopigmentation
Cediranib CIRCCa [72] NCT01229930	VEGFR-1/2/3	Metastatic or recurrent cervical cancer not amendable to surgery	II (69) —	Paclitaxel 175mg/m² + carboplatin AUC of 5 every 3 weeks for a maximum of 6 cycles + either cediranib 20mg or placebo daily	8.1 vs 6.7 months HR = 0.58; CI = 0.40–0.85; $P = 0.032$	13.6 vs 14.8 months HR = 0.94; CI = 0.65–1.36; $P = 0.42$	Grade 2–3 HTN (34% vs 11%). Grade≥ 3 diarrhea (16% vs 3%) Fatigue (13% vs 6%), neutropenia (31% vs 11%), and febrile neutropenia (16% vs 0%)

Epithelial Ovarian Cancer

In two phase III randomized controlled trials in women with chemotherapy-naïve EOC, the incorporation of bevacizumab with platinum and taxane-based therapy followed by maintenance bevacizumab improved PFS. In GOG 218, a PFS benefit was seen in women with newly diagnosed disease who received both concurrent and maintenance bevacizumab (14.1 vs 10.3 months; HR = 0.717; $P < 0.0001$), however, there was no difference in OS [50]. Similarly in ICON 7, a marginal benefit of PFS was noted in the bevacizumab arm (HR = 0.81; $P = 0.0041$), but in the entire cohort there was no difference in OS. A post hoc exploratory subgroup analysis revealed a significant OS improvement for women with stage IV or stage IIIC disease with >1 cm residual tumor and was later confirmed in the mature analysis (34.5 vs 39.3 months, $P = 0.03$) [49,76]. The optimal treatment duration of bevacizumab treatment (22 vs 44 cycles) is being explored in the BOOST trial (NCT01462890).

The addition of bevacizumab to chemotherapy was evaluated in women with platinum-sensitive and -resistant recurrent EOC. In the phase III OCEANS trial, patients who received concurrent and maintenance bevacizumab had an improved response rate (RR) (78.5% vs 57.4%, $P < 0.0001$), and 4-month improvement in PFS (12.4 vs 8.4 months; HR = 0.484; $P < 0.0001$), but no difference in OS [51]. The phase III GOG 213 trial demonstrated that bevacizumab plus chemotherapy resulted in a nonstatistically significant 5-month improvement in OS compared with chemotherapy alone for women with platinum-sensitive recurrent EOC (42.2 vs 37.3 months; HR = 0.83; $P = 0.056$) [57]. The randomized phase III AURELIA trial demonstrated that the addition of bevacizumab to chemotherapy in platinum-resistant recurrent disease improved median PFS (6.7 vs 3.4 months, HR = 0.48; $P < 0.001$) and RR (27.3% vs 11.8%; $P = 0.001$), and decreased frequency of paracentesis (17% vs 2%) [54]. Moreover, the proportion of patients achieving a 15% improvement in patient-reported abdominal symptoms during chemotherapy was increased in the bevacizumab group (21.9% vs 9.3%; $P = 0.002$) [77].

Ongoing combination trials include the STAC trial bevacizumab ± erlotinib consolidation chemotherapy after carboplatin, paclitaxel, and bevacizumab in advanced EOC (NCT00520013), and bevacizumab in combination with weekly paclitaxel in relapsed ovarian sex-cord stromal tumors in the ALIENOR trial (NCT01770301).

Endometrial Cancer

Two phase II randomized trials evaluating bevacizumab in combination with chemotherapy indicated promising results with improved clinical outcomes

(Table 5.3). GOG 86P, a 3-arm study, incorporated bevacizumab, temsirolimus, and ixabepilone with taxane and platinum-based chemotherapy in women with chemonaïve advanced or recurrent endometrial cancer. Patients were randomized to experimental arms and compared individually to historical control data. OS was statistically significantly increased with bevacizumab, paclitaxel, and carboplatin (34 vs 22.7 months; HR = 0.71; $P <$ 0.039) relative to historical control [53]. The MITO END-2 randomized phase II trial compared paclitaxel and carboplatin with and without bevacizumab in women with advanced or recurrent endometrial cancer. The addition of bevacizumab significantly increased PFS (13 vs 9.7 months; HR = 0.59; $P =$ 0.036) and was associated with nonsignificant increased RR (72.7 vs 54.3%; $P = 0.065$) and OS (23.5 vs 18 months; $P = 0.24$) [52]. Together these randomized phase II trials support further exploration of bevacizumab combination therapy in endometrial cancer in phase III trials.

Cervical Cancer

Based on single agent activity in patients with heavily pretreated, recurrent cervical cancer [78], the GOG 240 phase III trial was developed to compare taxane-based doublets as well as the addition of bevacizumab to chemotherapy [55]. The addition of bevacizumab to chemotherapy significantly improved RR (48% vs 36%; $P = 0.008$), median PFS (8.2 vs 5.9 months; HR = 0.67; $P = 0.002$), and median OS (HR = 0.71; $P = 0.004$) when compared with chemotherapy alone. The most notable bevacizumab-associated toxicity was increase in gastrointestinal perforations or genitourinary fistulas (6% vs 0%; $P = 0.002$) (Table 5.3) [55].

RECEPTOR KINASE INHIBITORS

Several small-molecule inhibitors of the VEGF receptors and other angiogenic pathways have been developed and completed evaluation in phase III clinical trials in gynecologic cancers. The receptor TKIs evaluated are nintedanib, cediranib, and pazopanib. Nintedanib targets VEGFR-1/2/3, PDGFR-α/β, and FGFR-1/2/3 and also has inhibitory activity against FLT3 and the v-src avian sarcoma viral oncogene homolog (Src) family [79]. Pazopanib predominantly targets VEGFR-1/2/3, PDGFR-α/β, and FGFR-1/3 [79]. Cediranib inhibits VEGFR-1/2/3 [80]. Antitumor activity with nintedanib [64], cediranib [80], and pazopanib [81] noted in phase II single agent studies in recurrent EOC prompted further evaluation of these compounds with carboplatin and paclitaxel. The phase III clinical trials have demonstrated a PFS benefit in women with chemonaïve newly diagnosed and platinum-sensitive recurrent ovarian cancer indicating a potentially important role for the use

TKIs targeting angiogenesis pathways in the treatment of EOC (Table 5.2) [65,66]. Pazopanib has also been evaluated in cervical cancer with improved PFS (HR = 0.66; P = 0.013) and OS (HR = 0.67; P = 0.045) compared to lapatinib [70]. Pazopanib is approved for uterine leiomyosarcoma (LMS) (Table 5.3) and the PazoDoble trial is evaluating the combination of gemcitabine + pazopanib in patients with recurrent or metastatic uterine LMS or carcinosarcoma (NCT02203760) [82]. Additional ongoing randomized phase II trials include carboplatin and paclitaxel ± nintedanib in cervix cancer (NCT02009579); and pazopanib combinations gemcitabine (NCT01610206) and topotecan (TOPAZ, NCT01600573) in recurrent EOC.

FUSION PEPTIBODY

Trebananib is a peptide-Fc fusion protein that targets the Ang1/Ang2/Tie2 pathway and inhibits angiogenesis by blocking the interaction between Ang1/Ang2 with the Tie2 receptor. A phase III trial of weekly paclitaxel with AMG 386 or placebo for patients with recurrent EOC, FTC, and PPC revealed marginal improved PFS (7.2 vs 5.4 months, HR = 0.66; P < 0.0001), but no difference in OS [68]. Continued studies are ongoing to further evaluate the activity of AMG 386 in women with recurrent and newly diagnosed ovarian cancer (TRINOVA-3 20101129/ENGOT-ov2, NCT01493505) and endometrial cancer (Table 5.2).

COMBINATION BIOLOGIC THERAPY

Developing rationale combination therapies including antiangiogenic agents with novel emerging agents is ongoing. Promising results using antiangiogenic agents in combination with PARP inhibitors have been reported and trials evaluating the incorporation of immunotherapies are ongoing. One of the most promising combinations reported is the use of olaparib and cediranib in women with recurrent EOC. In a randomized open-label phase II study, women with platinum-sensitive high-grade serous or endometrioid epithelial ovarian cancer, or those with germline BRCA1/2 mutations were randomized to either olaparib alone or in combination with cediranib [67]. Median PFS was significantly longer for combination therapy compared to olaparib alone (17.7 vs 9.0 months; P = 0.005) (Table 5.2). Phase II/III trials evaluating the combination of olaparib and cediranib to standard chemotherapy have been approved in women with platinum-sensitive and resistant recurrent EOC. Pazopanib in combination with fosbretabulin is being explored in recurrent EOC (PAZOFOS, NCT02055690).

RESISTANCE TO ANTIANGIOGENIC THERAPIES

Though cancers may have an initial response to antiangiogenesis agents, resistance to these drugs can ultimately develop. The mechanisms of resistance to anti-VEGF and other antiangiogenic agents are multifactorial and reflect the complexity and redundancy of tumor angiogenic pathways as well as other mechanisms of tumor neovascularization. This phenomenon is unlike the classic mechanisms for resistance against conventional chemotherapeutic agents. Tumor endothelial cells are not prone to mutation, they adapt to VEGF inhibition using secondary signaling pathways (i.e., PDGF, FGF) to recruit vasculature, and rescue tumor angiogenesis when VEGF is blocked, providing a mechanism for resistance to VEGF inhibitors [43]. The MITO-16, MANGO2b study (NCT01802749) is evaluating the continued use of bevacizumab beyond progression in platinum-sensitive EOC based on promising retrospective data regarding bevacizumab beyond progression in colorectal cancer and EOC [83,84]. A phase II study evaluating nintedanib in women with recurrent EOC previously treated with bevacizumab to determine if an agent that simultaneously inhibits both the VEGF pathway and ancillary angiogenic pathways (i.e., FGF, PDGF) may overcome resistance to VEGF blockade. The TAPAZ trial, a phase II trial will evaluate weekly paclitaxel $\pm$ pazopanib in platinum-resistant EOC patients who relapse during bevacizumab maintenance (NCT02383251). Cancers can also exploit other mechanisms of tumor vascularization including vessel cooption, vascular mimicry, intussusception or vessel splitting, endothelial cell differentiation, and vasculogenesis (Fig. 5.2) [2].

BIOMARKERS

There is a critical need to identify novel angiogenic targets for therapeutic development as well as the elaboration of antiangiogenic agents in tandem with biomarkers to direct therapy with the goal to rationally direct therapy and minimize cost and toxicity. Correlative translational studies have identified several promising tumor tissue-based, plasma-based, clinical, and radiologic biomarkers (Table 5.4). However, most of these biomarkers are prognostic and have not been proven to be predictive. Predictive biomarkers differentiate outcome to intervention whereas prognostic biomarkers differentiate overall outcome, such as PFS and OS within a population [12]. Further validation in phase III clinical biomarker-directed trials is needed to determine if promising biomarkers, molecular subtypes, and ascites may direct individualized treatment with antiangiogenic agents in gynecologic cancers.

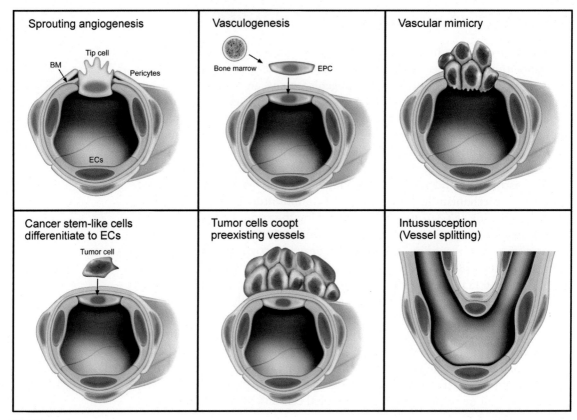

FIGURE 5.2

Other mechanisms of tumor vascularization include vessel cooption, vascular mimicry, intussusception or vessel splitting, endothelial cell differentiation, and vasculogenesis. Vessel cooption refers to the process where tumor cells arise near to or migrates to and grow around preexisting blood vessels. Intussusception occurs when preformed vessel splits into two daughter vessels. Vascular mimicry occurs when tumor cells replace endothelial cells. Vasculogenesis is the development of new blood vessel growth from endothelial progenitor cells (EPCs), which are derived from bone marrow-derived progenitor cells and in response to chemo attractants. The EPCs can differentiate into endothelial cells. Endothelial cell differentiation can also develop from cancer stem-like cells and these ECs are less sensitive to the effects of antiangiogenic therapy. *Adapted from Welti J, Loges S, Dimmeler S, Carmeliet P. Recent molecular discoveries in angiogenesis and antiangiogenic therapies in cancer. J Clin Invest 2013;123(8):3190–200.*

CONCLUSION

Current scientific research has led to an improved understanding of the complex regulation of tumor angiogenesis, development of resistance to antiangiogenic agents, and identification of promising targets and biomarkers. Currently, nine phase III randomized trials including antiangiogenic agents in the treatment of primary and recurrent epithelial ovarian cancers

Table 5.4 Biomarkers Predictive of Outcome to Antiangiogenic Therapy

Agents	Biomarkers	Studies	Results
Bevacizumab [85]	Molecularly defined proangiogenic vs immune subgroup	ICON7	Immune subgroup had worse PFS (HR = 1.73; CI = 1.12–2.68) and OS (HR = 2.00; CI = 1.11–3.61) when treated with BEV compared to chemotherapy alone. Proangiogenic group, there was a nonsignificant trend to improved PFS for the addition of BEV (median 17.4 vs 12.3 months)
Bevacizumab [86]	Molecularly defined subgroups: proliferative and mesenchymal subtypes compared with immunoreactive or differentiated subtypes	ICON7	Patients with serous proliferative subtype carcinomas had greatest benefit from BEV with an improvement of median PFS of 12.8 months ($P = 0.032$). In contrast, median PFS improvements with BEV were not significantly greater in the differentiated, immunoreactive, and mesenchymal subtypes. Patients with mesenchymal serous cancers had improvement in median OS with BEV (HR = 0.27, CI = 0.08–0.96, $P = 0.03$)
Bevacizumab [72]	Serum VEGFR3, α_1-acid glycoprotein, and mesothelin combined with CA125	ICON7	The signature-positive group demonstrated improved median PFS in the BEV arm compared with the control arm (17.9 vs 12.4 months; $P = 0.040$). The signature-negative group had an improved median PFS in the standard chemotherapy arm compared with the BEV arm (36.3 vs 20 months; $P = 0.006$)
Bevacizumab [87]	Serum Ang1 and Tie2	ICON7	Patients with high Ang1 and low Tie2 levels had a significant benefit from BEV (median PFS = 23.0 vs 16.2 months, $P = 0.006$). Conversely, among patients with high Ang1 and high Tie2 levels the median PFS was lower for BEV arm than for control arm (12.8 vs 28.5 months, $P=0.007$); patients with low Ang1 levels had no significant PFS differences associated with treatment regardless of Tie2 levels
Bevacizumab [88]	Tumor CD31 MVD, VEGF, VEGFR2, NRP1, or MET	GOG 218	When comparing BEV throughout vs control, higher CD31 MVD potential predictive value for PFS (>Q3 MVD, HR = 0.38, CI = 0.25–0.58; ≤Q3 MVD, HR = 0.68, CI = 0.54–0.86; $P = 0.018$) and OS (>Q3 MVD, HR = 0.57, CI = 0.39–0.83; ≤Q3 MVD, HR = 1.03, CI = 0.83–1.27; $P = 0.0069$) Tumor VEGF showed potential predictive value BEV throughout vs control with a Q3 cutoff for OS (>Q3 Tumor VEGF, HR = 0.62, CI = 0.43–0.91; ≤Q3 Tumor VEGF, HR = 1.01, CI = 0.82–1.25; $P = 0.023$) No prognostic or predictive association was seen for plasma VEGF or VEGFR2 or tumor VEGFR2, NRP1 or MET [89,90]
Bevacizumab [91]	Ascites	GOG 218	Patients with ascites treated with BEV has improved PFS (HR = 0.72, CI = 0.63–0.83; $P < 0.001$) and OS (HR = 0.82, CI = 0.7–0.96; $P = 0.01$). Patients without ascites did not demonstrate a difference in PFS (HR = 0.77; CI = 0.57–1.04; $P = 0.091$) or OS (HR = 0.88; CI = 0.61–1.28; $P = 0.5$)

Ang, angiopoietin; BEV, bevacizumab; CI, confidence interval; GOG, Gynecologic Oncology Group; HR, hazard ratio; ICON, International Collaborative Ovarian Neoplasm; MET, MNNG HOS transforming gene also known as hepatocyte growth factor receptor; MVD, microvessel density; NRP, neuropilin; OS, overall survival; PFS, progression-free survival; VEGF, vascular endothelial growth factor; VEGFR, vascular endothelial growth factor receptor.

have demonstrated improved PFS [49–51,54,57,61,66,68]. Furthermore, the pivotal phase III trial evaluating bevacizumab in women with advanced or recurrent cervical cancer demonstrated improved PFS and OS [55]. While the addition of antiangiogenic agents has demonstrated improved clinical outcomes in women with gynecologic cancer, identification of molecular biomarkers to rationally direct therapy is needed. Translational research data from pivotal trials have yielded important information about predictive and prognostic biomarkers and will be translated into integral biomarker-directed clinical trials in the future. The current landscape of antiangiogenic therapies continues to expand and randomized clinical trials have demonstrated that therapies targeting the tumor microenvironment can improve survival outcomes in women with select gynecologic cancers.

GLOSSARY

Angiogenesis The development of new blood vessels from preexisting vessels is a complex multifaceted process that is essential for continued primary tumor growth, facilitation of metastasis, subsequent support for metastatic tumor growth, and cancer progression.

Extracellular matrix A dynamic complex substance composed of over thousands of proteins including proteoglycans (collagen and elastin) and associated proteins produced and surrounding by the embedded cells.

Fusion peptibody A scientifically designed biologically active peptides grafted onto the part of an antibody known as the Fc domain.

Lymphangiogenesis Formation of lymphatic vessels from preexisting lymphatic vessels.

Monoclonal antibody A scientifically designed antibodies that specifically binds a certain antigen.

Tumor microenvironment The tumor's cellular environment including surrounding blood vessels, immune cells, connective tissue, signaling molecules, and the extracellular matrix.

Tyrosine kinase receptor Cell surface receptors that can bind to cytokines, growth factors, and hormones and have an important function in normal cellular processes and malignancies.

LIST OF ACRONYMS AND ABBREVIATIONS

AE   adverse event
AGO   Arbeitsgemeinschaft Gynaekologische Onkologie
Ang   angiopoietin
AUC   area under the curve
BEV   bevacizumab
BRCA   breast cancer
CI   confidence interval
CIN   cervical intraepithelial neoplasia
DFS   disease-free survival

DLL	Delta like ligand
EC	endothelial cells
ECM	extracellular matrix
EMEA	European Medicines Agency
EOC	epithelial ovarian cancer
Eph	Ephrin
FDA	Food and Drug Administration
FGF	fibroblast growth factor
FGFR	fibroblast growth factor receptor
FLT3	fms-related tyrosine kinase 3
FTC	fallopian tube
GIP	gastrointestinal perforation
GOG	Gynecologic Oncology Group
HR	hazard ratio
HTN	hypertension
ICON	International Collaborative Ovarian Neoplasm
Jag	Jagged
KIT	v-kit Hardy-Zuckerman 4 feline sarcoma viral oncogene homolog
LMS	leiomyosarcoma
LRF	locoregional failure
MET	MNNG HOS transforming gene also known as hepatocyte growth factor receptor
MITO	Multicenter Italian Trials in Ovarian cancer and gynecologic malignancies
MMP	matrix metalloproteinase
mRNA	messenger RNA
MVD	microvessel density
NRP	neuropilin
OS	overall survival
PARP	poly ADP ribose polymerase
PDGF	platelet-derived growth factor
PDGFR	platelet-derived growth factor receptor
PFS	progression-free survival
PlGF	placental growth factor
PPE	palm-plantar erythrodysaethesia syndrome
PPC	primary peritoneal cancer
RECK	Reversion-inducing cysteine-rich protein with Kazal motifs
Robo	roundabout receptors
RR	response rate
RTOG	Radiation Therapy Oncology Group
SEMA	semaphorin
Src	v-src avian sarcoma viral oncogene homolog
TEM	Temsirolimus
TGF-β	transforming growth factor-β
TKI	tyrosine kinase inhibitor
TKR	tyrosine kinase receptor
VEGF	vascular endothelial growth factor
VEGFR	vascular endothelial growth factor receptor
VE-PTP	vascular endothelial protein tyrosine phosphatase
WPRT	whole pelvic radiation therapy

References

[1] Burger RA. Overview of anti-angiogenic agents in development for ovarian cancer. Gynecol Oncol 2011;121(1):230–8.

[2] Welti J, Loges S, Dimmeler S, Carmeliet P. Recent molecular discoveries in angiogenesis and antiangiogenic therapies in cancer. J Clin Invest 2013;123(8):3190–200.

[3] Folkman J. What is the evidence that tumors are angiogenesis dependent. J Natl Cancer I 1990;82(1):4–6.

[4] Biselli-Chicote PM, Oliveira AR, Pavarino EC, Goloni-Bertollo EM. VEGF gene alternative splicing: pro- and anti-angiogenic isoforms in cancer. J Cancer Res Clin Oncol 2012;138(3):363–70.

[5] Rak J, Yu JL, Klement G, Kerbel RS. Oncogenes and angiogenesis: signaling three-dimensional tumor growth. J Invest Dermatol Symp Proc 2000;5(1):24–33.

[6] Kofler NM, Shawber CJ, Kangsamaksin T, Reed HO, Galatioto J, Kitajewski J. Notch signaling in developmental and tumor angiogenesis. Genes Cancer 2011;2(12):1106–16.

[7] Thurston G, Suri C, Smith K, et al. Leakage-resistant blood vessels in mice transgenically overexpressing angiopoietin-1. Science 1999;286(5449):2511–14.

[8] Yacyshyn OK, Lai PF, Forse K, Teichert-Kuliszewska K, Jurasz P, Stewart DJ. Tyrosine phosphatase beta regulates angiopoietin-Tie2 signaling in human endothelial cells. Angiogenesis 2009;12(1):25–33.

[9] Wu Q, Suo Z, Kristensen GB, Baekelandt M, Nesland JM. The prognostic impact of EphB2/B4 expression on patients with advanced ovarian carcinoma. Gynecol Oncol 2006;102(1):15–21.

[10] Haramis AP, Perrakis A. Selectivity and promiscuity in Eph receptors. Structure 2006;14(2):169–71.

[11] Ricard-Blum S, Vallet SD. Proteases decode the extracellular matrix cryptome. Biochimie 2015.

[12] Romano-Fitzgerald S, De Meritens AB, Secord AA, Kohn EC. Invasion, metastasis, and angiogenesis. In: Barakat R, Berchuck A, Markman M, Randall ME, editors. Principles and practice of gynecologic oncology. 6th ed.; 2013. p. 72–88.

[13] Alexius-Lindgren M, Andersson E, Lindstedt I, Engstrom W. The RECK gene and biological malignancy—its significance in angiogenesis and inhibition of matrix metalloproteinases. Anticancer Res 2014;34(8):3867–73.

[14] Rubatt JM, Darcy KM, Hutson A, et al. Independent prognostic relevance of microvessel density in advanced epithelial ovarian cancer and associations between CD31, CD105, p53 status, and angiogenic marker expression: a Gynecologic Oncology Group study. Gynecol Oncol 2009;112(3):469–74.

[15] Cantu De Leon D, Lopez-Graniel C, Frias Mendivil M, Chanona Vilchis G, Gomez C, De La Garza Salazar J. Significance of microvascular density (MVD) in cervical cancer recurrence. Int J Gynecol Cancer 2003;13(6):856–62.

[16] Ozalp S, Yalcin OT, Acikalin M, Tanir HM, Oner U, Akkoyunlu A. Microvessel density (MVD) as a prognosticator in endometrial carcinoma. Eur J Gynaecol Oncol 2003;24(3–4):305–8.

[17] Vong S, Kalluri R. The role of stromal myofibroblast and extracellular matrix in tumor angiogenesis. Genes Cancer 2011;2(12):1139–45.

[18] Bryant CS, Munkarah AR, Kumar S, et al. Reduction of hypoxia-induced angiogenesis in ovarian cancer cells by inhibition of HIF-1 alpha gene expression. Arch Gynecol Obstet 2010;282(6):677–83.

[19] Makker A, Goel MM. Tumor progression, metastasis and modulators of EMT in endometrioid endometrial carcinoma: an update. Endocr Relat Cancer 2015.

[20] Macciò A, Madeddu C. Inflammation and ovarian cancer. Cytokine 2012;58(2):133–47.

[21] Cooper BC, Ritchie JM, Broghammer CL, et al. Preoperative serum vascular endothelial growth factor levels: significance in ovarian cancer. Clin Cancer Res 2002;8(10):3193–7.

[22] Li L, Wang L, Zhang W, et al. Correlation of serum VEGF levels with clinical stage, therapy efficacy, tumor metastasis and patient survival in ovarian cancer. Anticancer Res 2004;24(3b):1973–9.

[23] Hefler LA, Zeillinger R, Grimm C, et al. Preoperative serum vascular endothelial growth factor as a prognostic parameter in ovarian cancer. Gynecol Oncol 2006;103(2):512–17.

[24] Han ES, Burger RA, Darcy KM, et al. Predictive and prognostic angiogenic markers in a gynecologic oncology group phase II trial of bevacizumab in recurrent and persistent ovarian or peritoneal cancer. Gynecol Oncol 2010;119(3):484–90.

[25] Kassim SK, El-Salahy EM, Fayed ST, et al. Vascular endothelial growth factor and interleukin-8 are associated with poor prognosis in epithelial ovarian cancer patients. Clin Biochem 2004;37(5):363–9.

[26] Dobrzycka B, Mackowiak-Matejczyk B, Kinalski M, Terlikowski SJ. Pretreatment serum levels of bFGF and VEGF and its clinical significance in endometrial carcinoma. Gynecol Oncol 2013;128(3):454–60.

[27] Dobbs SP, Hewett PW, Johnson IR, Carmichael J, Murray JC. Angiogenesis is associated with vascular endothelial growth factor expression in cervical intraepithelial neoplasia. Br J Cancer 1997;76(11):1410–15.

[28] Cheng WF, Chen CA, Lee CN, Chen TM, Hsieh FJ, Hsieh CY. Vascular endothelial growth factor in cervical carcinoma. Obstet Gynecol 1999;93(5):761–5.

[29] Sallinen H, Heikura T, Koponen J, et al. Serum angiopoietin-2 and soluble VEGFR-2 levels predict malignancy of ovarian neoplasm and poor prognosis in epithelial ovarian cancer. BMC Cancer 2014:14.

[30] Yokoyama Y, Charnock-Jones DS, Licence D, et al. Expression of vascular endothelial growth factor (VEGF)-D and its receptor, VEGF receptor 3, as a prognostic factor in endometrial carcinoma. Clin Cancer Res 2003;9(4):1361–9.

[31] Ben-Hamo R, Efroni S. Biomarker robustness reveals the PDGF network as driving disease outcome in ovarian cancer patients in multiple studies. BMC Syst Biol 2012;6:3.

[32] Madsen CV, Steffensen KD, Olsen DA, et al. Serial measurements of serum PDGF-AA, PDGF-BB, FGF2, and VEGF in multiresistant ovarian cancer patients treated with bevacizumab. J Ovarian Res 2012;5(1):23.

[33] Ding J, Li XM, Liu SL, Zhang Y, Li T. Overexpression of platelet-derived growth factor-D as a poor prognosticator in endometrial cancer. Asian Pacific J Cancer Prevent 2014;15(8):3741–5.

[34] Taja-Chayeb L, Chavez-Blanco A, Martinez-Tlahuel J, et al. Expression of platelet derived growth factor family members and the potential role of imatinib mesylate for cervical cancer. Cancer Cell Int 2006;6:22.

[35] Touat M, Ileana E, Postel-Vinay S, André F, Soria J-C. Targeting FGFR signaling in cancer. Clin Cancer Res 2015;21(12):2684–94.

[36] Fearon AE, Gould CR, Grose RP. FGFR signalling in women's cancers. Int J Biochem Cell Biol 2013;45(12):2832–42.

[37] Wang H, Huang X, Zhang J, et al. The expression of VEGF and Dll4/Notch pathway molecules in ovarian cancer. Clin Chim Acta 2014;436:243–8.

[38] Mitsuhashi Y, Horiuchi A, Miyamoto T, Kashima H, Suzuki A, Shiozawa T. Prognostic significance of Notch signalling molecules and their involvement in the invasiveness of endometrial carcinoma cells. Histopathology 2012;60(5):826–37.

[39] Yang S, Liu Y, Xia B, et al. DLL4 as a predictor of pelvic lymph node metastasis and a novel prognostic biomarker in patients with early-stage cervical cancer. Tumour Biol 2015.

[40] Hata K, Nakayama K, Fujiwaki R, Katabuchi H, Okamura H, Miyazaki K. Expression of the angopoietin-1, angopoietin-2, Tie2, and vascular endothelial growth factor gene in epithelial ovarian cancer. Gynecol Oncol 2004;93(1):215–22.

[41] Holland CM, Day K, Evans A, Smith SK. Expression of the VEGF and angiopoietin genes in endometrial atypical hyperplasia and endometrial cancer. Br J Cancer 2003;89(5):891–8.

[42] Kopczynska E, Makarewicz R, Biedka M, Kaczmarczyk A, Kardymowicz H, Tyrakowski T. Plasma concentration of angiopoietin-1, angiopoietin-2 and Tie-2 in cervical cancer. Eur J Gynaecol Oncol 2009;30(6):646–9.

[43] Siamakpour-Reihani S, Owzar K, Jiang C, et al. Prognostic significance of differential expression of angiogenic genes in women with high-grade serous ovarian carcinoma. Gynecol Oncol 2015;139(1):23–9.

[44] Alam SM, Fujimoto J, Jahan I, Sato E, Tamaya T. Overexpression of ephrinB2 and EphB4 in tumor advancement of uterine endometrial cancers. Ann Oncol 2007;18(3):485–90.

[45] Alam SM, Fujimoto J, Jahan I, Sato E, Tamaya T. Coexpression of EphB4 and ephrinB2 in tumor advancement of uterine cervical cancers. Gynecol Oncol 2009;114(1):84–8.

[46] Castellvi J, Garcia A, de la Torre J, et al. Ephrin B expression in epithelial ovarian neoplasms correlates with tumor differentiation and angiogenesis. Hum Pathol 2006;37(7):883–9.

[47] Kamat AA, Coffey D, Merritt WM, et al. EphA2 overexpression is associated with lack of hormone receptor expression and poor outcome in endometrial cancer. Cancer 2009;115(12):2684–92.

[48] Wu D, Suo Z, Kristensen GB, et al. Prognostic value of EphA2 and EphrinA-1 in squamous cell cervical carcinoma. Gynecol Oncol 2004;94(2):312–19.

[49] Perren TJ, Swart AM, Pfisterer J, et al. A phase 3 trial of bevacizumab in ovarian cancer. N Engl J Med 2011;365(26):2484–96.

[50] Burger RA, Brady MF, Bookman MA, et al. Incorporation of bevacizumab in the primary treatment of ovarian cancer. N Engl J Med 2011;365(26):2473–83.

[51] Aghajanian C, Blank SV, Goff BA, et al. OCEANS: a randomized, double-blind, placebo-controlled phase iii trial of chemotherapy with or without bevacizumab in patients with platinum-sensitive recurrent epithelial ovarian, primary peritoneal, or fallopian tube cancer. J Clin Oncol 2012;30(17):2039–45.

[52] Lorusso D, Ferrandina G, Colombo N, et al. Randomized phase II trial of carboplatin–paclitaxel (CP) compared to carboplatin–paclitaxel–bevacizumab (CP-B) in advanced (stage III–IV) or recurrent endometrial cancer: the MITO END-2 trial. J Clin Oncol 2015;33 suppl; abstr 5502.

[53] Aghajanian C, Filiaci VL, Dizon DS, et al. A randomized phase II study of paclitaxel/carboplatin/bevacizumab, paclitaxel/carboplatin/temsirolimus and ixabepilone/carboplatin/bevacizumab as initial therapy for measurable stage III or IVA, stage IVB or recurrent endometrial cancer, GOG-86P. J Clin Oncol 2015;33 suppl; abstr 5500.

[54] Pujade-Lauraine E, Hilpert F, Weber B, et al. Bevacizumab combined with chemotherapy for platinum-resistant recurrent ovarian cancer: The AURELIA Open-Label Randomized Phase III Trial. J Clin Oncol 2014;32(13):1302–8.

[55] Tewari KS, Sill MW, Long HJ, et al. Improved survival with bevacizumab in advanced cervical cancer. N Engl J Med 2014;370(8):734–43.

[56] Li X, Zhu S, Hong C, Cai H. Angiogenesis inhibitors for patients with ovarian cancer: a meta-analysis of 12 randomized controlled trials. Curr Med Res Opin 2015:1–21.

[57] Coleman RL BR, Herzog TJ, Brady MF, et al. A phase III randomized controlled clinical trial of carboplatin and paclitaxel alone or in combination with bevacizumab followed by bevacizumab and secondary cytoreductive surgery in platinum-sensitive, recurrent ovarian, peritoneal primary and fallopian tube cancer. In: Society of Gynecologic Oncology's Annual Meeting on Women's Cancer 2015; LBA 1.

[58] Tew WP, Sill M, McMeekin DS, et al. A randomized phase II trial of bevacizumab (BV) plus oral everolimus (EV) versus bevacizumab alone for recurrent or persistent epithelial ovarian (EOC), fallopian tube (FTC), or primary peritoneal cancer (PPC). J Clin Oncol 2014;32:15.

[59] Campos SM PR, Matulonis U, et al. STAC: a randomized phase II trial of avastin or avastin + erlotinib as first line consolidation chemotherapy after standard therapy. Br Gynecol Soc Meet 2008 abstract 93.

[60] Monk B, Sill M, Walker J, et al. Randomized phase 2 evaluation of bevacizumab versus bevacizumab/fosbretabulin in recurrent ovarian, tubal or peritoneal carcinoma: a Gynecologic Oncology Group Study. Int J Gynecol Cancer 2014;24(9):42–3.

[61] Du Bois A, Kristensen G, Ray-Coquard I, et al. Standard first-line chemotherapy with or without nintedanib for advanced ovarian cancer (AGO-OVAR 12): a randomised, double-blind, placebo-controlled phase 3 trial. Lancet Oncol 2015;17(1):78–89.

[62] Zang RY, Wu LY, Zhu JQ, et al. Pazopanib (Paz) monotherapy in Asian women who have not progressed after first-line chemotherapy for advanced ovarian, Fallopian tube, or primary peritoneal carcinoma. J Clin Oncol 2013:31. (suppl; abstr 5512).

[63] Pignata S, Lorusso D, Scambia G, et al. Pazopanib plus weekly paclitaxel versus weekly paclitaxel alone for platinum-resistant or platinum-refractory advanced ovarian cancer (MITO 11): a randomised, open-label, phase 2 trial. Lancet Oncol 2015;16(5):561–8.

[64] Ledermann JA, Hackshaw A, Kaye S, et al. Randomized phase II placebo-controlled trial of maintenance therapy using the oral triple angiokinase inhibitor BIBF 1120 after chemotherapy for relapsed ovarian cancer. J Clin Oncol 2011;29(28):3798–804.

[65] du Bois A, Kristensen G, Ray-Coquard I, et al. Standard first-line chemotherapy with or without nintedanib for advanced ovarian cancer (AGO-OVAR 12): a randomised, double-blind, placebo-controlled phase 3 trial. Lancet Oncol 2015;17(1):78–89.

[66] Ledermann JA, Perren TJ, Raja FA, et al. Randomised double-blind phase III trial of cediranib (AZD 2171) in relapsed platinum sensitive ovarian cancer: results of the ICON6 trial. Eur J Cancer 2013;49:S5–S6.

[67] Liu JF, Barry WT, Birrer M, et al. Combination cediranib and olaparib versus olaparib alone for women with recurrent platinum-sensitive ovarian cancer: a randomised phase 2 study. Lancet Oncol 2014;15(11):1207–14.

[68] Monk BJ, Poveda A, Vergote I, et al. Anti-angiopoietin therapy with trebananib for recurrent ovarian cancer (TRINOVA-1): a randomised, multicentre, double-blind, placebo-controlled phase 3 trial. Lancet Oncol 2014;15(8):799–808.

[69] Schefter T, Winter K, Kwon JS, et al. RTOG 0417: Efficacy of bevacizumab in combination with definitive radiation therapy and cisplatin chemotherapy in untreated patients with locally advanced cervical carcinoma. Int J Radiat Oncol 2014;88(1):101–5.

[70] Monk BJ, Mas Lopez L, Zarba JJ, et al. Phase II, open-label study of pazopanib or lapatinib monotherapy compared with pazopanib plus lapatinib combination therapy in patients with advanced and recurrent cervical cancer. J Clin Oncol 2010;28(22):3562–9.

[71] Van Der Graaf WTA, Blay JY, Chawla SP, et al. Pazopanib for metastatic soft-tissue sarcoma (PALETTE): a randomised, double-blind, placebo-controlled phase 3 trial. Lancet 2012;379(9829):1879–86.

[72] Collinson F, Hutchinson M, Craven RA, et al. Predicting response to bevacizumab in ovarian cancer: a panel of potential biomarkers informing treatment selection. Clin Cancer Res 2013;19(18):5227–39.

[73] Burger RA, Sill MW, Monk BJ, Greer BE, Sorosky JI. Phase II trial of bevacizumab in persistent or recurrent epithelial ovarian cancer or primary peritoneal cancer: a Gynecologic Oncology Group Study. J Clin Oncol 2007;25(33):5165–71.

[74] Monk BJ, Willmott LJ, Sumner DA. Anti-angiogenesis agents in metastatic or recurrent cervical cancer. Gynecol Oncol 2010;116(2):181–6.

[75] Aghajanian C, Sill MW, Darcy KM, et al. Phase II trial of bevacizumab in recurrent or persistent endometrial cancer: a Gynecologic Oncology Group study. J Clin Oncol 2011;29(16):2259–65.

[76] Oza AM, Cook AD, Pfisterer J, et al. Standard chemotherapy with or without bevacizumab for women with newly diagnosed ovarian cancer (ICON7): overall survival results of a phase 3 randomised trial. Lancet Oncol 2015;16(8):928–36.

[77] Stockler MR, Hilpert F, Friedlander M, et al. Patient-reported outcome results from the Open-Label Phase III AURELIA Trial evaluating bevacizumab-containing therapy for platinum-resistant ovarian cancer. J Clin Oncol 2014;32(13):1309–16.

[78] Monk BJ, Sill MW, Burger RA, Gray HJ, Buekers TE, Roman LD. Phase II trial of bevacizumab in the treatment of persistent or recurrent squamous cell carcinoma of the cervix: a Gynecologic Oncology Group Study. J Clin Oncol 2009;27(7):1069–74.

[79] Davidson BA, Secord AA. Profile of pazopanib and its potential in the treatment of epithelial ovarian cancer. Int J Women's Health 2014;6:289–300.

[80] Matulonis UA, Berlin S, Ivy P, et al. Cediranib, an oral inhibitor of vascular endothelial growth factor receptor kinases, is an active drug in recurrent epithelial ovarian, fallopian tube, and peritoneal cancer. J Clin Oncol 2009;27(33):5601–6.

[81] Friedlander M, Hancock KC, Rischin D, et al. A phase II, open-label study evaluating pazopanib in patients with recurrent ovarian cancer. Gynecol Oncol 2010;119(1):32–7.

[82] Gadducci A, Guerrieri ME. Pharmacological treatment for uterine leiomyosarcomas. Expert Opin Pharmacother 2015;16(3):335–46.

[83] Tsutsumi S, Ishibashi K, Uchida N, et al. Phase II trial of chemotherapy plus bevacizumab as second-line therapy for patients with metastatic colorectal cancer that progressed on bevacizumab with chemotherapy: the Gunma Clinical Oncology Group (GCOG) trial 001 SILK study. Oncology 2012;83(3):151–7.

[84] McCann GA, Smith B, Backes FJ, et al. Recurrent ovarian cancer: is there a role for re-treatment with bevacizumab after an initial complete response to a bevacizumab-containing regimen? Gynecol Oncol 2012;127(2):362–6.

[85] Gourley CMA, Perren T, Paul J, et al. Molecular subgroup of high-grade serous ovarian cancer (HGSOC) as a predictor of outcome following bevacizumab. J Clin Oncol 2014;32:5s. (suppl; abstr 5502).

[86] Winterhoff BKS, Oberg AL, Wang C, et al. Bevacizumab and improvement of progression-free survival (PFS) for patients with the mesenchymal molecular subtype of ovarian cancer. J Clin Oncol 2014;32:5s. (suppl; abstr 5509).

[87] Backen A, Renehan AG, Clamp AR, et al. The combination of circulating Ang1 and Tie2 levels predicts progression-free survival advantage in bevacizumab-treated patients with ovarian cancer. Clin Cancer Res 2014;20(17):4549–58.

[88] Birrer MJ, Choi Y, Brady MF, et al. Retrospective analysis of candidate predictive tumor biomarkers (BMs) for efficacy in the GOG-0218 trial evaluating front-line carboplatin-paclitaxel (CP) +/− bevacizumab (BEV) for epithelial ovarian cancer (EOC). J Clin Oncol 2015;33(15) (suppl; abstr 5505).

[89] Birrer MJ CY, Brady MF, Mannel RS, et al. Grp NOGO. Retrospective analysis of candidate predictive tumor biomarkers (BMs) for efficacy in the GOG-0218 trial evaluating front-line carboplatin–paclitaxel (CP) ± bevacizumab (BEV) for epithelial ovarian cancer (EOC). J Clin Oncol 2015:33. (suppl; abstr 5505).

[90] Birrer MJ, Lankes H, Burger RA, et al. Biomarker (Bm) results from Gog-0218, a phase 3 trial of front-line bevacizumab (Bv) plus chemotherapy (Ct) for ovarian cancer (Oc). Ann Oncol 2012;23:81–2.

[91] Ferriss JS, Java JJ, Bookman MA, et al. Ascites predicts treatment benefit of bevacizumab in front-line therapy of advanced epithelial ovarian, fallopian tube and peritoneal cancers: an NRG Oncology/GOG study. Gynecol Oncol 2015;139(1):17–22.

CHAPTER 6

Homologous Recombination and BRCA Genes in Ovarian Cancer: Clinical Perspective of Novel Therapeutics

J.F. Liu and P.A. Konstantinopoulos

Dana-Farber Cancer Institute, Boston, MA, United States

CONTENTS

INTRODUCTION

Homologous recombination (HR) is one of the two major pathways in eukaryotic cells for the repair of double-strand DNA breaks. While mutations in genes that encode for proteins that are critical for HR have been associated with the development of certain malignancies, including breast and ovarian cancers, recent discoveries have highlighted that cells deficient in HR are also vulnerable to therapies directed at other DNA repair pathways. These developments have led to novel therapies in ovarian cancer, including the approval of the poly(ADP-ribose) polymerase (PARP) inhibitor olaparib by the FDA for women with *BRCA*-mutated ovarian cancer who have received three or more prior therapies for their disease.

In this chapter, we will review the role of HR in gynecologic malignancies, particularly ovarian cancer, and the development of new therapeutic strategies targeting cells displaying an HR-deficient (HRD) phenotype.

HR AND DOUBLE-STRAND DNA BREAK REPAIR

Two major pathways have been classically described for the repair of double-strand DNA breaks in eukaryotic cells, HR and nonhomologous end joining (NHEJ), although additional pathways such as alternative end joining (alt-EJ) and single-strand annealing have been described more recently (reviewed in Ref. [1]). Of these pathways, HR is the most conservative in nature, with a lower tendency toward mutagenesis, as it typically utilizes a sister chromatid as the template for repair [2]. Detailed reviews of the HR process may be

Translational Advances in Gynecologic Cancers. DOI: http://dx.doi.org/10.1016/B978-0-12-803741-6.00006-9

found elsewhere [3,4]. In brief, during the HR process, the double-strand break is identified and end-resection occurs, whereby the DNA at the break is resected to single strands by the MRN complex, which includes proteins such as Mre11, RAD50, and NBS1. The exposed single-stranded DNA is then coated with RPA, and the RPA is subsequently replaced by RAD51. Once RAD51 has been loaded onto the DNA, this mediates strand invasion, where the RAD51 nucleoprotein filament can invade the sister chromatid, leading to initiation of DNA synthesis and repair using the sister chromatid as a template. A large number of proteins have been described to be critical to HR, including BRCA1, BRCA2, ATM, ATR, PALB2, RAD51B, RAD51C, and RAD51D [5].

HR AND EPITHELIAL OVARIAN CANCER

Hereditary Risk

BRCA1 and *BRCA2* were discovered in 1994 and 1995, respectively, and described to be linked to hereditary breast and ovarian cancers [6–8]. In epithelial ovarian cancers, germline *BRCA1* or *BRCA2* mutations occur in approximately 10–18% of women with the disease [9–11]. Given these findings, genetic risk evaluation is recommended by the National Comprehensive Cancer Network (NCCN) for all women diagnosed with an epithelial ovarian, fallopian tube, or primary peritoneal cancer [12]. While *BRCA1* and *BRCA2* are the predominant genes associated with a risk of hereditary ovarian cancer, other HR genes may also be associated with a hereditary ovarian cancer risk. A study conducted by Walsh et al. [11] used next-generation sequencing to assess for germline mutations in 21 tumor suppressor genes from the DNA in 360 women with ovarian, fallopian tube, or primary peritoneal cancer and found that 24% of women had germline loss of function mutations. While the majority (18%) of these mutations were in *BRCA1* or *BRCA2*, the other 6% occurred in other genes, including those involved in HR, such as *BARD1*, *BRIP1*, *CHEK2*, *MRE11A*, *PALB2*, *RAD50*, and *RAD51C*. Of note, over 30% of the women with inherited mutations did not have a family history of breast or ovarian carcinoma, and more than 35% were 60 years or older at the time of diagnosis.

HR Pathway Alterations in Epithelial Ovarian Cancer

In 2011, The Cancer Genome Atlas (TCGA) project published a comprehensive genomic and molecular analysis on 489 high-grade serous ovarian cancer tumors (HGSOCs) [10]. One finding of note was that approximately 50% of HGSOCs had a molecular alteration in genes associated with HR

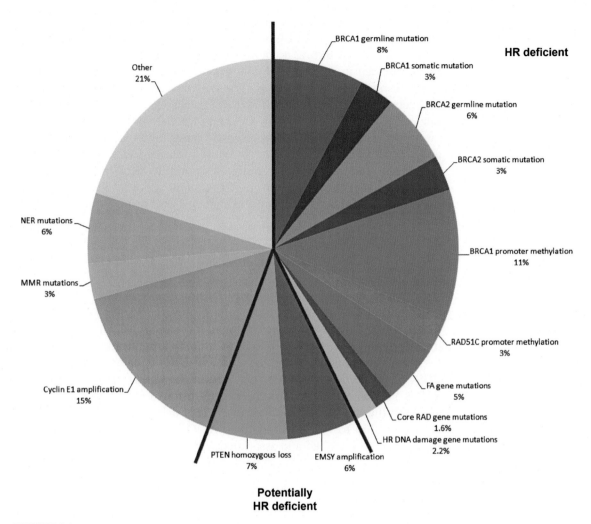

HR deficient

BRCA1 germline mutation
8%

BRCA1 somatic mutation
3%

BRCA2 germline mutation
6%

BRCA2 somatic mutation
3%

BRCA1 promoter methylation
11%

RAD51C promoter methylation
3%

FA gene mutations
5%

Core RAD gene mutations
1.6%

HR DNA damage gene mutations
2.2%

EMSY amplification
6%

PTEN homozygous loss
7%

Cyclin E1 amplification
15%

MMR mutations
3%

NER mutations
6%

Other
21%

**Potentially
HR deficient**

FIGURE 6.1

Alterations in genes involved in the HR pathway are present in approximately 50% of high-grade serous ovarian cancers.

(Fig. 6.1). In 316 cases where DNA could be isolated for whole exome sequencing from both the tumor and matched normal samples, a total of 17% cases had germline mutations in either *BRCA1* or *BRCA2*, consistent with other studies [9,11]. Additionally, another 6% of cases had somatic mutations in either *BRCA1* or *BRCA2*. Of note, in 81% of the cases with *BRCA1* mutation and in 72% of those with *BRCA2* mutation, there was heterozygous loss of the gene, indicating inactivation of both alleles and consistent with Knudson's two-hit hypothesis.

In addition to mutations in *BRCA1/2*, epigenetic silencing of *BRCA1* was observed in 11% of TCGA cases via promoter hypermethylation. While alterations to *BRCA1* or *BRCA2* accounted for over 30% of the HR pathway alterations observed in TCGA, additional genes were also implicated as potential mechanisms for HRD in HGSOC. These included mutations in Fanconi anemia genes such as *PALB2*, *FANCA*, *FANCC*, *FANCI*, and *FANCL* at an overall rate of 5%, core HR RAD genes including *RAD50* and *RAD54L* at an overall rate of 1.6%, and DNA damage response genes involved in HR such as *ATM* and *ATR* at an overall rate of 2.2%. While mutations in *RAD51C*, which has previously been described as a cancer susceptibility gene in breast and ovarian cancers [13], were not seen in TCGA, silencing of *RAD51C* by promoter hypermethylation was seen in approximately 3% of cases.

Further potential alterations in HR were observed in TCGA via amplification of *EMSY* (8%) or homozygous deletions of *PTEN* (7%). *EMSY* binds specifically to the transactivation domain in *BRCA2* and co-localizes with *BRCA2* at sites of DNA damage [14], suggesting that overexpression of *EMSY* may result in inhibition of *BRCA2* activity. Of note, however, *EMSY* amplification has been associated with worse outcomes in ovarian cancer [15] and in vitro functional studies have not demonstrated increased sensitivity to PARP inhibitors or platinum agents in cell lines with *EMSY* amplification [16]; therefore, the significance of *EMSY* amplification and HRD remains unclear. Similarly, *PTEN* has been described to be involved in the transcriptional regulation of *RAD51* [17] and *PTEN* loss has been demonstrated to compromise HR [18], although the functional significance of *PTEN* deletion in ovarian cancer and its relationship to HR remains unknown. Of interest, while HRD has been most robustly described and associated with HGSOC, alterations in HR have also been described in other epithelial ovarian cancer histologies, including clear cell, endometrioid, and carcinosarcomas [19].

Prognostic Impact of HRD in Ovarian Cancer

Several studies have now demonstrated that patients with ovarian cancer and germline *BRCA1* or *BRCA2* mutation have improved clinical outcomes compared to those without germline mutations [9,10,20,21]. Interestingly, this improvement in survival appears to be more pronounced in women with *BRCA2* mutations [22,23]. In part, these improved outcomes may be related to the prolonged platinum sensitivity observed in women with BRCA-related ovarian cancers [20]. Table 6.1 outlines some of the clinical characteristics of *BRCA1/2*-mutated ovarian cancers. The improved survival observed in BRCA-mutated ovarian cancers likely extends to ovarian cancers with other defects in HR; in a study of 390 cases, one group found that survival for women with non-*BRCA* HR mutations was similar to that of women with *BRCA1/2* mutations and improved compared to women without any HR mutation [19].

Table 6.1 Clinical Characteristics of BRCA1/2-Mutated Tumors

Characteristics	BRCA1/2-Mutated EOCs
Hereditary-breast ovarian syndrome	■ BRCA1 carriers have higher lifetime frequency of EOC ■ BRCA1 carriers develop EOC at an earlier age ■ Risk of different cancers in BRCA-1 vs BRCA-2 carriers
Pathology/stage	■ Association with serous tumors ■ Association with high-grade/undifferentiated tumors ■ Association with higher stage (stage III or IV) at presentation
Immunogenicity	■ BRCA1/2-mutated tumors are associated with higher CD3+ and CD8+ TILs
Overall survival	■ Improved survival for BRCA1 vs sporadic (HR = 0.73) ■ Improved survival for BRCA2 vs sporadic (HR = 0.49) ■ Improved survival for BRCA2 vs BRCA1 (HR = 0.64)
Response to chemotherapy	■ Improved response to platinum and PARP ■ Improved response to other double-strand DNA break inducing agents, i.e., PLD ■ BRCA2 tumors are more responsive to platinum and have greater genomic instability than BRCA1 tumors
Patterns of recurrence	■ More likely to develop visceral metastases (parenchymal lung, liver, spleen, adrenal, and brain metastases) ■ This effect seems more prominent for BRCA1 tumors

EOC, epithelial ovarian cancer; TILs, tumor-infiltrating lymphocytes.

TARGETING HRD IN OVARIAN CANCER

PARP Inhibitors: Preclinical Experience

PARP1 is the best characterized member of the 17-member PARP family of proteins and has been implicated in multiple aspects of DNA-repair, including recruitment to the site of DNA damage, where it synthesizes long-branching chains of poly(ADP-ribose) (reviewed in Ref. [24]). This polymer subsequently serves as a scaffold for DNA repair machinery to be recruited to the site of DNA damage, typically in the setting of base excision repair (BER). While PARP1's role in BER has been best described, PARP1 has also been postulated to facilitate HR, as the recruitment of MRE11 and ATM to the site of double-strand DNA breaks has been observed to be dependent upon poly(ADP-ribose) synthesis [25,26]. PARP1 has also been implicated in NHEJ, where PARP1 can compete with Ku70 and Ku80 for the repair of DNA double-strand breaks [27]. Of note, despite its multiple roles in DNA repair and cellular function, PARP1 is not essential in normal conditions, and *Parp1* knockout mice develop normally unless they are subject to DNA damage [28,29].

Based upon the critical role PARP1 plays in the repair of single-strand break repair, it was hypothesized that PARP inhibition could prove to be synthetically lethal with loss of HR, such as through deficiencies in BRCA1 or BRCA2 function, as unrepaired single-strand DNA breaks would convert into double-strand breaks and eventually prove lethal in cells that lacked intact HR. In 2005, two landmark papers demonstrated that cells deficient in BRCA2 were indeed vulnerable to the effects of PARP inhibitors [30,31]. Subsequent studies have demonstrated that deficiency in HRD by loss of other components of HRD also result in sensitivity to PARP inhibition [32]. Although the initial hypothesis for synthetic lethality between PARP inhibitors and HRD was based on PARP1's role in BER, some subsequent findings have suggested that this model may not provide a complete explanation for the activity of PARP inhibitors in HRD cells. For example, when expression of XRCC1, which operates downstream of PARP1 in BER, is knocked down, no synthetic lethality is observed between XRCC1 knockdown and HRD, suggesting that loss of BER activity alone is not lethal for cells that are HRD [33].

This observation has led to the investigation of how other functions of PARP1 might contribute to the sensitivity of HRD cells to PARP inhibition [34]. As PARP1 can potentially inhibit NHEJ, it has also been hypothesized that PARP inhibition therefore increases genomic instability by promoting increased NHEJ, which is particularly toxic to HRD cells [33]. This model has been supported by findings that inhibition of NHEJ reduces the activity of PARP inhibitors in HRD cells [27,35,36]. Another postulated mechanism of action for PARP inhibitors is the concept of "PARP trapping," whereby PARP1 that has been inactivated by the PARP inhibitor remains bound to the site of DNA damage, thereby inhibiting repair [37]; however, this model cannot explain why PARP1 knockdown also results in toxicity in cells deficient in BRCA1 or BRCA2 [30,31]. Other mechanisms have also been suggested to contribute to the activity of PARP inhibitors in HRD cells, including impairment of BRCA1 recruitment to the site of DNA damage in the setting of certain *BRCA1* mutations [38].

PARP Inhibitors: Clinical Experience

The activity of PARP inhibitors in epithelial ovarian cancer has now been extensively studied, and olaparib has been approved in the United States by the FDA in women with *BRCA*-mutated ovarian cancer who have received at least three prior lines of therapy for their disease and in Europe as a maintenance therapy following response to platinum therapy after platinum-sensitive recurrence. Multiple studies of PARP inhibitors in various clinical settings in ovarian cancer remain ongoing as of late 2015. Table 6.2 summarizes the results of several of the key Phase 2 trials with PARP inhibitors to date.

BRCA-Mutated Ovarian Cancers

The first activity of PARP inhibitors was described in a Phase 1 trial of the PARP inhibitor olaparib [48], where 9 of 19 evaluable patients who had

Table 6.2 Key Phase 2 Trials of PARP Inhibitors in Ovarian Cancer

Trial	Setting	# of Pts	Treatment Arms	PFS (months)	OS	P Value
Audeh et al. [39]	Relapsed BRCAmt ovarian cancer	57	Olaparib[a] 100 or 400 mg BID	1.9 (100 mg) 5.8 (400 mg)	NR	NA (nonrandomized trial)
Gelmon et al. [40]	Relapsed HGSOC	64	Olaparib 400 mg BID	7.4 (BRCAmt) 6.4 (BRCA non-mt)	NR	NA
Coleman et al. [41]	Relapsed BRCAmt ovarian cancer	50	Veliparib 400 mg BID	8.2	NR	NA
Kristeleit et al. [42]	Relapsed platinum-sensitive ovarian cancer	206	Rucaparib 600 mg BID	12.8 (BRCAmt) 5.7 (BRCA like) 5.3 (biomarker negative)	NR	<0.0001 (BRCAmt vs biomarker negative) 0.045 (BRCA like vs biomarker negative)
Kaye et al. [43]	Relapsed platinum-resistant or partially resistant (<12 months) BRCAmt ovarian cancer	97	Olaparib 200 mg BID Olaparib 400 mg BID PLD	6.5 8.8 7.1	NR	0.78 (Olaparib 200 mg vs PLD) 0.63 (Olaparib 400 mg vs PLD)
Oza et al. [44]	Relapsed platinum-sensitive HGSOC	173	Chemo[b] + olaparib 200 mg BID d1-10 and olaparib 400 mg BID maintenance Chemo[b] alone	12.2 9.6	33.8 37.6	PFS: 0.0012 OS: 0.44
Liu et al. [45]	Relapsed platinum-sensitive HGSOC	90	Olaparib 200 mg BID + cediranib 30 mg daily Olaparib 400 mg BID	17.7 9.0	NR	0.005
Ledermann et al. [46,47]	Maintenance therapy following platinum in relapsed platinum-sensitive HGSOC	265	Olaparib 400 mg BID Placebo	7.4 (11.2 for BRCAmt) 5.5 (4.3 for BRCAmt)	29.7 (37.1 for BRCAmt) 29.9 (37.6 for BRCAmt)	PFS: 0.54 (0.18 BRCAmt) OS: NS

[a]Olaparib dosed in capsule formulation in all trials.
[b]Chemo= carboplatin and paclitaxel, dosed AUC4 and $175 \, mg/m^2$ when administered with olaparib and AUC6 and $175 \, mg/m^2$ when dosed alone; NR, not reported; NA, not applicable; NS, not statistically significant; OS, overall survival.

BRCA-related ovarian, breast, or prostate cancer experienced a partial or complete response. A subsequent expansion of this trial into 50 women with *BRCA*-related ovarian cancer demonstrated a response rate of 40% [49]. Of interest, significant association was observed between platinum sensitivity and the likelihood of response, suggesting that mechanisms of platinum resistance and PARP inhibitor resistance might have at least partial overlap.

Subsequent clinical studies have confirmed the activity of various PARP inhibitors in women with *BRCA*-related ovarian cancer. A Phase 2 study of olaparib in women with *BRCA*-related ovarian cancer demonstrated a response rate of 33% [39]. Similar activity, with a response rate of 31%, was observed in women with ovarian cancer in a Phase 2 study of olaparib monotherapy in *BRCA*-mutated tumors across multiple tumor types [50]. Other PARP inhibitors have demonstrated similar single-agent activity: a Phase 2 study of veliparib found a response rate of 26% in women with *BRCA*-related ovarian cancer [41]. In the setting of platinum-sensitive disease, PARP inhibitor monotherapy is associated with high response rates in *BRCA*-related ovarian cancer, with a response rate of 69% to olaparib in platinum-sensitive patients in the Fong et al. study [49] and of 75% in the subset of patients harboring a *BRCA1/2* mutation in the Phase 2 ARIEL2 study of rucaparib monotherapy in platinum-sensitive recurrent ovarian cancer [42].

Despite the marked activity of PARP inhibitor monotherapy in these trials, PARP inhibitors have not been shown to have superior activity to chemotherapy, even in patients with *BRCA*-related ovarian cancer. A Phase 2 study comparing olaparib monotherapy to pegylated liposomal doxorubicin (PLD) in women with partially platinum-sensitive (recurrence within 12 months of last platinum) *BRCA*-mutated ovarian cancer demonstrated no difference in activity between olaparib and PLD [43]. Notably, the activity of PLD in this study was more robust than seen in prior studies, raising the possibility that the lack of difference in activity between PARP inhibitors and chemotherapy in this trial was because patients with *BRCA*-related ovarian cancer also have increased sensitivity to PLD. An ongoing trial, SOLO3, is currently exploring olaparib monotherapy compared to single-agent nonplatinum-based chemotherapy in women with platinum-sensitive relapsed *BRCA*-mutated ovarian cancer.

PARP inhibitors as maintenance therapy have also been widely explored in ovarian cancer. A Phase 2 study of olaparib monotherapy as maintenance therapy following platinum-based chemotherapy for platinum-sensitive recurrence significantly increased progression-free survival (PFS), especially in women with a *BRCA* mutation, where PFS increased from 4.3 to 11.2 months with olaparib maintenance [46,47]. Similarly, women with *BRCA*-mutated ovarian cancer in a Phase 3 study of niraparib maintenance therapy following platinum-based chemotherapy in platinum-sensitive recurrence (the NOVA trial) also derived significant PFS improvement, from 5.5 to 21.0 months, compared to placebo (Mirza et al., N Engl J Med, 2016). Phase 3 trials of olaparib maintenance therapy in *BRCA*-mutated ovarian cancer following initial surgery and chemotherapy for newly diagnosed disease (SOLO1) and following platinum-based chemotherapy for platinum-sensitive recurrence (SOLO2) completed accrual in 2015 and are awaiting data maturation.

BRCA Wild-Type Ovarian Cancers

While much of the early research on PARP inhibitors has focused on the *BRCA*-mutated population, growing evidence suggests that responses to PARP inhibitors will not be limited to *BRCA*-related tumors alone. As discussed earlier, data from TCGA suggests that approximately 50% of HGSOCs have an alteration in genes related to HR (TCGA), while another study of 390 ovarian cancers found germline and somatic mutations in HR genes in 31% of cases [19], and PARP inhibitors have demonstrated in vitro synergy with deficiency in multiple HR genes aside from *BRCA1* and *BRCA2* [32]. In fact, in an early Phase 2 trial in women with or without *BRCA* mutations and breast or ovarian cancer, the response rate to olaparib in women with ovarian cancer and without a *BRCA* mutation was 24% [40]. Numerous studies are now ongoing to explore the activity of PARP inhibitors in women with ovarian cancer who do not have a known deleterious *BRCA* mutation.

One of the challenges in assessing the activity of PARP inhibitors in women without a *BRCA* mutation is identifying a biomarker that might suggest HRD. Multiple approaches have emerged toward identifying a signature of HRD in ovarian cancer, including gene expression signatures of "BRCAness" [51], mutational profiling of HR and other DNA repair genes by targeted capture and massively parallel genomic sequencing [19], and genome-wide chromosomal aberrations that can occur in the setting of HR [52–54]. One caveat regarding using genome-wide chromosomal aberrations as a marker of HRD is that the changes that occur in the setting of HRD do not revert if HR is restored. Thus, a cell that has restored HR will retain any chromosomal aberrations that occurred when HR was lost and will appear in this assay to be "HRD" despite being HR proficient [52].

Current studies of PARP inhibitors in nonselected populations incorporate measures of HRD as secondary or exploratory biomarkers. For example, in ARIEL2, which is assessing the activity of rucaparib in women with recurrent ovarian cancer, genome-wide loss of heterozygosity (LOH) has been used to classify women with *BRCA* wild-type but HRD tumors. The first part of this study, which reported on the outcomes of 204 women with platinum-sensitive ovarian cancer receiving rucaparib monotherapy, found the response rate to be 75% in women with a *BRCA* mutation (germline or somatic), 36% in women whose tumors were *BRCA* wild type but who had evidence of genomic LOH (a "BRCA-like" population), and 16% in patients who were both *BRCA* wild type and did not have genomic LOH ("biomarker negative") [42]. The hazard ratio for PFS was 0.22 between the *BRCA*-mutated and biomarker negative groups ($p < 0.0001$) and 0.67 between the *BRCA*-like and biomarker negative groups ($p = 0.045$). Interestingly, while the overall PFS of *BRCA*-like patients was significantly less than that for *BRCA*-mutated patients (12.8 vs 5.7 months), when duration of response was examined in those patients who had responded to rucaparib, the duration of response was more similar between *BRCA*-mutated and *BRCA*-like patients compared to biomarker negative patients (9.5, 8.2, and 5.5 months, respectively). This observation may reflect the inability of the genomic LOH signature to differentiate between the active presence of HRD and the presence of HRD in the past.

Another trial (QUADRA) examining the activity of niraparib monotherapy in women who have received at least three prior lines of chemotherapy for ovarian cancer is ongoing. Like ARIEL2, this trial does not preselect patients with known *BRCA* mutations or HRD; however, responses will be analyzed by HRD status using the Myriad HRD Test (Myriad), which calculates the HRD score from three components reflecting different types of tumor rearrangements, including LOH of regions >15 Mb and <whole chromosome, large-scale state transitions (consisting of chromosome breaks in adjacent segments of DNA), and telomeric allelic imbalance [55]. Results of this trial are not yet available. However, in a Phase 3 study of maintenance niraparib in recurrent platinum-sensitive ovarian cancer (NOVA), activity was observed in women without a germline *BRCA* mutation, with an increase in PFS from 3.9 to 9.3 months with a hazard ratio of 0.45 (Mirza et al., N Engl J Med, 2016). Patients without a germline *BRCA* mutation but with a positive HRD score appeared to have slightly increased benefit with maintenance niraparib, with a PFS improvement from 3.8 to 12.9 months and a hazard ratio 0.38, suggesting that assays of HRD may help identify patients who are likely to derive the greatest benefit from PARP inhibitors. The results of a similarly-designed Phase 3 study of maintenance rucaparib in recurrent platinum sensitive ovarian cancer, are not yet available.

Mechanisms of PARP Inhibitor Resistance

Multiple mechanisms of resistance to PARP inhibitors have now been described. In the setting of *BRCA1/2* mutation, one of the more common methods of resistance is the development of "reversion" or "secondary somatic" mutations, a situation in which the tumor cell develops a secondary mutation in the mutated copy of *BRCA1/2* that results in restoration of the expression of a functional version of the protein, typically either by restoring the open reading frame (e.g., in settings where the original mutation resulted

Table 6.3 Potential Mechanisms of PARP Inhibitor and Platinum Resistance

Mechanism of Resistance	Predicted Effect on PARP Inhibitor Resistance	Predicted Effect on Platinum Resistance
Reversion mutation	Resistant	Resistant
Epigenetic reversion	Resistant	Resistant
53BP1 loss	Resistant (in BRCA1-deficient tumors)	Sensitive
Decreased PARP1 expression	Resistant	Sensitive
Increased PgP expression	Resistant	Sensitive
Stabilization of BRCA1	Resistant	Resistant

in a frameshift) or by fully reverting the original mutation to restore a wild-type version of the gene [56,57]. This form of PARP inhibitor resistance was first described in vitro and also results in resistance to platinum; subsequent studies from clinical samples of patients with platinum or PARP inhibitor resistance have demonstrated that secondary somatic mutations also occur in the clinical setting [58–62]. In one study, 46% of patients with *BRCA*-related ovarian cancer that had become resistant to platinum were found to have secondary somatic mutations [62].

Although not clinically validated in patient samples, other mechanisms of potential PARP inhibitor resistance have also been described (Table 6.3). These include loss of activity of the NHEJ factor 53BP1, decreased PARP1 expression, and elevated expression of ABC transporters, such as the P-glycoprotein efflux pump (PgP) [63]. Other groups have documented resistance to platinum in clinical samples via a combination of HSP90-mediated stabilization of mutant BRCA1 and decreased 53BP1 protein expression [64]. Understanding the exact mechanism of resistance to PARP inhibitors may have significant clinical implications; for example, secondary mutations will likely result in resistance to both PARP inhibitors and platinum agents, whereas decreased PARP1 expression or PgP upregulation should result in PARP inhibitor resistance but preserved platinum sensitivity. Similarly, loss of 53BP1 would only affect patients with BRCA1-deficient tumors, as *53bp1* deletion does not appear to have any effect on cells that are BRCA2 deficient.

Beyond PARP Inhibitors in HRD

Although PARP inhibitors have been the first targeted therapy in ovarian cancer to leverage vulnerabilities in cells that are HRD, other effective methods of targeting HRD tumors may exist. The role of the cell cycle checkpoints in allowing for appropriate DNA repair and the recent observation of synthetic lethality between PolΘ (i.e., POLQ) and HRD suggest future methods of attacking HRD tumors.

Cell Cycle and DNA Damage Checkpoints

Cell cycle checkpoints are critical to prevent the cell from progressing to the next phase of the cell cycle before the prior phase has been completed. Premature entry into the next phase of the cell cycle can result in catastrophic consequences for the cell and cell death. p53 is a key regulator of the cell cycle and plays a critical role in the function of the G_1 checkpoint. Thus, HGSOCs, in which *TP53* mutations are nearly universal (TCGA), have typically lost G_1 checkpoint control and thus are increasingly dependent upon the S/G_2 checkpoints.

One of the triggers for cell cycle arrest is DNA damage; for example, the DNA damage response proteins ATM and ATR can induce G_1 and S/G_2 arrest via the ATM-CHK2 and ATR-CHK1 signaling pathways, respectively [65]. Cells that are HRD may be particularly vulnerable to checkpoint inhibitors that prevent the cell cycle checkpoints from appropriately inducing cell arrest, especially as their ability to repair DNA damage is already impaired. For example, one study has demonstrated that Fanconi anemia–deficient tumor cells are sensitive to CHK1 inhibition; this sensitivity was further heightened when combined with cisplatin [66]. The combination of HRD and p53 loss may allow for increased sensitivity to agents that inactivate the ATR-CHK1-WEE1 pathway, either alone or in combination with chemotherapy agents that promote replication stress, and agents that target various elements of the pathway, including ATR, CHK1, and Wee1, are under active development. Of note, a Phase 1 trial of the Wee1 inhibitor AZD1775 demonstrated partial responses in two of nine patients with *BRCA1* or *BRCA2* mutations [67].

PolΘ and Alternative End Joining

HR and NHEJ have been classically described as the two major pathways by which double-stranded DNA breaks are repaired. However, two additional pathways have been more recently described, including alt-EJ, which has also been referred to as the error-prone microhomology-mediated end-joining pathway [1,68]. Two studies in 2015 described the protein PolΘ as being a critical player within the alt-EJ pathway and found that PolΘ depletion in HRD cells resulted in synthetic lethality [68,69]. These findings raise the question of whether PolΘ represents a novel druggable target, particularly in tumors that are HRD.

PROMOTING HR DEFICIENCY IN HR-PROFICIENT CELLS

While the effectiveness of PARP inhibitors in *BRCA*-related and other HRD ovarian cancers leverages preexisting HRD within the tumor cells to achieve synthetic lethality, additional research is now focusing on how HR-proficient cells can be induced into more HRD states, allowing for potential activity of drugs such as PARP inhibitors [70]. Such strategies could include combining PARP inhibitors with other targeted therapies that can result in inhibited BRCA function or expression, such as CDK1 inhibitors, which decrease BRCA1 phosphorylation [71]; PI3K or AKT inhibitors, where inhibition of PI3K/AKT signaling can result in suppression of *BRCA1/2* expression and Rad51 focus formation [72,73]; HDAC inhibitors, which can result in downregulation of HR [74]; or HSP90 inhibitors, as many HR proteins are also HSP90 clients [75].

Preclinical studies have supported the presence of potential synergy between inhibitors of these targets and PARP inhibitors or platinum, suggesting that HR deficiency resulting in sensitivity to PARP inhibitors or platinum may indeed be induced in HR-proficient cells by inhibiting these pathways and/ or proteins. One caveat to the approach of targeting cancer cells by combining an agent that induces HR deficiency with a PARP inhibitor or platinum is that this approach loses the selectivity of toxicity that PARP inhibitors carry with intrinsically HRD tumor cells. In theory, normal cells may also experience induction of HRD and therefore be similarly vulnerable to these combinations. Thus, the therapeutic window for such combinations may be much narrower than what has been observed to date with PARP inhibitor monotherapy.

CONCLUSION

The past decade has witnessed rapid advancement in the understanding of HR in ovarian cancer. With approximately 50% of HGSOC harboring alterations in the HR pathway, it has become clear that understanding the unique vulnerabilities of HRD cells may result in novel therapeutic strategies for ovarian cancer. Indeed, the discovery of synthetic lethality between PARP inhibitors and HRD and the approval of olaparib in BRCA-mutated ovarian cancer represent one of the first biomarker-directed therapies in gynecologic cancers. However, the complexities of HR require further research, and critical questions remain regarding biomarker identification in non-BRCA-mutated populations, inducing HRD and potential PARP inhibitor or platinum sensitivity, overcoming resistance to PARP inhibitors, and other mechanisms of targeting cells deficient in HR.

References

[1] Ceccaldi R, Rondinelli B, D'Andrea AD. Repair pathway choices and consequences at the double-strand break. Trends Cell Biol 2015. PubMed PMID: 26437586.

[2] Wyman C, Ristic D, Kanaar R. Homologous recombination-mediated double-strand break repair. DNA Repair (Amst) 2004;3(8–9):827–33. PubMed PMID: 15279767.

[3] Li X, Heyer WD. Homologous recombination in DNA repair and DNA damage tolerance. Cell Res 2008;18(1):99–113. PubMed PMID: 18166982. Pubmed Central PMCID: 3087377.

[4] San Filippo J, Sung P, Klein H. Mechanism of eukaryotic homologous recombination. Annu Rev Biochem 2008;77:229–57. PubMed PMID: 18275380.

[5] Walsh CS. Two decades beyond BRCA1/2: homologous recombination, hereditary cancer risk and a target for ovarian cancer therapy. Gynecol Oncol 2015;137(2):343–50. PubMed PMID: 25725131.

[6] Lancaster JM, Wooster R, Mangion J, Phelan CM, Cochran C, Gumbs C, et al. BRCA2 mutations in primary breast and ovarian cancers. Nat Genet 1996;13(2):238–40. PubMed PMID: 8640235.

[7] Miki Y, Swensen J, Shattuck-Eidens D, Futreal PA, Harshman K, Tavtigian S, et al. A strong candidate for the breast and ovarian cancer susceptibility gene BRCA1. Science 1994;266(5182):66–71. PubMed PMID: 7545954.

[8] Wooster R, Bignell G, Lancaster J, Swift S, Seal S, Mangion J, et al. Identification of the breast cancer susceptibility gene BRCA2. Nature 1995;378(6559):789–92. PubMed PMID: 8524414.

[9] Alsop K, Fereday S, Meldrum C, deFazio A, Emmanuel C, George J, et al. BRCA mutation frequency and patterns of treatment response in BRCA mutation-positive women with ovarian cancer: a report from the Australian Ovarian Cancer Study Group. J Clin Oncol Off J Am Soc Clin Oncol 2012;30(21):2654–63. PubMed PMID: 22711857. Pubmed Central PMCID: 3413277.

[10] Cancer Genome Atlas Research N Integrated genomic analyses of ovarian carcinoma. Nature 2011;474(7353):609–15. PubMed PMID: 21720365. Pubmed Central PMCID: 3163504.

[11] Walsh T, Casadei S, Lee MK, Pennil CC, Nord AS, Thornton AM, et al. Mutations in 12 genes for inherited ovarian, fallopian tube, and peritoneal carcinoma identified by massively parallel sequencing. Proc Natl Acad Sci USA 2011;108(44):18032–7. PubMed PMID: 22006311. Pubmed Central PMCID: 3207658.

[12] Network NCC. NCCN Guidelines [01/03/2014].

[13] Meindl A, Hellebrand H, Wiek C, Erven V, Wappenschmidt B, Niederacher D, et al. Germline mutations in breast and ovarian cancer pedigrees establish RAD51C as a human cancer susceptibility gene. Nat Genet 2010;42(5):410–14. PubMed PMID: 20400964.

[14] Hughes-Davies L, Huntsman D, Ruas M, Fuks F, Bye J, Chin SF, et al. EMSY links the BRCA2 pathway to sporadic breast and ovarian cancer. Cell 2003;115(5):523–35. PubMed PMID: 14651845.

[15] Brown LA, Kalloger SE, Miller MA, Shih Ie M, McKinney SE, Santos JL, et al. Amplification of 11q13 in ovarian carcinoma. Genes Chromosomes Cancer 2008;47(6):481–9. PubMed PMID: 18314909.

[16] Wilkerson PM, Dedes KJ, Wetterskog D, Mackay A, Lambros MB, Mansour M, et al. Functional characterization of EMSY gene amplification in human cancers. J Pathol 2011;225(1):29–42. PubMed PMID: 21735447.

[17] Shen WH, Balajee AS, Wang J, Wu H, Eng C, Pandolfi PP, et al. Essential role for nuclear PTEN in maintaining chromosomal integrity. Cell 2007;128(1):157–70. PubMed PMID: 17218262.

[18] McEllin B, Camacho CV, Mukherjee B, Hahm B, Tomimatsu N, Bachoo RM, et al. PTEN loss compromises homologous recombination repair in astrocytes: implications for glioblastoma therapy with temozolomide or poly(ADP-ribose) polymerase inhibitors. Cancer Res 2010;70(13):5457–64. PubMed PMID: 20530668. Pubmed Central PMCID: 2896430.

[19] Pennington KP, Walsh T, Harrell MI, Lee MK, Pennil CC, Rendi MH, et al. Germline and somatic mutations in homologous recombination genes predict platinum response and survival in ovarian, fallopian tube, and peritoneal carcinomas. Clin Cancer Res Off J Am

Assoc Cancer Res 2014;20(3):764–75. PubMed PMID: 24240112. Pubmed Central PMCID: 3944197.

[20] Tan DS, Rothermundt C, Thomas K, Bancroft E, Eeles R, Shanley S, et al. "BRCAness" syndrome in ovarian cancer: a case–control study describing the clinical features and outcome of patients with epithelial ovarian cancer associated with BRCA1 and BRCA2 mutations. J Clin Oncol Off J Am Soc Clin Oncol 2008;26(34):5530–6. PubMed PMID: 18955455.

[21] Yang D, Khan S, Sun Y, Hess K, Shmulevich I, Sood AK, et al. Association of BRCA1 and BRCA2 mutations with survival, chemotherapy sensitivity, and gene mutator phenotype in patients with ovarian cancer. JAMA 2011;306(14):1557–65. PubMed PMID: 21990299. Pubmed Central PMCID: 4159096.

[22] Bolton KL, Chenevix-Trench G, Goh C, Sadetzki S, Ramus SJ, Karlan BY, et al. Association between BRCA1 and BRCA2 mutations and survival in women with invasive epithelial ovarian cancer. JAMA 2012;307(4):382–90. PubMed PMID: 22274685. Pubmed Central PMCID: 3727895.

[23] Liu J, Cristea MC, Frankel P, Neuhausen SL, Steele L, Engelstaedter V, et al. Clinical characteristics and outcomes of BRCA-associated ovarian cancer: genotype and survival. Cancer Genet 2012;205(1–2):34–41. PubMed PMID: 22429596. Pubmed Central PMCID: 3337330.

[24] Rouleau M, Patel A, Hendzel MJ, Kaufmann SH, Poirier GG. PARP inhibition: PARP1 and beyond. Nat Rev Cancer 2010;10(4):293–301. PubMed PMID: 20200537. Pubmed Central PMCID: 2910902.

[25] Haince JF, Kozlov S, Dawson VL, Dawson TM, Hendzel MJ, Lavin MF, et al. Ataxia telangiectasia mutated (ATM) signaling network is modulated by a novel poly(ADP-ribose)-dependent pathway in the early response to DNA-damaging agents. J Biol Chem 2007;282(22):16441–53. PubMed PMID: 17428792.

[26] Haince JF, McDonald D, Rodrigue A, Dery U, Masson JY, Hendzel MJ, et al. PARP1-dependent kinetics of recruitment of MRE11 and NBS1 proteins to multiple DNA damage sites. J Biol Chem 2008;283(2):1197–208. PubMed PMID: 18025084.

[27] Wang M, Wu W, Wu W, Rosidi B, Zhang L, Wang H, et al. PARP-1 and Ku compete for repair of DNA double strand breaks by distinct NHEJ pathways. Nucleic Acids Res 2006;34(21):6170–82. PubMed PMID: 17088286. Pubmed Central PMCID: 1693894.

[28] de Murcia JM, Niedergang C, Trucco C, Ricoul M, Dutrillaux B, Mark M, et al. Requirement of poly(ADP-ribose) polymerase in recovery from DNA damage in mice and in cells. Proc Natl Acad Sci USA 1997;94(14):7303–7. PubMed PMID: 9207086. Pubmed Central PMCID: 23816.

[29] Wang ZQ, Auer B, Stingl L, Berghammer H, Haidacher D, Schweiger M, et al. Mice lacking ADPRT and poly(ADP-ribosyl)ation develop normally but are susceptible to skin disease. Genes Dev 1995;9(5):509–20. PubMed PMID: 7698643.

[30] Bryant HE, Schultz N, Thomas HD, Parker KM, Flower D, Lopez E, et al. Specific killing of BRCA2-deficient tumours with inhibitors of poly(ADP-ribose) polymerase. Nature 2005;434(7035):913–17. PubMed PMID: 15829966. eng.

[31] Farmer H, McCabe N, Lord CJ, Tutt AN, Johnson DA, Richardson TB, et al. Targeting the DNA repair defect in BRCA mutant cells as a therapeutic strategy. Nature 2005;434(7035):917–21. PubMed PMID: 15829967. eng.

[32] McCabe N, Turner NC, Lord CJ, Kluzek K, Bialkowska A, Swift S, et al. Deficiency in the repair of DNA damage by homologous recombination and sensitivity to poly(ADP-ribose) polymerase inhibition. Cancer Res 2006;66(16):8109–15. PubMed PMID: 16912188.

[33] Patel AG, Sarkaria JN, Kaufmann SH. Nonhomologous end joining drives poly(ADP-ribose) polymerase (PARP) inhibitor lethality in homologous recombination-deficient cells. Proc

Natl Acad Sci USA 2011;108(8):3406–11. PubMed PMID: 21300883. Pubmed Central PMCID: 3044391.

[34] Scott CL, Swisher EM, Kaufmann SH. Poly (ADP-ribose) polymerase inhibitors: recent advances and future development. J Clin Oncol Off J Am Soc Clin Oncol 2015;33(12):1397–406. PubMed PMID: 25779564. Pubmed Central PMCID: 4517072.

[35] Hochegger H, Dejsuphong D, Fukushima T, Morrison C, Sonoda E, Schreiber V, et al. Parp-1 protects homologous recombination from interference by Ku and Ligase IV in vertebrate cells. EMBO J 2006;25(6):1305–14. PubMed PMID: 16498404. Pubmed Central PMCID: 1422167.

[36] Paddock MN, Bauman AT, Higdon R, Kolker E, Takeda S, Scharenberg AM. Competition between PARP-1 and Ku70 control the decision between high-fidelity and mutagenic DNA repair. DNA Repair (Amst) 2011;10(3):338–43. PubMed PMID: 21256093. Pubmed Central PMCID: 4079052.

[37] Murai J, Huang SY, Das BB, Renaud A, Zhang Y, Doroshow JH, et al. Trapping of PARP1 and PARP2 by clinical PARP inhibitors. Cancer Res 2012;72(21):5588–99. PubMed PMID: 23118055. Pubmed Central PMCID: 3528345.

[38] Li M, Yu X. Function of BRCA1 in the DNA damage response is mediated by ADP-ribosylation. Cancer Cell 2013;23(5):693–704. PubMed PMID: 23680151. Pubmed Central PMCID: 3759356.

[39] Audeh MW, Carmichael J, Penson RT, Friedlander M, Powell B, Bell-McGuinn KM, et al. Oral poly(ADP-ribose) polymerase inhibitor olaparib in patients with BRCA1 or BRCA2 mutations and recurrent ovarian cancer: a proof-of-concept trial. Lancet 2010;376(9737):245–51. PubMed PMID: 20609468. eng.

[40] Gelmon KA, Tischkowitz M, Mackay H, Swenerton K, Robidoux A, Tonkin K, et al. Olaparib in patients with recurrent high-grade serous or poorly differentiated ovarian carcinoma or triple-negative breast cancer: a phase 2, multicentre, open-label, non-randomised study. Lancet Oncol 2011;12(9):852–61. PubMed PMID: 21862407. eng.

[41] Coleman RL, Sill MW, Bell-McGuinn K, Aghajanian C, Gray HJ, Tewari KS, et al. A phase II evaluation of the potent, highly selective PARP inhibitor veliparib in the treatment of persistent or recurrent epithelial ovarian, fallopian tube, or primary peritoneal cancer in patients who carry a germline BRCA1 or BRCA2 mutation—an NRG Oncology/Gynecologic Oncology Group study. Gynecol Oncol 2015;137(3):386–91. PubMed PMID: 25818403. Pubmed Central PMCID: 4447525.

[42] Kristeleit R, Swisher EM, Oza A, Coleman RL, Scott C, Konecny G, et al. Final results of ARIEL2 (Part 1): a phase 2 trial to prospectively identify ovarian cancer (OC) responders to rucaparib using tumor genetic analysis. Presented at ECCO 2015; 2015.

[43] Kaye SB, Lubinski J, Matulonis U, Ang JE, Gourley C, Karlan BY, et al. Phase II, open-label, randomized, multicenter study comparing the efficacy and safety of olaparib, a poly (ADP-ribose) polymerase inhibitor, and pegylated liposomal doxorubicin in patients with BRCA1 or BRCA2 mutations and recurrent ovarian cancer. J Clin Oncol Off J Am Soc Clin Oncol 2012;30(4):372–9. PubMed PMID: 22203755.

[44] Oza AM, Cibula D, Benzaquen AO, Poole C, Mathijssen RH, Sonke GS, et al. Olaparib combined with chemotherapy for recurrent platinum-sensitive ovarian cancer: a randomised phase 2 trial. Lancet Oncol 2015;16(1):87–97. PubMed PMID: 25481791.

[45] Liu JF, Barry WT, Birrer M, Lee JM, Buckanovich RJ, Fleming GF, et al. Combination cediranib and olaparib versus olaparib alone for women with recurrent platinum-sensitive ovarian cancer: a randomised phase 2 study. Lancet Oncol 2014;15(11):1207–14. PubMed PMID: 25218906. Pubmed Central PMCID: 4294183.

[46] Ledermann J, Harter P, Gourley C, Friedlander M, Vergote I, Rustin G, et al. Olaparib maintenance therapy in platinum-sensitive relapsed ovarian cancer. N Engl J Med 2012;366(15):1382–92. PubMed PMID: 22452356.

[47] Ledermann J, Harter P, Gourley C, Friedlander M, Vergote I, Rustin G, et al. Olaparib maintenance therapy in patients with platinum-sensitive relapsed serous ovarian cancer: a preplanned retrospective analysis of outcomes by BRCA status in a randomised phase 2 trial. Lancet Oncol 2014 PubMed PMID: 24882434.

[48] Fong PC, Boss DS, Yap TA, Tutt A, Wu P, Mergui-Roelvink M, et al. Inhibition of poly(ADP-ribose) polymerase in tumors from BRCA mutation carriers. N Engl J Med 2009;361(2):123–34. PubMed PMID: 19553641. eng.

[49] Fong PC, Yap TA, Boss DS, Carden CP, Mergui-Roelvink M, Gourley C, et al. Poly(ADP)-ribose polymerase inhibition: frequent durable responses in BRCA carrier ovarian cancer correlating with platinum-free interval. J Clin Oncol Off J Am Soc Clin Oncol 2010;28(15):2512–19. PubMed PMID: 20406929.

[50] Kaufman B, Shapira-Frommer R, Schmutzler RK, Audeh MW, Friedlander M, Balmana J, et al. Olaparib monotherapy in patients with advanced cancer and a germline BRCA1/2 mutation. J Clin Oncol Off J Am Soc Clin Oncol 2015;33(3):244–50. PubMed PMID: 25366685.

[51] Konstantinopoulos PA, Spentzos D, Karlan BY, Taniguchi T, Fountzilas E, Francoeur N, et al. Gene expression profile of BRCAness that correlates with responsiveness to chemotherapy and with outcome in patients with epithelial ovarian cancer. J Clin Oncol Off J Am Soc Clin Oncol 2010;28(22):3555–61. PubMed PMID: 20547991. Pubmed Central PMCID: 2917311.

[52] Abkevich V, Timms KM, Hennessy BT, Potter J, Carey MS, Meyer LA, et al. Patterns of genomic loss of heterozygosity predict homologous recombination repair defects in epithelial ovarian cancer. Br J Cancer 2012;107(10):1776–82. PubMed PMID: 23047548. Pubmed Central PMCID: 3493866.

[53] Birkbak NJ, Wang ZC, Kim JY, Eklund AC, Li Q, Tian R, et al. Telomeric allelic imbalance indicates defective DNA repair and sensitivity to DNA-damaging agents. Cancer Discov 2012;2(4):366–75. PubMed PMID: 22576213. Pubmed Central PMCID: 3806629.

[54] Popova T, Manie E, Rieunier G, Caux-Moncoutier V, Tirapo C, Dubois T, et al. Ploidy and large-scale genomic instability consistently identify basal-like breast carcinomas with BRCA1/2 inactivation. Cancer Res 2012;72(21):5454–62. PubMed PMID: 22933060.

[55] Wilcoxen KM, Becker M, Neff C, Abkevich V, Jones JT, Hou X, et al. Use of homologous recombination deficiency (HRD) score to enrich for niraparib sensitive high grade ovarian tumors. J Clin Oncol Off J Am Soc Clin Oncol 2015:33. [abstract 5532].

[56] Edwards SL, Brough R, Lord CJ, Natrajan R, Vatcheva R, Levine DA, et al. Resistance to therapy caused by intragenic deletion in BRCA2. Nature 2008;451(7182):1111–15. PubMed PMID: 18264088.

[57] Sakai W, Swisher EM, Karlan BY, Agarwal MK, Higgins J, Friedman C, et al. Secondary mutations as a mechanism of cisplatin resistance in BRCA2-mutated cancers. Nature 2008;451(7182):1116–20. PubMed PMID: 18264087. Pubmed Central PMCID: 2577037.

[58] Barber LJ, Sandhu S, Chen L, Campbell J, Kozarewa I, Fenwick K, et al. Secondary mutations in BRCA2 associated with clinical resistance to a PARP inhibitor. J Pathol 2013;229(3):422–9. PubMed PMID: 23165508.

[59] Patch AM, Christie EL, Etemadmoghadam D, Garsed DW, George J, Fereday S, et al. Whole-genome characterization of chemoresistant ovarian cancer. Nature 2015;521(7553):489–94. PubMed PMID: 26017449.

[60] Sakai W, Swisher EM, Jacquemont C, Chandramohan KV, Couch FJ, Langdon SP, et al. Functional restoration of BRCA2 protein by secondary BRCA2 mutations in BRCA2-mutated ovarian carcinoma. Cancer Res 2009;69(16):6381–6. PubMed PMID: 19654294. Pubmed Central PMCID: 2754824.

[61] Swisher EM, Sakai W, Karlan BY, Wurz K, Urban N, Taniguchi T. Secondary BRCA1 mutations in BRCA1-mutated ovarian carcinomas with platinum resistance. Cancer Res 2008;68(8):2581–6. PubMed PMID: 18413725. Pubmed Central PMCID: 2674369.

[62] Norquist B, Wurz KA, Pennil CC, Garcia R, Gross J, Sakai W, et al. Secondary somatic mutations restoring BRCA1/2 predict chemotherapy resistance in hereditary ovarian carcinomas. J Clin Oncol Off J Am Soc Clin Oncol 2011;29(22):3008–15. PubMed PMID: 21709188. Pubmed Central PMCID: 3157963.

[63] Lord CJ, Ashworth A. Mechanisms of resistance to therapies targeting BRCA-mutant cancers. Nat Med 2013;19(11):1381–8. PubMed PMID: 24202391.

[64] Johnson N, Johnson SF, Yao W, Li YC, Choi YE, Bernhardy AJ, et al. Stabilization of mutant BRCA1 protein confers PARP inhibitor and platinum resistance. Proc Natl Acad Sci USA 2013;110(42):17041–6. PubMed PMID: 24085845. Pubmed Central PMCID: 3801063.

[65] Curtin NJ. DNA repair dysregulation from cancer driver to therapeutic target. Nat Rev Cancer 2012;12(12):801–17. PubMed PMID: 23175119.

[66] Chen CC, Kennedy RD, Sidi S, Look AT, D'Andrea A. CHK1 inhibition as a strategy for targeting Fanconi anemia (FA) DNA repair pathway deficient tumors. Mol Cancer 2009;8:24. PubMed PMID: 19371427. Pubmed Central PMCID: 2672921.

[67] Do K, Wilsker D, Ji J, Zlott J, Freshwater T, Kinders RJ, et al. Phase I study of single-agent AZD1775 (MK-1775), a wee1 kinase inhibitor, in patients with refractory solid tumors. J Clin Oncol Off J Am Soc Clin Oncol 2015;33(30):3409–15. PubMed PMID: 25964244. Pubmed Central PMCID: 4606059.

[68] Ceccaldi R, Liu JC, Amunugama R, Hajdu I, Primack B, Petalcorin MI, et al. Homologous-recombination-deficient tumours are dependent on Poltheta-mediated repair. Nature 2015;518(7538):258–62. PubMed PMID: 25642963. Pubmed Central PMCID: 4415602.

[69] Mateos-Gomez PA, Gong F, Nair N, Miller KM, Lazzerini-Denchi E, Sfeir A. Mammalian polymerase theta promotes alternative NHEJ and suppresses recombination. Nature 2015;518(7538):254–7. PubMed PMID: 25642960.

[70] Konstantinopoulos PA, Ceccaldi R, Shapiro GI, D'Andrea AD. Homologous recombination deficiency: exploiting the fundamental vulnerability of ovarian cancer. Cancer Discov 2015;5(11):1137–54. PubMed PMID: 26463832. Pubmed Central PMCID: 4631624.

[71] Johnson N, Li YC, Walton ZE, Cheng KA, Li D, Rodig SJ, et al. Compromised CDK1 activity sensitizes BRCA-proficient cancers to PARP inhibition. Nat Med 2011;17(7):875–82. PubMed PMID: 21706030. Pubmed Central PMCID: 3272302.

[72] Ibrahim YH, Garcia-Garcia C, Serra V, He L, Torres-Lockhart K, Prat A, et al. PI3K inhibition impairs BRCA1/2 expression and sensitizes BRCA-proficient triple-negative breast cancer to PARP inhibition. Cancer Discov 2012;2(11):1036–47. PubMed PMID: 22915752.

[73] Juvekar A, Burga LN, Hu H, Lunsford EP, Ibrahim YH, Balmana J, et al. Combining a PI3K inhibitor with a PARP inhibitor provides an effective therapy for BRCA1-related breast cancer. Cancer Discov 2012;2(11):1048–63. PubMed PMID: 22915751. Pubmed Central PMCID: 3733368.

[74] Konstantinopoulos PA, Wilson AJ, Saskowski J, Wass E, Khabele D. Suberoylanilide hydroxamic acid (SAHA) enhances olaparib activity by targeting homologous recombination DNA repair in ovarian cancer. Gynecol Oncol 2014;133(3):599–606. PubMed PMID: 24631446. Pubmed Central PMCID: 4347923.

[75] Choi YE, Battelli C, Watson J, et al. Sublethal concentrations of 17-AAG suppress homologous recombination DNA repair and enhance sensitivity to carboplatin and olaparib in HR proficient ovarian cancer cells. Oncotarget 2014;5(9):2678–87. PubMed PMID: 24798692. Pubmed Central PMCID: 4058036.

CHAPTER 7

Molecular Basis of PARP Inhibition and Future Opportunities in Ovarian Cancer Therapy

B.L. Collins, A.N. Gonzalez, A. Hanbury, L. Ceppi and R.T. Penson
Massachusetts General Hospital, Boston, MA, United States

CONTENTS

DNA DAMAGE RESPONSE

DNA is the cardinal repository of the code of life and is guarded by a multi-layered and complex system of maintenance and repair. There are estimated to be 10,000–30,000 episodes of DNA damage each day in humans that require high fidelity repair to ensure biologic integrity and preservation of unique characteristics in subsequent generations [1].

The DNA damage response is hugely complex with >700 human proteins phosphorylated at sites recognized by ataxia telangiectasia mutated (ATM) and ataxia telangiectasia and Rad3-related (ATR) kinase substrates [2]. These proteins (a) sense DNA damage, (b) regulate cell division, and (c) repair DNA. Many of the proteins are encoded by tumor suppressor genes and, when mutated, result in cancer predisposition syndromes. Elledge described the process by which a cell duplicates itself as on the scale of duplicating a small city, each cell containing a detailed blueprint for the entire process: DNA. Key elements of DNA repair are illustrated in Fig. 7.1 [3].

Mutations in DNA damage–response genes cause a wide range of human diseases. Those that lead to cancer fall into two broad categories: signaling mutations and repair mutations. DNA damage signaling is impaired with mutations in *TP53* and *CHEK2*, and specifically with respect to double-strand DNA breaks in *ATM* and *NBS1* mutations. DNA repair is impaired with mutations on the tumor suppressor genes involved in homologous recombination (HR) (*BRCA1*, *BRCA2*, *PALB2*, *RAD51C*, and *RAD51D*), interstrand crosslink repair (*FANCs*), mismatch repair (*MLH1*, *MSH2*, *MSH6*, and *PMS2*) associated with colorectal cancer, nucleotide excision repair (*XPsA-F*) associated

Translational Advances in Gynecologic Cancers. DOI: http://dx.doi.org/10.1016/B978-0-12-803741-6.00007-0

with skin cancer, and effects in translesional synthesis (*POLH*) and helicases vgous end joining (DNA-PKcs), and DNA repair (AGT) [4].

Although initially thought to be the primary target of polyadenosine 5′-diphosphoribose polymerase (PARP) inhibitors, base-excision repair involving glycosylase, lyase, APE1, PAR, and DNA-PolΘ such as XRCC1 and Lig3 is no longer considered the only target [3].

BRCA GENE MUTATIONS

Evaluation of high-risk gene mutations has become a key part of care for ovarian cancer patients [5]. *BRCA1* and *BRCA2* are tumor suppressor genes that contribute key elements to DNA repair [6]. Mutations are DNA errors that can form and propagate in cell replication. They are "spelling errors" in the code for the structure and function of our biology. Every day, humans make 1000s of DNA mutations. The DNA repair apparatus detects and corrects these, or suicides the cell in question. If DNA repair is compromised, such as in those with a *BRCA* mutation, mutations accumulate and can lead to breast and ovarian cancer.

Approximately 15% of patients with ovarian cancer inherit a *BRCA1* or *BRCA2* mutation. *BRCA* mutations occur in approximately 90% of families with both breast and ovarian cancer and confer approximately 10–30 times the risk of cancer [7]. These mutations are present in all populations, more common in certain population groups: Ashkenazi Jews, Icelandic populations, French Canadians, and white races. For example, Ashkenazi Jewish women carry a 10-fold higher risk of carrying a *BRCA* mutation when compared to US women population. *BRCA1* mutations appear to be more common and potentially higher risk than *BRCA2* mutations, corresponding with a younger age of diagnosis, higher risk of triple-negative breast cancer, and greater risk of ovarian cancer [8].

BRCA genes are autosomal dominant, highly penetrant mutations. While women in the general population have a lifetime incidence of breast cancer of 12%, the incidence of ovarian cancer is much lower at 1/70 (1.4%), translating in an increased lifetime risk for breast cancer of 50–80%. *BRCA1* gene on chromosome 17q21, and *BRCA2* at chromosome 13q12–13 are involved in pleiotropic DNA repair of double-strand breaks, commonly called HR DNA damage repair, classified as tumor suppressors, they largely contribute to chromosome stability during replication, checkpoint activation, and transcription regulation. Their absence plays an important role in the pathogenesis of ovarian cancer [9].

This association has been characterized as hereditary breast and ovarian cancer syndrome. A *gBRCA1* mutation is associated with a 59% lifetime risk of epithelial ovarian cancer while *gBRCA2* has a 16.5% lifetime risk [10]. *gBRCA1/2* mutated ovarian cancers typically present with high-grade serous histology and advanced stage, but are associated with longer overall survival than their sporadic counterparts [11]. Individuals with *BRCA1/2* mutations also have an elevated risk of other malignancies such as pancreatic, cervical, and prostate cancer [12].

BRCA GENOTYPE-PHENOTYPE ASSOCIATIONS

Early on, it was observed that patients with BRCA-associated ovarian cancer had a better prognosis [13]. In the largest study to date that pooled analyses of 26 studies (including 304 *BRCA2* carriers, 909 *BRCA1* carriers, and 2666 patients with wild-type ovarian cancers), *BRCA2* carriers were notably older with more advanced-stage tumors when compared with *BRCA1* carriers ($p < 0.001$). Despite this, five year overall survival for *BRCA2m* carriers was 52%, compared with 44% in *BRCA1m* carriers, and only 36% in noncarriers [14]. In general, the conclusion has been that BRCA mutations predict for response to platinum and perhaps to anthracyclines [15].

Linking the structural and functional domains of the BRCA genes was anticipated to reveal genotype-phenotype associations, so that risk could be further predicted by specific mutations, that for example disrupt the RING finger, *Rad51* binding, or *ATM* phosphorylation [16]. However, this has not yet been apparent, other than in specific situations where the BRCA mutations confer a very different phenotype, such as certain mutations, such as *BRCA1* C61G, that may lead to poorer outcomes [17]. There has been debate about whether other mutations may confer platinum resistance [18,19].

OTHER OVARIAN CANCER ASSOCIATED SYNDROMES

Other syndromic genes associated with ovarian cancer are the Fanconi Anemia (FA) genes, Lynch II syndrome (HNPCC or hereditary nonpolyposis colorectal cancer), mismatch repair (MMR) family genes, including *MSH2*, *MLH1*, *MSH6*, and *PMS2*, basal cell nevus (Gorlin) syndrome with *PTCH1* mutations, and Werner's syndrome with *MEN1* mutations.

FA-BRCA PATHWAY

Other key components of HR are being elaborated in what is now being called the FA-BRCA pathway. Mutations in these genes can confer significant cancer risk [20]. *TP53*, *PTEN* (Mutated in Cowden's syndrome), and *STK11* (Peutz-Jeghers) are most proximal to DNA damage and then activate the *FA*

Core complex or ATM. *ATMs* have a penetrance of 15% and relative risk (RR) 2–4× for breast and pancreatic cancers, though not ovarian cancer [21]. *ATM* activated *CHEK2* has a 1100delC mutation carrier prevalence of 1% in the Dutch and a RR 2–5× of breast, ovarian, and other cancers [22]. *Partner and localizer BRCA2* (*PALB2*) has a mutation carrier rate of 0.08% and a 40% chance of triple negative breast, ovarian, and pancreatic cancers by 70 years of age [23].

MORE THAN JUST *BRCA*: STRONG PENETRANCE GENES

Since the discovery of the *BRCA1/2* genes, there has been a growing appreciation for the complexity of DNA repair. While mutations in genes other than *BRCA1/2* are independently rare, together they make up a substantial proportion of families with breast and ovarian cancer. Despite this, however, roughly 50% of familial breast cancer remains unresolved by gene testing [5]. While the relative weight of *BRCA1/2* in discussions about breast cancer genetics is slowly decreasing, screening for them is mainstream. As new genes are identified and pathways are elucidated, more diagnostic data, potential for screening and prophylaxis, and more effective therapies will become available. While *BRCA1/2* are classically high penetrance genes for breast and ovarian cancer, other high penetrance genes include *TP53*, *STK11*, and *PTEN*.

Li-Fraumeni syndrome is autosomal dominant condition associated with the development of early onset cancers, caused by *TP53* germline mutation and is a significant risk factor for the development of breast (49% risk by age 60) and ovarian malignancy, with one-third of the breast cancers diagnosed before age 30 [24,25].

Peutz-Jeghers syndrome is caused by a mutation in *STK11*, a serine/threonine kinase tumor suppressor gene, with development of hamartomas around the lips, hands, and genitals, as well as polyps throughout the GI tract. The syndrome is associated with cancers of the GI, pancreas, lung, uterus, breast, and ovary. These patients have a RR of 15 (CI 7.2–27), with an average age of diagnosis at 37 [26].

Cowden syndrome is autosomal dominant condition characterized by the development of multiple hamartomas on the skin and mucous membranes, and increased risk of the breast, thyroid, endometrium, and ovarian malignancies, due to a mutation in the tumor suppressor gene *PTEN*. The lifetime risk for breast cancer is 85.2% (95% CI 71.4–99.1) [27].

These syndromes have theoretical risks for ovarian cancer but no clear association has been demonstrated yet.

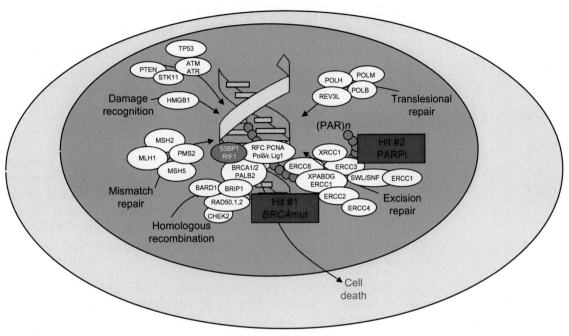

FIGURE 7.1
DNA repair genes and high-grade serous ovarian cancer.

FA-RELATED MODERATE PENETRANCE GENES

FA is an X-linked, rare inherited blood disorder that manifests with childhood bone marrow failure and increased susceptibility to malignancies, most commonly lymphoma. FA as a disease consists of 13 different complementation groups, each manifesting with a particular pattern. The FA-BRCA pathway is critical for maintaining genetic integrity through DNA repair by HR, and broadened our understanding of the development of breast and ovarian carcinoma. While the pathway is still being investigated, it has been shown that eight FA proteins form a complex that is activated by *FANCD2* ubiquitination. The modified *FANCD2* translocates to damaged nuclear centers containing *BRCA1*, *BRCA2*, and *RAD51*, allowing recognition and DNA repair [28]. Disruption of this process impairs the ability of cells to repair damaged DNA before replication due to less accurate nonhomologous end-joining repair, with accumulation of mutations resulting in genomic instability and apoptosis [29]. These mutations are related to medulloblastoma, leukemia, breast, and ovarian cancer. The phenotype of patients with *FANCD1* mutations is similar to those with *gBRCA2mut*.

Few others genes of FA-BRCA pathway are related to ovarian cancer risk, with moderate penetrance, such as *CHEK2*, *PALB2*, and *ATM*.

CHEK2 is part of the FA-BRCA pathway and involved in repair pathways, and checkpoint function. *CHEK2* mutation 1100delC has been specially identified as a risk factor for breast cancer. Studies estimate that this mutation is found in 1–2% of the population [30], and 5% of individuals with *BRCA1/2* mutation-negative breast cancer and 14% of individuals from families with male breast cancer [28].

PALB2 is another member of the FA-BRCA pathway. *PALB2* protein intimately binds to *BRCA2*, stabilizing it and allowing it to perform its reparative functions, essential for *BRCA2* tumor suppression and damage control activity. Biallelic mutation of *PALB2* results in FA. *PALB2* is associated with eightfold higher risk for breast malignancy in women older than 40 [23]. There is very likely an association of ovarian cancer risk even though no clear association has been demonstrated yet.

Ataxia telangiectasia is an autosomal recessive neurodegenerative disease that causes cerebellar dysfunction, immune deficiency, and increased risk of breast cancer, due to inability to repair broken DNA. *ATM* protein is an early initiator in the FA-BRCA pathway as an important cell cycle checkpoint kinase that phosphorylates *BRCA1*.

Heterozygous *ATM* mutations carry a RR of breast cancer of ≥ 2 (95% CI 1.90–12.9), which increases to fivefold in those below age 50 [31].

LOW PENETRANCE GENES

There are also a number of genes that are not as well characterized, with only preliminary data suggesting an increased risk of both breast and ovarian cancer. Among these are *BRIP1*, *RAD51C*, and *RAD51D*. These genes act as coeffectors with *BRCA1* and *BRCA2*, and were found in individuals with *mBRCA* breast cancer families with increased risk of disease when present [32–35].

BRCA TESTING FOR OVARIAN CANCER

An increasing number of targeted therapies are designed for cancer, so the genetic testing has become an essential part of the evaluation of ovarian cancer.

In the past, clinicians only tested women with ovarian cancer thought to be at high risk of carrying a *BRCA* mutation, indicated by a family history of two or more first degree relatives (mother, sister, daughter) with breast or ovarian cancer, with at least one relative diagnosed under 50 years of age. However, it has been demonstrated that family history and age at diagnosis are poor predictors of *BRCA* status in ovarian cancer patients. Guidelines from organizations such as National Comprehensive Cancer Network (NCCN), American Society of Clinical Oncology, and Society of Gynecologic Oncology now recommend that all patients with epithelial ovarian cancer be tested for g*BRCA*mut. Approximately 15% of women with ovarian cancer have a deleterious *BRCA* mutation [36,37]. Almost half (47%) of *BRCA*-positive ovarian cancer patients have no significant family history of ovarian or breast cancer [38]. Over two-thirds (71%) of *BRCA*-positive ovarian cancer patients are aged 50 or older [39]. Using next-generation sequencing, a Japanese group in a single hospital evaluated 95 unselected women with ovarian cancer between 2013 and 2015. Twelve of the 95 patients (13%) had deleterious mutations. Among the 36 cases with a family history, 6 (17%) had *BRCA* mutations, and 6 of the 59 cases (10%) without a family history also had *BRCA* germline mutations ($p = 0.36$) indicating that *BRCA1/2* genetic testing should be performed for all patients with ovarian cancers, not just those that fall within the previously favored limitations [40]. The incidence of homologous recombination deficiency (HRD) mutations and epigenetic events that might contribute to PARP inhibitor sensitivity is illustrated in Fig. 7.2 based on the Cancer Genome Atlas [41].

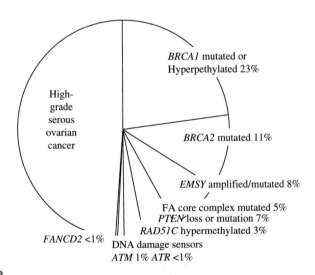

FIGURE 7.2

PARPi mechanism of action. *From Peng G, Lin SY. Exploiting the homologous recombination DNA repair network for targeted cancer therapy. World J Clin Oncol. 2011;2(2):73–9; Wiegand KC, Shah SP, Al-Agha OM, et al. ARID1A mutations in endometriosis-associated ovarian carcinomas. N Engl J Med. 2010;363(16):1532–43; Pearce CL, Templeman C, Rossing MA, et al. Association between endometriosis and risk of histological subtypes of ovarian cancer: a pooled analysis of case-control studies. Lancet Oncol. 2012;13(4):385–94; Pennington KP, Swisher EM. Hereditary ovarian cancer: beyond the usual suspects. Gynecol Oncol. 2012;124(2):347–53; Lord CJ, Ashworth A. Mechanisms of resistance to therapies targeting BRCA-mutant cancers. Nat Med 2013;19(11):1381–8.*

Next-generation sequencing has enabled low cost multiplex genetic testing, with the identification of other, less penetrant genes that also confer a risk of ovarian cancer. Remarkably, these screening techniques are often cheaper than *BRCA* testing, and potentially cover HRD-related gene mutations that may have clinical significance.

SCREENING AND PROPHYLACTIC SURGERY

Organizations including the American Cancer Society and NCCN have established clear guidelines for highly penetrant genes that have a well-established link to breast and ovarian cancer, such as *BRCA1/2*, *TP53*, and *PTEN* [42]. Screening methods from 25 years include annual mammography and breast MRI, CA125 and TVUS surveillance is recommended during the child bearing years, and risk-reducing surgery is advocated at age 40, or 10 years earlier than the earliest age at which a relative was diagnosed. The prophylactic use of selective estrogen receptor modulators and/or aromatase inhibitors has a growing evidence base. No recommendations are present for moderate to low penetrance gene mutations.

POLY (ADP-RIBOSE) POLYMERASE

PARP was discovered in 1962 and was initially anticipated to facilitate tissue protection and repair. However, it was not until the 1990s that its role in DNA repair began to be fully understood and exploited. The PARP family of poly(ADP-ribosyl)ating proteins contains 17 members [43]. They all use NAD^+ as substrate to polymerize adenosine diphosphate to make (ADP-ribose)$_n$. PARP-1 is the most studied, founding member family and probably most important member, but generally PARP inhibitors lack enzyme specificity. PARP-1 has a DNA-binding domain, automodification domain, and PARP homology catalytic domain. PARPs are nuclear proteins with DNA-binding domains that localize PARP to the site of DNA damage servicing as DNA damage sensors and signaling molecules for repair. PARP activity is essential for repair of single-strand DNA breaks through NHEJ, BER, and other mechanisms. Knocking out PARP-1 is sufficient to significantly impair DNA repair after radiation or cytotoxic insult. PARP-1 is also involved in innate immunity and inflammation, stress responses, cellular and organismal metabolism, pluripotency, differentiation, and hormonal signaling. Higher expression of PARP in cancer cells leads to general drug resistance.

PARP is part of the DNA repair mechanism, forming a polymer scaffold for other important enzymes. If this system is inhibited, DNA repair cannot happen if other important DNA repair systems are compromised, as when a patient carries a *BRCA1* or *2* mutation.

PARP INHIBITOR MECHANISM AND EARLY CLINICAL DEVELOPMENT HISTORY

PARP inhibitors first emerged as an approach of interest in cancer after data demonstrated a potential therapeutic effect in tumor cell lines with *BRCA* mutations [44]. In the original studies the novel PARP inhibitor, AG14361, restored sensitivity to temozolomide in mismatch repair-deficient cells [45]. Pharmacology has manipulated this concept in order to selectively target cancer cells, allowing for minimal side effect [46]. A significant unsuccessful result in the development of PARP inhibitors was the drug iniparib (BSI-201), which failed in the clinical setting [47], yet providing molecular basis of the PARP inhibitory activity [48]. In 1997, Steve Jackson was the founder and first developer of AZ2281, aka olaparib. First tested in *gBRCA*mut patients however after clinical success, it is clear that the range of treated patients may become broader [49]. PARP inhibitors prevent the normal compensation for HRDs, interfering with tumor DNA repair [50]. Although HR is the most important high-fidelity DNA repair mechanism, other repair mechanisms can compensate for loss of HR repair. However, HRD sets up an intrinsic vulnerability, a "synthetic lethality" or "Achilles' Heel" and inhibiting these other DNA repair mechanisms with a PARP inhibitor can trigger apoptosis [51]. Inhibition of PARP leads to the formation of double-strand breaks, as single-strand breaks become double-strand breaks during replication [52]. PARP-1 Trapping may be an important element to PARP inhibition [53]. This PARP-1 Trapping appears to be most potent for Biomarin's BMN-673, aka talazoparib [54]. Synthetic lethality and the mechanism of PARP inhibition is illustrated in Fig. 7.2.

CLINICAL STUDIES OF PARP INHIBITORS IN OVARIAN CANCER: OLAPARIB

Early clinical studies with olaparib demonstrated durable responses in *BRCA* carrier ovarian cancer tumors [55]. In the Phase I clinical trial, a clinical response was demonstrated in 63% of the 19 *BRCA* carriers who had ovarian, breast, or prostate cancers on olaparib 200 mg bid [56]. In the Phase II study (ICEBERG), the olaparib 400 mg bid cohort performed substantially better than the 100 mg bid cohort with a 33% response rate [57]. Furthermore, a Canadian study in breast and ovarian cancer patients observed an overall response rate in ovarian cancer patients of 29% (18/63) with no responses in *gBRCA*mut breast cancer (0/8) and triple negative breast cancer (0/15) [58]. Notably, the response rate in platinum-resistant *gBRCA*mut points was 33% (4/12), and the response rate was similar in weight *BRCA* platinum-sensitive tumors (50% (10/20)) and *gBRCA*mut platinum-sensitive tumors (60% (3/5)).

The race to a robust efficacy signal has been challenging. A small randomized Phase II trial (Study 12), which compared olaparib 200 mg bid, 400 mg bid, and pegylated liposomal doxorubicin 50 mg/m^2 in 97 women with *BRCA*-mutated tumors that had recurred in <12 months after platinum, could not demonstrate any difference in response or survival [59]. Concrete evidence of benefit instead came from studies of maintenance olaparib [60,61] (please refer to Chapter 6).

GO-NO-GO DECISION

The clinical development of the PARP inhibitors has been a dramatic roller-coaster. Pharmaceutical companies typically have a "LIP" when they commit to further development at the Late-stage Investment Point (LIP). Anticipating a blockbuster drug, they may set a high bar for Phase II requiring a significant signal of efficacy, given the Food and Drug Administration (FDA) requires for an improvement in survival or clinical benefit. Given the demonstrated smaller effect, on December 20, 2012, AstraZeneca announced that olaparib would not progress to Phase III: the challenges were that patients with *BRCA1/2*-associated cancer represent a small target population and so limit financial return, iniparib had failed to demonstrate efficacy in triple-negative breast cancer, and olaparib did not have orphan drug status because of the broad options to recurrent ovarian cancer patients [62]. A reversion in olaparib investigation after a secondary reinvestment, announced in September 4, 2013 [63], that brought to successful clinical trials [60,61] (see further detail in chapter: Ovarian Cancer: New Targets and Future Directions).

FDA approved olaparib in the United States as palliative therapy for patients with three or more prior lines of chemotherapy, where it was approximately twice as effective as chemotherapy in *gBRCAmut* patients with a response rate of 31% and a median PFS of 7 months [64]. It was coapproved with a companion diagnostic (CDx), Myriad's *BRCA* mutation assay, which identifies those patients most likely to benefit from further disruption of DNA repair (synthetic lethality).

European Medicines Agency approval, in contrast, did reflect the data from the maintenance study, (Study 19) with a 7-month prolongation in median PFS [60].

Given the large margin of benefit observed with maintenance therapy, two additional trials were opened: Study of OLaparib in Ovarian Cancer (SOLO)-1 after first line chemotherapy and SOLO-2 in second or subsequent remission of platinum-sensitive recurrent ovarian cancer [65].

REGISTRATION STRATEGIES

The FDA approves new agents with the convincing demonstration of a survival advantage or clinical benefit. The approval of bevacizumab and olaparib in 2014 marked the first new drugs approved in ovarian cancer for 8 years. The perceived and actual barriers in pathways to an approved indication were mainly focused on the issue of using PFS as a surrogate endpoint. This was compounded by a clinical setting in which multiple palliative options were available, and patients enjoyed a relatively long survival after recurrence. Different approaches are illustrated in Fig. 7.3. The best strategy remains controversial, particularly with the accelerated approval given to olaparib based on pooled Phase II data. Many clinical trial sponsors remain concerned about maintenance approaches as patients' health-related quality of life will be negatively impacted by toxicity while in remission.

COMBINATION THERAPY

Because of their hematologic toxicity PARP inhibitors are difficult to combine with myelosuppressive chemotherapy. In fact, adding olaparib to carboplatin required a 33% reduction in the platinum dose (from AUC 6 to 4) [61]. Combination with other biologics without overlapping toxicity has more potential. Liu et al. combined the oral antiangiogenic, cediranib with olaparib in women with recurrent platinum-sensitive ovarian cancer in a high-impact randomized phase 2 study [66]. Participants had measurable platinum-sensitive, relapsed, high-grade serous or endometrioid ovarian, fallopian tube, or primary peritoneal cancer, or deleterious germline *BRCA1/2* mutations and were randomly allocated to olaparib capsules 400 mg bid ($n = 46$) or 200 mg bid with cediranib 30 mg orally qd ($n = 44$). Median PFS was 17·at 7 months (95% CI 14.7–not reached) for the women treated with cediranib plus olaparib compared with 9 months (95% CI 5.7–16.5) for those treated with olaparib monotherapy (HR 0.42, 95% CI 0.23–0.76; $p = 0.005$). Grade 3 and 4 adverse events were more common with combination therapy than with monotherapy, including fatigue (12 vs 5 patients), diarrhea (10 vs 0), and hypertension (18 vs 0). Phase III trials have opened in both platinum-sensitive and -resistant disease, and the combination with bevacizumab is also in Phase III.

In contrast, veliparib can be combined with full dose carboplatin AUC 6 and weekly paclitaxel (Bell-McGuinn, personal communication—see Fig. 7.3). This is likely because the veliparib PARPi Ki is 5 nM in contrast to olaparib's (1.1 nM) and rucaparib (0.8 nM). In combination with oral cyclophosphamide, veliparib did not perform well, very likely because of a wholly inadequate dose (60 mg PO qd), although this was the Phase I defined dose for the

(A) Study design: concurrent and maintenance

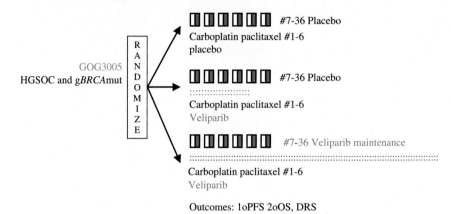

GOG3005
HGSOC and g*BRCA*mut

RANDOMIZE

#7-36 Placebo
Carboplatin paclitaxel #1-6
placebo

#7-36 Placebo
Carboplatin paclitaxel #1-6
Veliparib

#7-36 Veliparib maintenance
Carboplatin paclitaxel #1-6
Veliparib

Outcomes: 1oPFS 2oOS, DRS

(B) Study design: maintenance

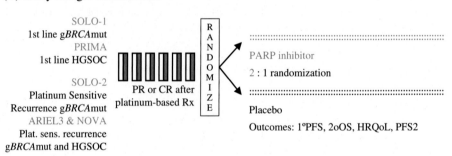

SOLO-1
1st line g*BRCA*mut
PRIMA
1st line HGSOC

SOLO-2
Platinum Sensitive
Recurrence g*BRCA*mut
ARIEL3 & NOVA
Plat. sens. recurrence
g*BRCA*mut and HGSOC

PR or CR after
platinum-based Rx

RANDOMIZE

PARP inhibitor
2 : 1 randomization

Placebo
Outcomes: 1°PFS, 2oOS, HRQoL, PFS2

(C) Study design: treatment

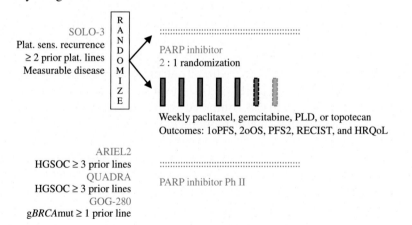

SOLO-3
Plat. sens. recurrence
≥ 2 prior plat. lines
Measurable disease

RANDOMIZE

PARP inhibitor
2 : 1 randomization

Weekly paclitaxel, gemcitabine, PLD, or topotecan
Outcomes: 1oPFS, 2oOS, PFS2, RECIST, and HRQoL

ARIEL2
HGSOC ≥ 3 prior lines
QUADRA
HGSOC ≥ 3 prior lines
GOG-280
g*BRCA*mut ≥ 1 prior line

PARP inhibitor Ph II

FIGURE 7.3

Registration strategies. SOLO-1 (Studies of OLaparib in Ovarian Cancer NCT01844986); NRG-GOG3005 (NCT02470585); SOLO-2 (NCT01874353); ARIEL3 (NCT01968213); NOVA (Niraparib in 2nd Line (Recurrent) Ovarian Cancer Maintenance NCT01847274); SOLO-3 (NCT02282020); QUADRA (NCT02354586); PFS2 includes analysis of time to first subsequent therapy (TFST) and time to second subsequent therapy (TSST); HGSOC (high-grade serous ovarian cancer); Disease-Related Symptom (DRS) scores.

combination. Rucaparib encountered similar problems to olaparib in Phase I with carboplatin.

OTHER PARP INHIBITORS

PARP has received a lot of interest with a rapidly expanding portfolio of drugs (Table 7.1). Rucaparib is the only other PARP inhibitor with clinical data that has been presented. ARIEL2, the Phase II trial of rucaparib, also had a major goal to prospectively predictive testing using tumor genetic analysis [67]. Certain HRD-related genotypes such as *RAD51C* mutation were as predictive

Table 7.1 PARP Inhibitors

PARPi	Phase	Company	Route	Studies
BMN-673 Talazoparib	I	Medivation	PO	Single agent Stalled
CEP-8983 (Prodrug CEP-9722)	I	Cephalon	PO	Single agent and combination with temozolomide Bought by Teva
E7016	I	Eisai	PO	Combination with temozolomide Stalled
GPI-21016	Preclinic	MGI-pharma	PO/IV	Ameliorates cisplatin nephrotoxicity Stalled
Iniparib BSI-201	II–III	BiPar/Sanofi	IV	Single agent and in combination with: carboplatin, gemcitabine, topotecan—breast ovary Failed
INO-1001	I	Inotek Genentech	IV	Shelved
Niraparib MK4827	III	Merck Tesaro	PO	Single agent and combination with carboplatin/paclitaxel and temozolomide
Olaparib AG14361 KU0058684 AZD-2281	III	Agouron KuDOS AstraZeneca	PO	Single agent (RPhII) and combo with cediranib, cisplatin, carbo, cyclo., lip. dox, carboTax, and other small molecules
Rucaparib AG014699 CO-338	III	Agouron Pfizer Clovis	IV	Single agent Original 3-aminobenzamide derivative
Veliparib ABT-888	III	Abbott	PO	Single agent (Japan) and combo Cyclo., temozolomide, carboT, lipo. dox, gem ± carbo, topo, irinotecan, mitomycin, and RT

of response as *BRCA*mut others, for example, *ATM* were not. The side effects were very similar to olaparib (nausea and fatigue) but there was more liver function abnormality and no myelodysplasia. It also caused more anemia but less neutropenia and 35% needed dose reductions. The last contrast to olaparib is that rucaparib also inhibits Tankyrase (PARP-5 and -7), although it is not known if this is an important target. In the *BRCA*mut group ($n = 35$) PFS was 9.4 months and the RR was 69% (RECIST).

The efficacy of Niraparib versus placebo as maintenance treatment for patients with platinum-sensitive, recurrent ovarian cancer was recently demonstrated in ENGOT-OV16/NOVA trial [XX]. In the trial patients were stratified based on the presence of a *gBRCA* mutation, and all treatment cohorts had a significantly longer median duration of PFS than did those in the placebo group: 21.0 vs. 5.5 months in the *gBRCA* cohort (hazard ratio, 0.27; 95% confidence interval [CI], 0.17 to 0.41); 12.9 months vs. 3.8 months in the non-*gBRCA* cohort (tumors with homologous recombination deficiency) (hazard ratio, 0.38; 95% CI, 0.24 to 0.59) and 9.3 months vs. 3.9 months in the overall non-*gBRCA* cohort (hazard ratio, 0.45; 95% CI, 0.34 to 0.61; $P < 0.001$ for all three comparisons). Moderate bone marrow toxicity was present: grade 3 or 4 adverse events were observed in the treatment group with thrombocytopenia in one third of the cases, anemia in one fourth, and neutropenia in one fifth [68].

PREDICTING RESPONSE TO PARP INHIBITOR THERAPY

Predictive markers may help avoid the use of ineffective therapy, but more importantly it increases the potential for benefit in other patients' subgroups and disease. Swisher et al. have developed a "DNA scar" test to prospectively identify ovarian cancer patients who respond to rucaparib [69]. Evaluating tumor LOH as a marker of HRD status (DNA "scarring") was predictive of benefit. *t*BRCALOH was associated with a HR of 0.61, and the *BRCA*wt/LOHhigh group ($n = 56$) had a PFS of 7.1 months compared with 3.7 months in the *BRCA*wt/LOHlow group ($n = 44$). Response rates were 30%, and 13%, respectively [67]. This predictive biomarker appears promising for all PARP inhibitors [70].

Abkevich, Mills, and Lanchbury have also developed a HRD score which reflects the number of long LOH regions [71]. They do not revert over time, and so are truly "scarred," and these patterns of heterozygous genomic loss predict HR repair defects in epithelial ovarian cancer. A pressing goal is to expand this sort of evaluation of DNA repair to identify the potential for benefit from PARP inhibitors more broadly [72].

COMBATING MECHANISMS OF PARP INHIBITOR RESISTANCE

An important clinical problem with PARP inhibitor treatment is the development of resistance. Classical mechanisms of drug resistance and drug-specific escape pathways, such as upregulation and loss of PARP-1 expression, have been demonstrated [73]. Other mechanisms are being explored, including efflux of the drug through overexpression of P-glycoprotein transporter [74], and increase in PARP-1 and other PARP expression [75]. Interestingly, the increase in genomic plasticity because of impaired DNA repair increases clonal variation, known as "Darwinian Escape" [76].

53BP1 and RIF1 have opposing activity to BRCA1 and prevent DNA resection and HR, and so confer a greater need for NHEJ. This and other counterbalancing DNA repair mechanisms have not as yet been described as resistance mechanisms, but likely contribute.

Perhaps the most elegant resistance mechanism is the occurrence of secondary mutations in BRCA alleles which restore wild-type BRCA activity [77]. They do this through frame shift proximal to a truncation mutation, that then results is a distorted, but full length transcript. These sorts of secondary mutations also confer platinum resistance, and the interaction of the effects of tumor genetic instability is an important research focus [18].

OTHER POTENTIAL OPPORTUNITIES

Perhaps the key and immediate challenge is to define other groups of patients who may benefit from PARP inhibition. The latest high impact study was recently published in the New England Journal of Medicine, and evaluated olaparib in a randomized phase II trial in patients with metastatic, castration-resistant prostate cancer [78] where 33% (95% CI, 20–48) had response. A total of 88% of patients with overall response had mutations in DNA repair genes. Also gastric cancer with low levels of ATM were demonstrated sensitive to PARP inhibitor olaparib in a Phase II trial where combination treatment with paclitaxel and maintenance with olaparib demonstrated improved overall survival versus placebo in both the overall population (HR, 0.56; 80% CI, 0.41–0.75; $p = 0.005$; median OS, 13.1 vs 8.3 months) and the ATM-low population (HR, 0.35; 80% CI, 0.22–0.56; $p = 0.002$; median OS, not reached vs 8.2 months).

The role of PTEN loss or mutation as a HRD marker remains controversial. One provocative case report in a woman with metastatic endometrioid endometrial adenocarcinoma with profound sensitivity to platinum, and notable benefit of olaparib in a phase I trial (DFI 8 months) [79].

Interestingly, PD-L1 upregulation occurs in high-grade serous ovarian cancer and the new frontier is to explore the combination of PARP inhibitors and check-point inhibitors [80].

OTHER DNA REPAIR TARGETED THERAPIES

Many tumors harbor mutations in the tumor suppressor gene, *TP53*, allowing for unregulated cellular growth and proliferation, and this remains a major target, with no therapeutic agent as yet. New hope arrived with the formulation of compound *APR-246*, part of the new quinuclidinones chemical class. *APR-246* is able to induce mutated, nonfunctional *TP53* to form a more stable, functional molecule. This "reactivated" p53 is able to perform its tumor suppressor functions, and eliminate the tumor cell [54]. Indeed, testing is already underway to examine the *APR-246*'s action against tumors like Ewing sarcoma [55]. The Phase Ib/II trial is with *APR-246* combination with carboplatin (AUC 5) and pegylated doxorubicin (30 mg/m^2), a popular second line standard of care chemotherapy for relapsed platinum-sensitive high-grade serous ovarian cancer, the PiSARRO trial.

Other researchers have shown that PARP inhibitors may be beneficial when used in cells with non-*BRCA* mutations that affect HR, namely genes in the FA-BRCA pathway [81]. For instance, researchers have shown that cells deficient in *RAD51D* have increased sensitivity to PARP inhibitors, highlighting the future potential of targeted therapy in patients with identifiable mutations in genes like *CHEK2*, *ATM*, and *RAD51D* [35].

CH(E)K1/wee1 inhibition can trigger a G2/M stop and lead to catastrophic mitotic death [82]. In acute leukemia AZ1775 has been combined with olaparib, and separately they are cytostatic but together appear to be cytotoxic.

VX970 is a promising inhibitor of DNA repair, and the *ATR* inhibitors VE-821 and VX-970 sensitize cancer cells to topoisomerase I inhibitors by disabling DNA replication initiation and fork elongation responses [83].

As previously discussed, the *CHK1* inhibitor UCN-01 has been shown to increase the sensitivity of *TP53*-mutated cells to ionizing radiation at the G2 checkpoint, and looks to be a potential agent with synergy with PARP inhibitors [30]. *CHK2* inhibitors (PF47736, AZD7762) similarly show potential. *CHK2*, along with *ATM*, is involved in communicating DNA damage signals to p53 [84]. Indeed, numerous studies have shown *CHK2* inhibitors to induce chemotherapy-sensitivity [85,86]. *ATM* (CP466722, KU55933) inhibitors are in also development, and have been shown to potentiate the effect of ionizing radiation as well, causing a loss of damage-induced cell cycle arrest [87]. One can image a time when genotyping is sophisticated enough to recommend a cocktail of specific targeted inhibitors in real-time, directed at each unique genotypic background [88].

CONCLUSION

Since the completion of the cancer genome atlas, we have been putting together a better understanding of genetics and the molecular biology of cancer. The PARPi story is one of the best examples of both the rewards of following the science through translational developments that benefit patients and the multiple challenges that obstruct progress. A deeper appreciation of the complexity of human biology has not slowed the pace of new drug development, and the FDA approval of olaparib in 2014 marked a new season of hope for effecting treatment of these dreadful diseases.

References

[1] Hoeijmakers JH. Genome maintenance mechanisms for preventing cancer. Nature 2001;411(6835):366–74.

[2] Matsuoka S, Ballif BA, Smogorzewska A, McDonald 3rd ER, Hurov KE, Luo J, et al. ATM and ATR substrate analysis reveals extensive protein networks responsive to DNA damage. Science 2007;316(5828):1160–6.

[3] Jasin M. Accolades for the DNA damage response. N Engl J Med 2015;373(16):1492–5.

[4] Ding J, Miao ZH, Meng LH, Geng MY. Emerging cancer therapeutic opportunities target DNA-repair systems. Trends Pharmacol Sci 2006;27(6):338–44.

[5] Walsh T, King MC. Ten genes for inherited breast cancer. Cancer Cell 2007;11(2):103–5.

[6] Yang D, Khan S, Sun Y, Hess K, Shmulevich I, Sood AK, et al. Association of BRCA1 and BRCA2 mutations with survival, chemotherapy sensitivity, and gene mutator phenotype in patients with ovarian cancer. JAMA 2011;306(14):1557–65.

[7] Lindor NM, McMaster ML, Lindor CJ, Greene MH, National Cancer Institute DoCPCO, Prevention Trials Research Group Concise handbook of familial cancer susceptibility syndromes - second edition. J Natl Cancer Inst Monogr 2008;38:1–93.

[8] Fortini P, Dogliotti E. Base damage and single-strand break repair: mechanisms and functional significance of short- and long-patch repair subpathways. DNA Repair (Amst) 2007;6(4):398–409.

[9] Somasundaram K. Breast cancer gene 1 (BRCA1): role in cell cycle regulation and DNA repair—perhaps through transcription. J Cell Biochem 2003;88(6):1084–91.

[10] Mavaddat N, Peock S, Frost D, Ellis S, Platte R, Fineberg E, et al. Cancer risks for BRCA1 and BRCA2 mutation carriers: results from prospective analysis of EMBRACE. J Natl Cancer Inst 2013;105(11):812–22.

[11] Chetrit A, Hirsh-Yechezkel G, Ben-David Y, Lubin F, Friedman E, Sadetzki S. Effect of BRCA1/2 mutations on long-term survival of patients with invasive ovarian cancer: the national Israeli study of ovarian cancer. J Clin Oncol 2008;26(1):20–5.

[12] Thompson D, Easton DF, Breast Cancer Linkage C Cancer Incidence in BRCA1 mutation carriers. J Natl Cancer Inst 2002;94(18):1358–65.

[13] Boyd J, Sonoda Y, Federici MG, Bogomolniy F, Rhei E, Maresco DL, et al. Clinicopathologic features of BRCA-linked and sporadic ovarian cancer. JAMA 2000;283(17):2260–5.

[14] Bolton KL, Chenevix-Trench G, Goh C, Sadetzki S, Ramus SJ, Karlan BY, et al. Association between BRCA1 and BRCA2 mutations and survival in women with invasive epithelial ovarian cancer. JAMA 2012;307(4):382–90.

[15] Safra T, Lai W, Borgato L, Nicoletto M, Berman T, Reich E, et al. BRCA mutations and outcome in epithelial ovarian cancer (EOC): experience in ethnically diverse groups. Ann Oncol 2013;24(Suppl. 8):viii, 63–8.

[16] Gibson BA, Kraus WL. New insights into the molecular and cellular functions of poly(ADP-ribose) and PARPs. Nat Rev Mol Cell Biol 2012;13(7):411–24.

[17] Drost R, Bouwman P, Rottenberg S, Boon U, Schut E, Klarenbeek S, et al. BRCA1 RING function is essential for tumor suppression but dispensable for therapy resistance. Cancer Cell 2011;20(6):797–809.

[18] Sakai W, Swisher EM, Karlan BY, Agarwal MK, Higgins J, Friedman C, et al. Secondary mutations as a mechanism of cisplatin resistance in BRCA2-mutated cancers. Nature 2008;451(7182):1116–20.

[19] Fojo T, Bates S. Mechanisms of resistance to PARP inhibitors—three and counting. Cancer Discov 2013;3(1):20–3.

[20] Pennington KP, Swisher EM. Hereditary ovarian cancer: beyond the usual suspects. Gynecol Oncol 2012;124(2):347–53.

[21] Roberts NJ, Jiao Y, Yu J, Kopelovich L, Petersen GM, Bondy ML, et al. ATM mutations in patients with hereditary pancreatic cancer. Cancer Discov 2012;2(1):41–6.

[22] Huzarski T, Cybulski C, Wokolorczyk D, Jakubowska A, Byrski T, Gronwald J, et al. Survival from breast cancer in patients with CHEK2 mutations. Breast Cancer Res Treat 2014;144(2):397–403.

[23] Antoniou AC, Casadei S, Heikkinen T, Barrowdale D, Pylkas K, Roberts J, et al. Breast-cancer risk in families with mutations in PALB2. N Engl J Med 2014;371(6):497–506.

[24] Hwang SJ, Lozano G, Amos CI, Strong LC. Germline p53 mutations in a cohort with childhood sarcoma: sex differences in cancer risk. Am J Hum Genet 2003;72(4):975–83.

[25] Birch JM, Hartley AL, Tricker KJ, Prosser J, Condie A, Kelsey AM, et al. Prevalence and diversity of constitutional mutations in the p53 gene among 21 Li-Fraumeni families. Cancer Res 1994;54(5):1298–304.

[26] Giardiello FM, Brensinger JD, Tersmette AC, Goodman SN, Petersen GM, Booker SV, et al. Very high risk of cancer in familial Peutz-Jeghers syndrome. Gastroenterology 2000;119(6):1447–53.

[27] Tan MH, Mester JL, Ngeow J, Rybicki LA, Orloff MS, Eng C. Lifetime cancer risks in individuals with germline PTEN mutations. Clin Cancer Res 2012;18(2):400–7.

[28] Meijers-Heijboer H, van den Ouweland A, Klijn J, Wasielewski M, de Snoo A, Oldenburg R, et al. Low-penetrance susceptibility to breast cancer due to CHEK2(*)1100delC in noncarriers of BRCA1 or BRCA2 mutations. Nat Genet 2002;31(1):55–9.

[29] Shimamura A, Montes de Oca R, Svenson JL, Haining N, Moreau LA, Nathan DG, et al. A novel diagnostic screen for defects in the Fanconi anemia pathway. Blood 2002;100(13):4649–54.

[30] Graves PR, Yu L, Schwarz JK, Gales J, Sausville EA, O'Connor PM, et al. The Chk1 protein kinase and the Cdc25C regulatory pathways are targets of the anticancer agent UCN-01. J Biol Chem 2000;275(8):5600–5.

[31] Renwick A, Thompson D, Seal S, Kelly P, Chagtai T, Ahmed M, et al. ATM mutations that cause ataxia-telangiectasia are breast cancer susceptibility alleles. Nat Genet 2006;38(8):873–5.

[32] Seal S, Thompson D, Renwick A, Elliott A, Kelly P, Barfoot R, et al. Truncating mutations in the Fanconi anemia J gene BRIP1 are low-penetrance breast cancer susceptibility alleles. Nat Genet 2006;38(11):1239–41.

[33] Shin DS, Pellegrini L, Daniels DS, Yelent B, Craig L, Bates D, et al. Full-length archaeal Rad51 structure and mutants: mechanisms for RAD51 assembly and control by BRCA2. EMBO J 2003;22(17):4566–76.

[34] Meindl A, Hellebrand H, Wiek C, Erven V, Wappenschmidt B, Niederacher D, et al. Germline mutations in breast and ovarian cancer pedigrees establish RAD51C as a human cancer susceptibility gene. Nat Genet 2010;42(5):410–14.

[35] Loveday C, Turnbull C, Ramsay E, Hughes D, Ruark E, Frankum JR, et al. Germline mutations in RAD51D confer susceptibility to ovarian cancer. Nat Genet 2011;43(9):879–82.

[36] Pal T, Permuth-Wey J, Betts JA, Krischer JP, Fiorica J, Arango H, et al. BRCA1 and BRCA2 mutations account for a large proportion of ovarian carcinoma cases. Cancer 2005;104(12):2807–16.

[37] Alsop K, Fereday S, Meldrum C, deFazio A, Emmanuel C, George J, et al. BRCA mutation frequency and patterns of treatment response in BRCA mutation-positive women with ovarian cancer: a report from the Australian Ovarian Cancer Study Group. J Clin Oncol 2012;30(21):2654–63.

[38] Song H, Cicek MS, Dicks E, Harrington P, Ramus SJ, Cunningham JM, et al. The contribution of deleterious germline mutations in BRCA1, BRCA2 and the mismatch repair genes to ovarian cancer in the population. Hum Mol Genet 2014;23(17):4703–9.

[39] Abkevich V, Timms KM, Hennessy BT, Potter J, Carey MS, Meyer LA, et al. Patterns of genomic loss of heterozygosity predict homologous recombination repair defects in epithelial ovarian cancer. Br J Cancer 2012;107(10):1776–82.

[40] Ikuko Sakamoto I, et al. BRCA1 and BRCA2 mutations in Japanese patients with ovarian, fallopian tube, and primary peritoneal cancer. Cancer Biol Ther 2015 Published online: <http://dx.doi.org/10.1002/cncr.29707>.

[41] Cancer Genome Atlas Research Network Integrated genomic analyses of ovarian carcinoma. Nature 2011;474(7353):609–15.

[42] Saslow D, Boetes C, Burke W, Harms S, Leach MO, Lehman CD, et al. American Cancer Society guidelines for breast screening with MRI as an adjunct to mammography. CA Cancer J Clin 2007;57(2):75–89.

[43] Hakme A, Wong HK, Dantzer F, Schreiber V. The expanding field of poly(ADP-ribosyl)ation reactions. 'Protein Modifications: Beyond the Usual Suspects' Review Series. EMBO Rep 2008;9(11):1094–100.

[44] Farmer H, McCabe N, Lord CJ, Tutt AN, Johnson DA, Richardson TB, et al. Targeting the DNA repair defect in BRCA mutant cells as a therapeutic strategy. Nature 2005;434(7035):917–21.

[45] Curtin NJ, Wang LZ, Yiakouvaki A, Kyle S, Arris CA, Canan-Koch S, et al. Novel poly(ADP-ribose) polymerase-1 inhibitor, AG14361, restores sensitivity to temozolomide in mismatch repair-deficient cells. Clin Cancer Res 2004;10(3):881–9.

[46] Peng G, Lin SY. Exploiting the homologous recombination DNA repair network for targeted cancer therapy. World J Clin Oncol 2011;2(2):73–9.

[47] Mateo J, Ong M, Tan DS, Gonzalez MA, de Bono JS. Appraising iniparib, the PARP inhibitor that never was—what must we learn? Nat Rev Clin Oncol 2013;10(12):688–96.

[48] O'Shaughnessy J, Osborne C, Pippen JE, Yoffe M, Patt D, Rocha C, et al. Iniparib plus chemotherapy in metastatic triple-negative breast cancer. N Engl J Med 2011;364(3):205–14.

[49] Hennessy BT, Timms KM, Carey MS, Gutin A, Meyer LA, Flake II DD, et al. Somatic mutations in BRCA1 and BRCA2 could expand the number of patients that benefit from poly (ADP ribose) polymerase inhibitors in ovarian cancer. J Clin Oncol 2010;28(22):3570–6.

[50] Patel AG, Sarkaria JN, Kaufmann SH. Nonhomologous end joining drives poly(ADP-ribose) polymerase (PARP) inhibitor lethality in homologous recombination-deficient cells. Proc Natl Acad Sci U S A 2011;108(8):3406–11.

[51] Iglehart JD, Silver DP. Synthetic lethality—a new direction in cancer-drug development. N Engl J Med 2009;361(2):189–91.

[52] Shrivastav M, De Haro LP, Nickoloff JA. Regulation of DNA double-strand break repair pathway choice. Cell Res 2008;18(1):134–47.

[53] Murai J, Huang SY, Das BB, Renaud A, Zhang Y, Doroshow JH, et al. Trapping of PARP1 and PARP2 by clinical PARP inhibitors. Cancer Res 2012;72(21):5588–99.

[54] Murai J, Huang SY, Renaud A, Zhang Y, Ji J, Takeda S, et al. Stereospecific PARP trapping by BMN 673 and comparison with olaparib and rucaparib. Mol Cancer Ther 2014;13(2):433–43.

[55] Fong PC, Yap TA, Boss DS, Carden CP, Mergui-Roelvink M, Gourley C, et al. Poly(ADP)-ribose polymerase inhibition: frequent durable responses in BRCA carrier ovarian cancer correlating with platinum-free interval. J Clin Oncol 2010;28(15):2512–19.

[56] Fong PC, Boss DS, Yap TA, Tutt A, Wu P, Mergui-Roelvink M, et al. Inhibition of poly(ADP-ribose) polymerase in tumors from BRCA mutation carriers. N Engl J Med 2009;361(2):123–34.

[57] Audeh MW, Carmichael J, Penson RT, Friedlander M, Powell B, Bell-McGuinn KM, et al. Oral poly(ADP-ribose) polymerase inhibitor olaparib in patients with BRCA1 or BRCA2 mutations and recurrent ovarian cancer: a proof-of-concept trial. Lancet 2010;376(9737):245–51.

[58] Gelmon KA, Tischkowitz M, Mackay H, Swenerton K, Robidoux A, Tonkin K, et al. Olaparib in patients with recurrent high-grade serous or poorly differentiated ovarian carcinoma or triple-negative breast cancer: a phase 2, multicentre, open-label, non-randomised study. Lancet Oncol 2011;12(9):852–61.

[59] Kaye S, Kaufman B, Lubinski J, Matulonis U, Gourley C, Karlan B, et al. Phase II study of the oral PARP inhibitor olaparib (AZD2281) versus liposomal doxorubicin in ovarian cancer patients with BRCA1 and/ or BRCA2 mutations [abstract]. Ann Oncol 2010;21(Suppl. 8): viii, 304–13.

[60] Ledermann J, Harter P, Gourley C, Friedlander M, Vergote I, Rustin G, et al. Olaparib maintenance therapy in platinum-sensitive relapsed ovarian cancer. N Engl J Med 2012;366(15):1382–92.

[61] Oza AM, Cibula D, Benzaquen AO, Poole C, Mathijssen RH, Sonke GS, et al. Olaparib combined with chemotherapy for recurrent platinum-sensitive ovarian cancer: a randomised phase 2 trial. Lancet Oncol 2015;16(1):87–97.

[62] Domchek SM, Mitchell G, Lindeman GJ, Tung NM, Balmana J, Isakoff SJ, et al. Challenges to the development of new agents for molecularly defined patient subsets: lessons from BRCA1/2-associated breast cancer. J Clin Oncol 2011;29(32):4224–6.

[63] <http://uk.reuters.com> [accessed 09.04.13].

[64] Kaufman B, Shapira-Frommer R, Schmutzler RK, Audeh MW, Friedlander M, Balmana J, et al. Olaparib monotherapy in patients with advanced cancer and a germline BRCA1/2 mutation. J Clin Oncol 2015;33(3):244–50.

[65] FDA Briefing Document. <http://www.fda.gov/downloads/AdvisoryCommittees/Committees MeetingMaterials/Drugs/OncologicDrugsAdvisoryCommittee/UCM402207.pdf>.

[66] Liu JF, Barry WT, Birrer M, Lee JM, Buckanovich RJ, Fleming GF, et al. Combination cediranib and olaparib versus olaparib alone for women with recurrent platinum-sensitive ovarian cancer: a randomised phase 2 study. Lancet Oncol 2014;15(11):1207–14.

[67] McNeish I, Oza A, Coleman R, editors. A phase 2 trial to prospectively identify ovarian cancer patients likely to respond to rucaparib using tumor genetic analysis. Chicago (IL): ASCO; 2015.

[68] Mirza MR, Monk BJ, Herrstedt J, Oza AM, Mahner S, Redondo A, et al. Niraparib maintenance therapy in platinum-sensitive, recurrent ovarian cancer. N Engl J Med 2016;375(10):2154–64.

[69] ARIEL2: a phase 2 study to prospectively identify ovarian cancer patients likely to respond to rucaparibSwisher E, editor. 26th EORTC-NCI-AACR Symposium on Molecular Targets and Cancer Therapeutics. Barcelona: AACR; 2015.

[70] Michels J, Vitale I, Saparbaev M, Castedo M, Kroemer G. Predictive biomarkers for cancer therapy with PARP inhibitors. Oncogene 2014;33(30):3894–907.

[71] Abkevich V, Timms KM, Hennessy BT, Potter J, Carey MS, Meyer LA, et al. Patterns of genomic loss of heterozygosity predict homologous recombination repair defects in epithelial ovarian cancer. Br J Cancer 2012;107(10):1776–82.

[72] McLornan DP, List A, Mufti GJ. Applying synthetic lethality for the selective targeting of cancer. N Engl J Med 2014;371(18):1725–35.

[73] Ashworth A. Drug resistance caused by reversion mutation. Cancer Res 2008;68(24): 10021–3.

[74] Rottenberg S, Jaspers JE, Kersbergen A, van der Burg E, Nygren AO, Zander SA, et al. High sensitivity of BRCA1-deficient mammary tumors to the PARP inhibitor AZD2281 alone and in combination with platinum drugs. Proc Natl Acad Sci U S A 2008;105(44):17079–84.

[75] Lord CJ, Ashworth A. Mechanisms of resistance to therapies targeting BRCA-mutant cancers. Nat Med 2013;19(11):1381–8.

[76] Greaves M, Maley CC. Clonal evolution in cancer. Nature 2012;481(7381):306–13.

[77] Norquist B, Wurz KA, Pennil CC, Garcia R, Gross J, Sakai W, et al. Secondary somatic mutations restoring BRCA1/2 predict chemotherapy resistance in hereditary ovarian carcinomas. J Clin Oncol 2011;29(22):3008–15.

[78] Mateo J, Carreira S, Sandhu S, et al. DNA-Repair Defects and Olaparib in Metastatic Prostate Cancer. N Engl J Med 2015;373:1697–708.

[79] Forster MD, Dedes KJ, Sandhu S, Frentzas S, Kristeleit R, Ashworth A, et al. Treatment with olaparib in a patient with PTEN-deficient endometrioid endometrial cancer. Nat Rev Clin Oncol 2011;8(5):302–6.

[80] Mantia-Smaldone G, Ronner L, Blair A, Gamerman V, Morse C, Orsulic S, et al. The immunomodulatory effects of pegylated liposomal doxorubicin are amplified in BRCA1—deficient ovarian tumors and can be exploited to improve treatment response in a mouse model. Gynecol Oncol 2014;133(3):584–90.

[81] McCabe N, Turner NC, Lord CJ, Kluzek K, Bialkowska A, Swift S, et al. Deficiency in the repair of DNA damage by homologous recombination and sensitivity to poly(ADP-ribose) polymerase inhibition. Cancer Res 2006;66(16):8109–15.

[82] Morgan MA, Parsels LA, Maybaum J, Lawrence TS. Improving the efficacy of chemoradiation with targeted agents. Cancer Discov 2014;4(3):280–91.

[83] Josse R, Martin SE, Guha R, Ormanoglu P, Pfister TD, Reaper PM, et al. ATR inhibitors VE-821 and VX-970 sensitize cancer cells to topoisomerase i inhibitors by disabling DNA replication initiation and fork elongation responses. Cancer Res 2014;74(23):6968–79.

[84] Kawabe T. G2 checkpoint abrogators as anticancer drugs. Mol Cancer Ther 2004;3(4):513–19.

[85] Carlessi L, Buscemi G, Larson G, Hong Z, Wu JZ, Delia D. Biochemical and cellular characterization of VRX0466617, a novel and selective inhibitor for the checkpoint kinase Chk2. Mol Cancer Ther 2007;6(3):935–44.

[86] Arienti KL, Brunmark A, Axe FU, McClure K, Lee A, Blevitt J, et al. Checkpoint kinase inhibitors: SAR and radioprotective properties of a series of 2-arylbenzimidazoles. J Med Chem 2005;48(6):1873–85.

[87] Hickson I, Zhao Y, Richardson CJ, Green SJ, Martin NM, Orr AI, et al. Identification and characterization of a novel and specific inhibitor of the ataxia-telangiectasia mutated kinase ATM. Cancer Res 2004;64(24):9152–9.

[88] Lapenna S, Giordano A. Cell cycle kinases as therapeutic targets for cancer. Nat Rev Drug Discov 2009;8(7):547–66.

CHAPTER 8

Ovarian Cancer: New Targets and Future Directions

L.P. Martin[1] and R.J. Schilder[2]

[1]Fox Chase Cancer Center, Philadelphia, PA, United States [2]Thomas Jefferson University, Philadelphia, PA, United States

CONTENTS

INTRODUCTION

Advancing knowledge about the molecular makeup of cancer cells, including genetic and epigenetic alterations, cellular surface receptors, aberrant pathway activation, and an understanding of the tumor microenvironment, has led to an unprecedented number of new targeted therapeutics currently under study in the preclinical and clinical arena across a broad range of cancers. Ovarian cancer (OC) remains the leading cause of death among the gynecologic malignancies and while 5-year survival has improved over the last decade, only about half of women with recurrent disease will survive 5 years [1]. While many women with OC respond well to traditional cytotoxic chemotherapy at diagnosis, most will develop recurrent disease, which ultimately will become resistant to the available cytotoxic regimens. Further, at recurrence, most women will remain on treatment to control their disease for prolonged periods, and these agents may have significant risk and toxicity. While advances have been made in optimizing results from standard cytotoxic chemotherapeutics in the last decade by adjusting the schedule or route of administration, these modifications have not resulted in an improvement in the rate of cure. The hope for improving outcomes for women with OC rests, in part, on developing novel treatments for this disease. Targeted therapeutics may offer the opportunity to treat disease with fewer side effects, improving quality of life, and if the right targets are identified, to improve outcomes.

Translational Advances in Gynecologic Cancers. DOI: http://dx.doi.org/10.1016/B978-0-12-803741-6.00008-2

IMMUNOTHERAPY

See Table 8.1 for quick reference of agents discussed here.

One of the most exciting areas of research in recent years has been in the area of immunotherapy. Success resulting in FDA-approved therapeutics has been seen in melanoma, prostate cancer, renal cell carcinoma (RCC), lymphoma, and non–small cell lung cancer (NSCLC). There are now multiple clinical trials evaluating various immune therapeutic approaches in OC. Evidence of OC immunogenicity has been present for over a decade. The presence of CD3+ tumor-infiltrating T lymphocytes (TILs) has been shown to correlate with improved survival [2]. Further work in this area confirmed that the presence of CD3+ intraepithelial T-cells, and perhaps more importantly, CD8+ intraepithelial T-cells correlates with survival, and that the improvement in survival may further correlate with the number of T-cells present [3]. Further, the presence of regulatory T-cells in the tumor environment, such as CD4+ CD25+ T-cells, which suppress immune response and maintain tolerance to self-antigens, appears to confer a worse survival [3]. Numerous tumor-associated antigens expressed by OC cells have been identified. Novel immunotherapeutic approaches in the treatment of OC include adoptive cell therapy, vaccines, and immune checkpoint inhibitors.

T-cell activity is regulated by co-stimulatory and inhibitory signals that allow the immune system to react to antigens, while maintaining self-tolerance. The inhibitory signals include a variety of immune checkpoints that have

Table 8.1 Immunotherapy			
Agents	**Types of Agents**	**Targets**	**Current Phase of Trials Which Include OC**
Checkpoint Inhibitors			
Ipilimumab	Monoclonal antibody	CTLA-4	Phases I and II in combinations with other agents
Nivolumab	Monoclonal antibody	PD-1	Phases I and II in combinations with other agents
IDO Inhibitors			
Indoximiid	Small molecule	IDO	Phase I
Epacadostat	Small molecule	IDO	Phases I and II in combination with other agents
NLG919	Small molecule	IDO	
TCR			
NY-ESO-1c259	TCR	NY-ESO antigen	Phase I/II

been studied and found to be overexpressed in some tumor cells, antigen-presenting cells (APCs), T-cells, or the tumor microenvironment. Additional immune inhibition can occur through the presence of metabolic enzymes which inhibit immune response against tumor cells by depleting the tumor microenvironment of essential amino acids that are required for lymphocyte function, such as indolamine-2,3-dioxygenase. (For a figure illustrating checkpoint interactions, see Ref. [4].)

Cytotoxic T-lymphocyte antigen 4 (CTLA-4) is a negative regulator of early T-cell activation and expansion and is expressed on the surface of T-cells [5]. Its co-stimulatory counterpart, CD28, facilitates T-cell activation in response to antigen presentation in the lymphatic system. When CTLA-4 is inhibited, T-cells are stimulated, both in the tumor environment and in other areas of the body, resulting in antitumor effect as well as autoimmune side effects in some patients. Ipilimumab is an anti-CTLA-4 monoclonal antibody (mAb) that has demonstrated improvement in survival and durable response in patients with metastatic melanoma, with 18% of patients surviving more than 2 years [6]. Side effects have included autoimmune-related rashes, hypothyroidism, colitis, and hypophysitis. Efforts have been made to enhance the anticancer effects and minimize the toxicity related to CTLA-4 inhibition through the use of combination regimens (Fig. 8.1).

The programmed death (PD)-1 receptor–ligand interaction is a major pathway of immune control. The normal function of PD-1 is to down-modulate unwanted or excessive immune responses, including autoimmune reactions in the peripheral tissues [7]. The receptor is expressed on the cell surface of activated T-cells, T regulatory cells, activated B–cells, and NK-cells. PD-L1 and PD-L2, the ligands for PD-1, are constitutively expressed or can be induced in some tumors. Binding of either of these ligands to PD-1 inhibits T-cell activation. PD-L1 is expressed at low levels on various nonhematopoietic tissues, including the vascular endothelium, whereas PD-L2 is only expressed on APCs found in lymphoid tissue or chronic inflammatory environments. PD-L2 is thought to control immune T-cell activation in lymphoid organs, while PD-L1 affects T-cell function in peripheral tissues. High expression of PD-L1 on tumor cells (and to a lesser extent of PD-L2) has been found to correlate with poor prognosis and survival in various cancer types, including RCC [8], pancreatic carcinoma [9], and ovarian carcinoma [10]. The correlation of clinical prognosis with PD-L1 expression in multiple cancers suggests that the PD-1/PD-L1 pathway plays a critical role in tumor immune evasion and is an attractive target for therapeutic intervention. Clinical trials of PD-1 inhibitors have been undertaken in women with recurrent OC. Studies with the PD-1 inhibitor, pembrolizumab, and the PD-L1 inhibitor, avelumab, have demonstrated evidence of efficacy in heavily pretreated women with recurrent OC [11,12]. A study in patients with recurrent, platinum-resistant

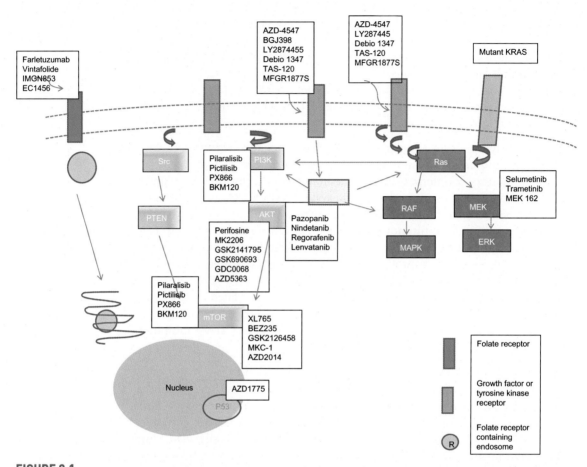

FIGURE 8.1

The network signaling. The signaling pathways are abbreviated to highlight the specific signaling molecules discussed in this chapter. Text boxes are superimposed upon the signaling molecules that are targeted by the drugs within.

disease evaluating nivolumab demonstrated two completed responses (CRs) and one partial response (PR) as well as four patients with stable disease in 20 patients; these responses were durable for some responders [13,14]. There are currently studies under way to evaluate combinations of checkpoint inhibitors in OC including NRG-GY003 which is a Phase II randomized trial evaluating the combination of ipilimumab and nivolumab vs nivolumab in women with recurrent OC (NCT02498600).

Indoleamine 2,3-dioxygenase-1(IDO1) is the first and rate-limiting enzyme involved in metabolizing the essential amino acid, tryptophan [15]. The expression of IDO1 in tumor cells is associated with poor prognosis and is associated with low numbers of TILs [16–18]. This enzyme is typically inactive in immune cells, until it is induced by cytokines. Upon activation, it depletes tryptophan from the local environment with a corresponding accumulation of immunosuppressive metabolites such as kynurenine, so that T-cells in the region are unable to undergo activation and become arrested [19]. There are three IDO inhibitors currently undergoing clinical evaluation, indoximod (1-methyl-D-tryptophan) [20], epacadostat (INCB24360), and NLG919 (NCT02048709) [21,22]. Objective responses have been low in patients participating in trials of IDO inhibitors to date, although prolonged periods of stable disease have been noted [20,21] and there is interest and ongoing trials evaluating these agents in combination with immune checkpoint inhibitors.

Adoptive T-cell therapy refers to a process in which T-cells are collected and isolated from a patient and grown and expanded ex vivo after which they are reinfused into the recipient patient in an effort to stimulate T-cell attack of tumor cells [23]. The main approaches currently being studied in solid tumors include treatment with expanded populations of TILs, or with genetically modified T-cells that have been altered by the introduction of chimeric antigen receptors (CARs) and genetically engineered T-cell receptors (TCRs). TILs are obtained from tumors and lymph nodes that have been surgically removed. Initial studies grew the collected T-cells in culture, but later studies evaluated these cells for the presence of specific tumor recognition antigens and selected this subset of cells for expansion. The expanded T-cells were then infused into the patient, who typically had undergone a conditioning regimen of chemotherapy geared to deplete lymphocytes. The majority of studies evaluating TIL therapy have been in melanoma, although this approach has also been evaluated in OC [24,25]. In the first study, Fujita et al. [24] treated 13 patients in the adjuvant setting who had achieved a CR after surgical cytoreduction and cisplatin chemotherapy. Eleven patients served as matched controls and underwent standard surgery and chemotherapy. After a median follow-up of 36 (range, 23–44) months in the TIL group and 33 (range, 14–48) months in the control group, 3-year overall survival (OS) for the TIL group and in the control group was 100% and 67.5%, respectively ($p < 0.01$). Additionally, the estimated 3-year disease-free survival of the patients in the TIL group and the control group was 82.1% and 54.5%, respectively ($p < 0.05$). Aoki et al. [25] isolated TILs from fresh ovarian tumors which were expanded with recombinant interleukin-2 and treated seven patients with advanced or recurrent OC, using adoptive transfer following a single i.v. injection of cyclophosphamide. One CR and four cases of greater than 50% decrease of tumor were noted (14.3% and 57.1%, respectively). Duration of response ranged from 3 to 5 months. In the same study, 10 patients were treated alternately with cisplatin-containing chemotherapy and the adoptive transfer of TILs. Seven patients experienced a CR and two patients noted a PR. Four of the seven patients with a CR had no recurrence for greater than 15 months of follow-up. This approach has been limited by the need for surgery and the limited availability of sufficient tumor-specific T lymphocytes [26].

To overcome these limitations, normal peripheral blood lymphocytes have been altered by the introduction of either antigen-specific TCRs or CARs. TCRs are comprised of an α chain and a β chain that recognize a specific antigen being presented by a patient's own APCs, and CARs are engineered receptors that are comprised of antibody heavy and light chains linked with intracellular signaling chains and can recognize antigens on any cell surface, and are not limited to MHC-specific APCs. These engineered cells are created by the insertion of genetic sequences encoding TCR α/β heterodimers or CARs into the genome of T-cells that were previously collected from the peripheral blood. One of the challenges of these strategies is identification of tumor-specific antigens that are limited to tumor cells and not expressed on normal cells, in order to limit toxicity to normal cells. A current Phase I/IIa clinical trial in solid tumors including OC will evaluate TCR-engineered autologous cells targeting NY-ESO-1c259 in women with recurrent OC, the HLA-A201 allele, and whose tumor expresses the NY-ESO-1 tumor antigen (NCT01567891). During this trial, patients will undergo a cyclophosphamide conditioning regimen 7 days prior to receiving T-cells to potentiate the immunotherapy. There are also ongoing studies evaluating CAR T-cells targeting mesothelin that are enrolling patients with OC (NCT01583686 and NCT02159716).

Therapeutic vaccines have been developed with a goal to enhancing the adaptive and innate immune response to target and eradicate cancer cells, and prevent recurrence of disease. Dendritic cell vaccines, peptide vaccines, and recombinant viral vaccines have been studied in OC. Dendritic cells are a class of APCs that process antigens and present them to naïve T-cells, B–cells, and NK-cells [27]. Dendritic cell vaccines have been studied in multiple diseases, and a vaccine-targeting mucin-1 demonstrated a potential progression-free survival (PFS) benefit in patients with recurrent OC who had achieved remission after treatment with chemotherapy [28].

FOLATE RECEPTOR-TARGETED THERAPY

Folate is a member of the B class of vitamins and plays a key role in DNA replication during cell division [29]. Folate can be transported into normal cells by the reduced folate carrier, the proton-coupled folate transporter, or by the folate receptor (FR) [30]. There are four known isoforms of the FR, the well-studied of which is the FRα. It is most prevalent during embryonic development and has limited expression on some of the epithelial cells of the proximal kidney tubules, the choroid plexus, the genital organs, the retina, and submandibular salivary glands. It is found on the luminal surface of the epithelial cells in these areas and had limited exposure to circulating folic acid. However, FRα is overexpressed in multiple cancers, including OC, where it has been found to be overexpressed in greater that 70% of OC and to maintain overexpression even after chemotherapy [31]. In patients with serous OC, overexpression has been identified in over 80% of samples, while the overexpression is slightly lower in patients with clear cell or endometrioid histology [31,32]. While chemotherapeutics have been developed that target folate metabolism in an effort to control cancer growth, more recently, FRα has been an attractive target for anticancer therapies because of its limited expression in normal cells and overexpression on tumor cells [33]. Two main approaches have been utilized to date for therapeutic targeting of FRα. mAbs and folate-conjugated drugs have been developed to target FRα. An example of the former, farletuzumab, was developed as a humanized monoclonal IgG kappa antibody that demonstrated a strong affinity to FRα, with limited binding to normal tissue [34]. Upon binding to FRα, tumor cell death was triggered by antibody-dependent cell-mediated cytotoxicity (ADCC) and complement-dependent cytotoxicity. In a Phase I trial in patients with recurrent, platinum-resistant or refractory OC, no responses were seen, but 36% of patients experienced stable disease, with a trend toward greater benefit at higher doses [34]. A follow-up Phase II trial was performed by Armstrong and colleagues [35], evaluating farletuzumab as a single agent in women with asymptomatic CA125 relapse or farletuzumab with carboplatin and taxane-based therapy in patients with symptomatic recurrent, platinum-sensitive disease; patients who started with farletuzumab monotherapy went on to receive farletuzumab with chemotherapy. While no responses were noted in the monotherapy arm, in the combination arm 75% of patients experienced a CR or PR, and 81% of patients were found to have a normalized CA125. It was noted that 20% of patients receiving the combination enjoyed a PFS interval that exceeded their initial PFS after front-line therapy. An international Phase III trial evaluated carboplatin and paclitaxel with or without farletuzumab in patients with platinum-sensitive first relapse and found no statistical difference in median PFS [36].

More recently, efforts have focused on FR conjugate drugs. Vintafolide is a small-molecule drug consisting of folate conjugated with a potent vinca alkaloid, desacetyl vinblastine hydrazine [37,38]. This agent was found to be well tolerated in a Phase I trial, and a Phase II trial in patients with recurrent OC demonstrated PFS of 15.4 weeks in a heavily pretreated group of patients with recurrent disease [39,40]. Given that this was comparable to prior trials of active agents in recurrent OC, a randomized Phase II trial (PRECEDENT) compared pegylated liposomal doxorubicin alone or in combination with vintafolide and noted a PFS of 5.0 months for the combination therapy vs 2.7 months for the monotherapy arm [41]. A Phase III trial (PROCEED) evaluating this combination is currently ongoing (NCT01170650).

Additional conjugate molecules targeting FRα are being developed, and Phase I studies are ongoing. EC1456 is a potent folic acid–tubulysin B hydrazide (TubBH) small-molecule drug conjugate that inhibits the polymerization of tubulin into microtubules (analogous to taxanes) and arrests cells in metaphase [42]. EC1456 acts as a cell-specific cytotoxic agent and preferentially targets TubBH to cancer cells that express FR. IMGN853 (mirvetuximab soravtansine) is an FRα-targeting agent that comprises an FRα-binding antibody conjugated with the potent maytansinoid, DM4 [43]. Early data presented from a Phase I trial of IMGN853 in an expansion cohort of heavily pretreated patients with platinum-resistant OC and FRα-positive tumor as assessed by IHC in archival tissue noted clinical benefit with an objective response rate of 40% and clinical benefit rate of 50% [44].

FIBROBLAST GROWTH FACTOR TARGETING

See Table 8.2 for quick reference of agents discussed here.

Fibroblast growth factors (FGFs) and their receptors (FGFRs) are involved in cell survival and proliferation with mutations and amplifications of FGFRs noted in a variety of cancers, including OC [45]. There are five known FGFRs, FGFR1–FGFR4 and FGRL1. FGFR1–FGFR4 are transmembrane receptors, with an extracellular immunoglobulin-like domain and intracellular tyrosine kinase domain, while FGRL1 does not have a tyrosine kinase domain. These receptors can be bound by any of the known 18 ligands, resulting in the activation of any of the key signaling pathways, including MAPK, PI3K/AKT, PLCΥ, or STAT. Amplification or mutation of the FGFR and FGF can be seen across a wide variety of cancers, and FGFR1 mutations and amplifications have been identified in approximately 5% of OC [46]. While this represents a small subset of patients with OC, the availability of new agents that target FGFR1 provides a potentially viable treatment option for this rare subset of women with such a mutation that is driving proliferation and propagation of her cancer, analogous to the finding of ALK mutation in NSCLC, in whom treatment with crizotinib and other, newer-generation ALK inhibitors has resulted in improvement in outcomes [47]. There are currently several agents that were developed as antiangiogenic therapies and target multiple tyrosine kinases, including FGFR1. For example, pazopanib is a multi-tyrosine kinase inhibitor (TKI) that targets VEGFR1-3, PDGFR-α, -β, FGFR-1, -3, and KIT [48]. du Bois et al. [49] performed a randomized trial of pazopanib or placebo for

Table 8.2 FGFR-Targeting Agents

Agents	Types of Agents	Targets	Current Phase of Trials Which Include OC
Pazopanib	Small-molecule TKI	VEGFR1-3, PDGFR-α, and -β, FGFR-1 and -3, and KIT	Phase II
Nintedanib	Small-molecule TKI	PDGFR-α/β, FGFR-1/3, and VEGFR1–3	Phase III in combination with other agents nation
Regorafenib	Small-molecule TKI	KIT, RET, RAF1, BRAF, VEGFR1–3, TIE2, PDGFR, FGFR	Phase II
Lenvatinib	Small-molecule TKI	VEGFR1–3, FGFR1–4, PDGFRα, RET, and KIT	No current studies
AZD4547	Small-molecule TKI	FGFR1–3	No current studies
BGJ398	Small-molecule TKI	FGFR1–3	Phase I as monotherapy and Phase Ib in combination with other agents
LY2874455	Small-molecule TKI	FGFR1–4	No current studies
Debio 1347	Small-molecule TKI	FGFR1–3	Phase I
TAS-120	Small-molecule TKI	FGFR1–4	Phase I
MFGR1877S	Monoclonal antibody	FGFR3	No current studies
FP-1039	TRAP antibody	FGF ligands	No current studies

up to 24 months in the maintenance setting in 940 patients with no evidence of progression after primary therapy and met is primary endpoint, demonstrating a PFS benefit of 5.6 months ($p = 0.0021$). However, 33% of patients receiving pazopanib stopped treatment for adverse events, most in the first 12 weeks and the most common reasons for cessation of treatment were hypertension (8%), diarrhea (2.9%), AST (2.5%), or ALT (2.3%) increase, neutropenia (2.3%), and palmar-plantar erythrodysesthesia (1.7%), and no OS benefit was noted.

Similarly, nintedanib [50], regorafenib [51], and lenvatinib [52] are multi-TKIs that include FGFR. Inhibition of FGFR signaling was not the primary goal during development of these agents, and there has been recent focus on development of small molecules that more specifically inhibit FGFR signaling. AZD4547 is a selective inhibitor of FGFR1, -2, and -3 that has been studied in a Phase I trial in patients with advanced solid tumors and found to be well tolerated [53]. Biomarker analysis noted increased clinical benefit in a later study in patients with tumors selected for amplification of FGFR1 or

-2 [54]. BGJ398 is a selective pan-FGFR inhibitor that was found to be well tolerated in a Phase I trial that enrolled patients whose tumors were found to have FGFR mutations at their local laboratory or on central testing [55]. Of the 94 patients enrolled, 8 patients with FGFR amplification or mutation were found to have tumor responses or stabilization of disease for greater than 16 weeks. Additional small molecules selectively targeting FGFR signaling that have recently begun Phase I clinical testing include LY2874455, Debio 1347, and TAS-120. In addition to the TKIs, MFGR1877S is an mAb-targeting FGFR3 showing some stabilization of disease in patients with bladder cancer, and this agent will undergo additional testing [56]. This may be of particular interest in clear cell carcinoma of the ovary, as data suggests an important role for FGFR3 in patients with this histology [57]. Finally, there is some drug development–targeting FGF ligands. FP-1039 is a soluble fusion protein that sequesters FGF ligands and prevents their binding to FGFR1, and this agent is being studied in combination with chemotherapy in patients with NSCLC [58,59].

PI3K/AKT/mTOR SIGNALING

See Table 8.3 for quick reference of agents discussed here.

The phosphatidylinositol 3-kinase (PI3K)/AKT (also known as protein kinase B)/mammalian target of rapamycin (mTOR) signaling pathway is dysregulated in many cancer types, and appears to be involved in malignant transformation, tumor growth and metastasis, and chemotherapy resistance. Dysregulation can occur through a variety of mechanisms, including overexpression or activation of AKT, loss of the negative regulator gene phosphatase and tensin homolog PTEN, mutation or amplification resulting in activation of PI3K, or loss or downregulation of the tuberous sclerosis tumor suppressor complex [60]. PI3K activation is driven by any of a number of receptor tyrosine kinases, including EGFR, HER2, IGF-1R, and PDGFR. Overexpression or activation of any of these receptors can result in activation of PI3K, and mutations have been found in the PIK3CA and PIK3R1 genes that result in overactivation, leading to the activation of AKT, which can drive cancer cell migration, invasion, proliferation, and epithelial to mesenchymal transition [61,62]. Additionally, AKT mutation or amplification of the one of the AKT genes can result in activation of the pathway. mTOR acts through mTORC1 (mTOR complex 1, rapamycin sensitive) or mTORC2 (mTOR complex 2, rapamycin insensitive), two multiprotein signaling complexes. mTOR1 is activated by PI3K/AKT signaling, while the mechanism of activation of mTORC2 is unknown, but it is thought to play a role in cell survival. Tumors with activation of the PI3K/AKT/mTOR pathway have a worse prognosis and have been found to have enhanced chemotherapy resistance [63,64]. There are

Table 8.3 PI3K/AKT/mTOR

Agents	Types of Agents	Targets	Current Phase of Trials Which Include OC
Pilaralisib	Small-molecule TKI	PI3Kα, -β, -γ, and -δ isoforms	No current studies
Pictilisib	Small-molecule TKI	PI3Kα, -β, -γ, and -δ isoforms	Phase Ib
PX866	Small-molecule TKI	PI3Kα, -β, -γ, and -δ isoforms	No current studies
BKM120	Small-molecule TKI	PI3Kα, -β, -γ, and -δ isoforms	Phases I and Ib in combination with other agents
Perifosine	Small-molecule TKI	AKT	No current studies
MK2206	Small-molecule TKI	AKT	No current studies
GSK2141795	Small-molecule TKI	AKT	Phase I/II in combination with other agents
GSK690693	Small-molecule TKI	AKT	No current studies
GDC0068	Small-molecule TKI	AKT	Phase I in combination with other agents
XL765	Small-molecule TKI	PI3Kα, -β, -γ, and -δ and mTOR1/2	No current studies
BEZ235	Small-molecule TKI	PI3Kα, -β, -γ, and -δ and mTOR	No current studies
GSK2126458	Small-molecule TKI	PI3K-α, -β, -γ, and -δ and mTOR1/2	No current studies
MKC-1	Small-molecule TKI	Importin-β, tubulin, and mTOR2	No current studies
AZD2014	Small-molecule TKI	mTOR1/2	Phase II
AZD5363	Small-molecule TKI	AKT	Phases I and Ib

multiple agents targeting components of this signaling pathway in development. Agents targeting PI3K include the pan-PI3K inhibitors pilaralisib (XL147) [65], pictilisib (GDC0941) [66], PX 866 [67], and BKM 120 [68]. As a central regulator of the PI3K/AKT/mTOR signaling pathway, agents targeting one or more isoforms of AKT have been developed and are in early phase testing, including perifosine, MK2206 [69], GSK2141795, GSK690693, and GDC0068. While there have been preliminary signals of activity in OC and other solid tumors noted in the early trials, work is being done to identify potential markers for response to these agents. mTOR inhibitors have been FDA approved for the treatment of RCC, breast cancer, and pancreatic neuroendocrine tumors. mTOR inhibition has been tested in OC, with a Phase II trial of temsirolimus demonstrating modest activity, with 24% of patients achieving PFS greater than 6 months, but with few objective responses (9.3%)

[70]. Everolimus, an oral mTOR inhibitor, is currently undergoing study in combination with bevacizumab in recurrent advanced gynecologic cancers.

Preclinical work has shown that mTOR inhibition can result in a positive feedback loop that leads to increased AKT expression [71]. In light of the modest responses and rapid development of resistance noted in trials targeting PI3K or mTOR, dual PI3K and mTOR inhibitors have been developed and are currently in early phase testing, including XL765 [72], BEZ235, GDC 0980, and GSK2126458. MKC-1 is a dual mTOR/AKT inhibitor which had limited single agent activity in a Phase II trial in women with recurrent gynecologic cancers [73]. Future strategies include combinations of PI3K/AKT/mTOR pathway inhibitors with chemotherapy or other targeted agents. AZD2014, an mTORC1/2 inhibitor, and AZD5363, an AKT inhibitor, are two agents currently being studied in combination with olaparib in a Phase I trial in patients with OC and other cancers (NCT02208375). Further, biomarkers that can be used to identify those patients most likely to benefit from these agents will be critical to select these subsets. Preclinical study suggests mTOR can be a promising therapeutic target in patients with clear cell OC [74], and PIK3CA mutations have been identified in up to 12% of patients with serous OC, but are more common (~33%) in patients with clear cell cancer of the ovary but this is a relatively rare entity and it is therefore difficult to perform trials limited to this histology [62,75]. Additionally, while expression of activated AKT has been identified in a high percentage of OC, the cause of activation may result in differential responses to agents targeting segments of this pathway, and further research of mechanisms of resistance to pathway inhibition is ongoing.

RAS/RAF/MEK/ERK PATHWAY INHIBITION

This signaling pathway regulates cellular processes of proliferation survival, apoptosis, differentiation, and metabolism [76]. The RAF kinases are ubiquitous serine/protein kinases that are activated by a variety of extracellular signals, and mutations or activation of this signaling pathway and RAS genes are among the most frequently mutated oncogenes in human cancer [77]. Vemurafenib is an oral inhibitor of BRAF and has been approved for the treatment of malignant metastatic melanoma with BRAF V600E mutation [78]. BRAF and KRAS mutations have been identified in patients with low-grade serous carcinoma, and efforts to target this mutation through inhibition of its downstream effector, MEK, has shown some success in patients with this subtype of OC [79]. Selumetinib (AZD6244) is a selective oral small-molecule inhibitor of MEK1 and MEK2 and has shown efficacy in a Phase II trial of women with recurrent low-grade serous cancer [80]. Other agents under development include trametinib, a MEK1 and MEK2 inhibitor and MEK162, which is currently being studied in a Phase III trial in women with recurrent low-grade serous OC [81]. A Phase III trial in melanoma combining trametinib with dabrafenib, a BRAF inhibitor, demonstrated superior RR, PFS, and OS without increased toxicity in patients with recurrent metastatic melanoma with a BRAF mutation, suggesting that dual inhibition of BRAF and MEK1 and MEK2 may add benefit in patients whose tumors are dependent upon this pathway [82].

p53

TP53 is a tumor suppressor gene that codes for the p53 protein and is the most frequently mutated gene in human cancers, with inactivation the most common outcome of these mutations [83]. High-grade serous OC is noted to have near universal aberration in p53 [84]. p53 plays an important role in multiple cellular activities, including cell cycle arrest, apoptosis, cellular metabolism, and autophagy [85]. Because *TP53* mutation typically results in loss of function, it has been difficult to identify methods for targeting mutant p53 [86]. Furthermore, several different mutations of p53 have been identified, and different mutations can result in different effects in human cancers, thus a single approach to restoring function of p53 is unlikely to have broad effects [87]. The heat-shock protein (HSP) 90 chaperones multiple mutant p53 client proteins and prevents their degradation [88,89]. Preclinical evaluation of HSP90 inhibition has demonstrated activity of HSP90 inhibition in controlling growth of cancer [90,91]. The HSP 90 inhibitor, ganetespib, is currently undergoing evaluation in patients with recurrent OC (NCT01962948). p53 is also important in cell cycle regulation and triggers G1 arrest in the face of DNA damage [92]. Given that these cells become more dependent upon later checkpoints in the cell cycle for arrest and DNA damage repair, inhibition of additional checkpoints may result in synthetic lethality in p53 mutant cells. For example, WEE1 is a nuclear tyrosine kinase required for regulation of the G2 checkpoint. AZD1775 (formerly MK 1775) is a first-in-class, pyrazolopyrimidine derivative and potent small-molecule inhibitor of WEE1 kinase that inhibits WEE1 activity through inhibition of its substrate and can result in apoptosis in cells [93–95]. WEE1 inhibition may also enhance the effects of cytotoxic chemotherapy [94,96]. AZD1775 is currently undergoing evaluation in two Phase II trials in combination with various chemotherapeutic regimens in women with recurrent OC (NCT02272790 and NCT02101775). Additional approaches may ultimately include immunotherapeutic agents such as vaccines-targeting mutant p53. One such vaccine demonstrated p53-specific T-cell responses in women with recurrent OC; however, the vaccine did not appear to provide significant disease control either at the time of initial dosing or upon subsequent therapy, in spite of persistence of T-cells [97,98]. While specific targeting of mutant p53 remains a challenge, its presence is restricted to tumor cells and it remains a target of interest in this disease.

CONCLUSION

At present, cytotoxic chemotherapy remains the predominant modality in the management of initial and recurrent OC; however, most women continue to die of advanced, recurrent disease, and new treatments are needed. This chapter highlights some of the new directions being evaluated in the treatment of solid tumors that have relevance to, or are undergoing active study in OC. This is by no means an exhaustive profile of targeted therapeutics that may be of interest in this disease. Some of these agents may have broad activity, while others may serve small subpopulations of women with OC. The challenge for the future will be to develop predictive markers to identify the best treatments for individuals with OC. Further, as we learn more about cancer cells and their microenvironments, new targets of importance will be identified. Additionally, combinations of new therapeutics may result in synergistic effects, and identifying the most effective combinations will require more knowledge about the pathways in these cancers, and the specific actions of the agents we are using to try to control them.

References

[1] Siegel RL, Miller KD, Jemal A. Cancer statistics, 2015. CA Cancer J Clin 2015;65(1):5–29.

[2] Zhang L, Conejo-Garcia JR, Katsaros D, Gimotty PA, Massobrio M, Regnani G, et al. Intratumoral T cells, recurrence, and survival in epithelial ovarian cancer. New EnglJ Med 2003;348(3):203–13.

[3] Curiel TJ, Coukos G, Zou L, Alvarez X, Cheng P, Mottram P, et al. Specific recruitment of regulatory T cells in ovarian carcinoma fosters immune privilege and predicts reduced survival. Nat Med 2004;10(9):942–9.

[4] Freeman GJ, Sharpe AH. A new therapeutic strategy for malaria: targeting T cell exhaustion. Nat Immunol 2012;13(2):113–15.

[5] Egen JG, Kuhns MS, Allison JP. CTLA-4: new insights into its biological function and use in tumor immunotherapy. Nat Immunol 2002;3(7):611–18.

[6] Hodi FS, O'Day SJ, McDermott DF, Weber RW, Sosman JA, Haanen JB, et al. Improved survival with ipilimumab in patients with metastatic melanoma. New Engl J Med 2010;363(8):711–23.

[7] Francisco LM, Sage PT, Sharpe AH. The PD-1 pathway in tolerance and autoimmunity. Immunol Rev 2010;236:219–42.

[8] Thompson RH, Dong H, Lohse CM, Leibovich BC, Blute ML, Cheville JC, et al. PD-1 is expressed by tumor-infiltrating immune cells and is associated with poor outcome for patients with renal cell carcinoma. Clin Cancer Res 2007;13(6):1757–61.

[9] Nomi T, Sho M, Akahori T, Hamada K, Kubo A, Kanehiro H, et al. Clinical significance and therapeutic potential of the programmed death-1 ligand/programmed death-1 pathway in human pancreatic cancer. Clin Cancer Res 2007;13(7):2151–7.

[10] Hamanishi J, Mandai M, Iwasaki M, Okazaki T, Tanaka Y, Yamaguchi K, et al. Programmed cell death 1 ligand 1 and tumor-infiltrating CD8+ T lymphocytes are prognostic factors of human ovarian cancer. Proc Natl Acad Sci 2007;104(9):3360–5.

[11] Varga A, Piha-Paul SA, Ott PA, Mehnert JM, Berton-Rigaud D, Johnson EA, et al. Antitumor activity and safety of pembrolizumab in patients (pts) with PD-L1 positive advanced ovarian cancer: interim results from a phase Ib study. J Clin Oncol 2015;33(15 Suppl.) [abstract 5510].

[12] Disis ML, Patel MR, Pant S, Infante JR, Lockhart AC, Kelly K, et al. Avelumab (MSB0010718C), an anti-PD-L1 antibody, in patients with previously treated, recurrent or refractory ovarian cancer: a phase Ib, open-label expansion trial. J Clin Oncol. 2015;33(15 Suppl.) [abstract 5509].

[13] Hamanishi J, Mandai M, Ikeda T, Minami M, Kawaguchi A, Matsumura N, et al. Efficacy and safety of anti-PD-1 antibody (nivolumab: BMS-936558, ONO-4538) in patients with platinum-resistant ovarian cancer. J Clin Oncol. 2014;32(15 Suppl.) [abstract 5511].

[14] Hamanishi J, Mandai M, Ikeda T, Minami M, Kawaguchi A, Murayama T, et al. Safety and antitumor activity of anti–PD-1 antibody, nivolumab, in patients with platinum-resistant ovarian cancer. J Clin Oncol 2015.

[15] Taylor MW, Feng GS. Relationship between interferon-gamma, indoleamine 2,3-dioxygenase, and tryptophan catabolism. FASEB J 1991;5(11):2516–22.

[16] de Jong RA, Kema IP, Boerma A, Boezen HM, der Want JJLv Gooden MJM, et al. Prognostic role of indoleamine 2,3-dioxygenase in endometrial carcinoma. Gynecol Oncol 2012;126(3):474–80.

[17] Godin-Ethier J, Hanafi L-A, Piccirillo CA, Lapointe R. Indoleamine 2,3-dioxygenase expression in human cancers: clinical and immunologic perspectives. Clin Cancer Res 2011;17(22):6985–91.

[18] Brandacher G, Perathoner A, Ladurner R, Schneeberger S, Obrist P, Winkler C, et al. Prognostic value of indoleamine 2,3-dioxygenase expression in colorectal cancer: effect on tumor-infiltrating T cells. Clin Cancer Res 2006;12(4):1144–51.

[19] Munn DH, Shafizadeh E, Attwood JT, Bondarev I, Pashine A, Mellor AL. Inhibition of T cell proliferation by macrophage tryptophan catabolism. J Exp Med 1999;189(9):1363–72.

[20] Soliman HH, Neuger A, Noyes D, Vahanian NN, Link CJ, Munn D, et al. A phase I study of 1-methyl-D-tryptophan in patients with advanced malignancies. J clin Oncol. 2012;30(15 Suppl.) [abstract 2501].

[21] Newton RC, Scherle PA, Bowman K, Liu X, Beatty GL, O'Dwyer PJ, et al. Pharmacodynamic assessment of INCB024360, an inhibitor of indoleamine 2,3-dioxygenase 1 (IDO1), in advanced cancer patients. J Clin Oncol. 2012;30(15 Suppl.) [abstract 2500].

[22] Khleif S, Munn D, Nyak-Kapoor A, Mautino MR, Kennedy E, Vahanian NN, et al. First-in-human phase 1 study of the novel indoleamine-2,3-dioxygenase (IDO) inhibitor NLG-919. J Clin Oncol. 2014;32(15 Suppl.) [abstract TPS3121].

[23] Feldman SA, Assadipour Y, Kriley I, Goff SL, Rosenberg SA. Adoptive cell therapy—tumor-infiltrating lymphocytes, T-cell receptors, and chimeric antigen receptors. Semin Oncol 2015;42(4):626–39.

[24] Fujita K, Ikarashi H, Takakuwa K, Kodama S, Tokunaga A, Takahashi T, et al. Prolonged disease-free period in patients with advanced epithelial ovarian cancer after adoptive transfer of tumor-infiltrating lymphocytes. Clin Cancer Res 1995;1(5):501–7.

[25] Aoki Y, Takakuwa K, Kodama S, Tanaka K, Takahashi M, Tokunaga A, et al. Use of adoptive transfer of tumor-infiltrating lymphocytes alone or in combination with cisplatin-containing chemotherapy in patients with epithelial ovarian cancer. Cancer Res 1991;51(7):1934–9.

[26] Yannelli JR, Hyatt C, McConnell S, Hines K, Jacknin L, Parker L, et al. Growth of tumor-infiltrating lymphocytes from human solid cancers: summary of a 5-year experience. Int J Cancer 1996;65(4):413–21.

[27] Vasaturo A, Di Blasio S, Peeters DGA, De Koning CCH, De Vries J, Figdor C, et al. Clinical implications of co-inhibitory molecule expression in the tumor microenvironment for DC vaccination: a game of stop and go. Front Immunol 2013;4:417.

[28] Gray HJ, Gargosky SE, CAN-003 Study Team. Progression-free survival in ovarian cancer patients in second remission with mucin-1 autologous dendritic cell therapy. J Clin Oncol. 2014;32(15 Suppl.) [abstract 5504].

[29] Stover PJ. Physiology of folate and vitamin B12 in health and disease. Nutr Rev 2004;62:S3–S12.

[30] Salazar MA, Ratnam M. The folate receptor: What does it promise in tissue-targeted therapeutics? Cancer Metastasis Rev 2007;26(1):141–52.

[31] Kalli KR, Oberg AL, Keeney GL, Christianson TJH, Low PS, Knutson KL, et al. Folate receptor alpha as a tumor target in epithelial ovarian cancer. Gynecol Oncol 2008;108(3):619–26.

[32] Toffoli G, Cernigoi C, Russo A, Gallo A, Bagnoli M, Boiocchi M. Overexpression of folate binding protein in ovarian cancers. Int. J Cancer 1997;74(2):193–8.

[33] Parker N, Turk MJ, Westrick E, Lewis JD, Low PS, Leamon CP. Folate receptor expression in carcinomas and normal tissues determined by a quantitative radioligand binding assay. Anal Biochem 2005;338(2):284–93.

[34] Konner JA, Bell-McGuinn KM, Sabbatini P, Hensley ML, Tew WP, Pandit-Taskar N, et al. Farletuzumab, a humanized monoclonal antibody against folate receptor α, in epithelial ovarian cancer: a phase I study. Clin Cancer Res 2010;16(21):5288–95.

[35] Armstrong DK, White AJ, Weil SC, Phillips M, Coleman RL. Farletuzumab (a monoclonal antibody against folate receptor alpha) in relapsed platinum-sensitive ovarian cancer. Gynecol Oncol 2013;129(3):452–8.

[36] Vergote I, Armstrong D, Scambia G, Fujiwara K, Gorbunova V, Schweizer C, et al. Phase III double-blind, placebo-controlled study of weekly farletuzumab with carboplatin/taxane in subjects with platinum-sensitive ovarian cancer in first relapse. European Soceity of Gynaecological Oncology 18th International Meeting; October 19–22, 2013; Liverpool, UK Int J Gynecol Cancer 2013;23(8 Suppl. 1):11.

[37] LoRusso PM, Edelman MJ, Bever SL, Forman KM, Pilat M, Quinn MF, et al. Phase I study of folate conjugate EC145 (vintafolide) in patients with refractory solid tumors. J Clin Oncol 2012;30(32):4011–16.

[38] Vlahov IR, Santhapuram HKR, Kleindl PJ, Howard SJ, Stanford KM, Leamon CP. Design and regioselective synthesis of a new generation of targeted chemotherapeutics. Part 1: EC145, a folic acid conjugate of desacetylvinblastine monohydrazide. Bioorg Med Chem Lett 2006;16(19):5093–6.

[39] Naumann RW, Morris R, Harb W, et al. Protocol EC-FV-02: a Phase II study of EC145 in patients with advanced ovarian cancer. International Meeting of the European Society of Gynaecological Oncology Belgrade, Serbia, 11–14 October 2009. Int J Gynecol Cancer. 2009;19(Suppl. 2) [abstract 1181].

[40] Morris RT, Joyrich RN, Naumann RW, Shah NP, Maurer AH, Strauss HW, et al. Phase II study of treatment of advanced ovarian cancer with folate-receptor-targeted therapeutic (vintafolide) and companion SPECT-based imaging agent (99mTc-etarfolatide). Ann Oncol 2014;25(4):852–8.

[41] Naumann RW, Coleman RL, Burger RA, Sausville EA, Kutarska E, Ghamande SA, et al. PRECEDENT: A randomized phase II trial comparing vintafolide (EC145) and pegylated liposomal doxorubicin (PLD) in combination versus PLD alone in patients with platinum-resistant ovarian cancer. J Clin Oncol 2013;31(35):4400–6.

[42] Harb WA, Ramanathan RK, Matei DE, Nguyen B, Sausville EA. A phase 1 dose-escalation study of EC1456, a folic acid-tubulysin small-molecule drug conjugate, in adult patients (pts) with advanced solid tumors. J clin Oncol. 2014;32(15 Suppl.) [abstract TPS2630].

[43] Ab O, Whiteman KR, Bartle LM, Sun X, Singh R, Tavares D, et al. IMGN853, a folate receptor-α (FRα)–targeting antibody–drug conjugate, exhibits potent targeted antitumor activity against FRα-expressing tumors. Mol Cancer Ther 2015;14(7):1605–13.

[44] Moore KN, Martin LP, Seward SM, Bauer TM, O'Malley DM, Perez RP, et al. Preliminary single agent activity of IMGN853, a folate receptor alpha (FR{alpha})-targeting antibody–drug conjugate (ADC), in platinum-resistant epithelial ovarian cancer (EOC) patients (pts): phase I trial. J clin Oncol. 2015;33(15 Suppl.) [abstract 5518].

[45] Helsten T, Schwaederle M, Kurzrock R. Fibroblast growth factor receptor signaling in hereditary and neoplastic disease: biologic and clinical implications. Cancer Metastasis Rev 2015;34(3):479–96.

[46] Gorringe KL, Jacobs S, Thompson ER, Sridhar A, Qiu W, Choong DYH, et al. High-resolution single nucleotide polymorphism array analysis of epithelial ovarian cancer reveals numerous microdeletions and amplifications. Clin Cancer Res 2007;13(16):4731–9.

[47] Shaw AT, Kim D-W, Nakagawa K, Seto T, Crinó L, Ahn M-J, et al. Crizotinib versus chemotherapy in advanced ALK-positive lung cancer. New Engl J Med 2013;368(25):2385–94.

[48] Hurwitz HI, Dowlati A, Saini S, Savage S, Suttle AB, Gibson DM, et al. Phase I trial of pazopanib in patients with advanced cancer. Clin Cancer Res Off J Am Assoc Cancer Res 2009;15(12):4220–7.

[49] du Bois A, Floquet A, Kim J-W, Rau J, del Campo JM, Friedlander M, et al. Incorporation of pazopanib in maintenance therapy of ovarian cancer. J Clin Oncol 2014.

[50] Okamoto I, Kaneda H, Satoh T, Okamoto W, Miyazaki M, Morinaga R, et al. Phase I safety, pharmacokinetic, and biomarker study of BIBF 1120, an oral triple tyrosine kinase inhibitor in patients with advanced solid tumors. Mol Cancer Ther 2010;9(10):2825–33.

[51] Mross K, Frost A, Steinbild S, Hedbom S, Büchert M, Fasol U, et al. A phase I dose-escalation study of regorafenib (BAY 73-4506), an inhibitor of oncogenic, angiogenic, and stromal kinases, in patients with advanced solid tumors. Clin Cancer Res 2012;18(9):2658–67.

[52] Boss DS, Glen H, Beijnen JH, Keesen M, Morrison R, Tait B, et al. A phase I study of E7080, a multitargeted tyrosine kinase inhibitor, in patients with advanced solid tumours. Br J Cancer 2012;106(10):1598–604.

[53] Kilgour E, Ferry D, Saggese M, Arkenau H-T, Rooney C, Smith NR, et al. Exploratory biomarker analysis of a phase I study of AZD4547, an inhibitor of fibroblast growth factor receptor (FGFR), in patients with advanced solid tumors. J Clin Oncol. 2014;32(15 Suppl.) [abstract 11010].

[54] Smyth EC, Turner NC, Peckitt C, Pearson A, Brown G, Chua S, et al. Phase II multicenter proof of concept study of AZD4547 in FGFR amplified tumours. J Clin Oncol. 2015;33(15 Suppl.) [abstract 2508].

[55] Sequist LV, Cassier P, Varga A, Tabernero J, Schellens JH, Delord J-P, et al. Abstract CT326: phase I study of BGJ398, a selective pan-FGFR inhibitor in genetically preselected advanced solid tumors. Cancer Res 2014;74(19 Suppl.):CT326.

[56] O'Donnell P, Goldman JW, Gordon MS, Shih K, Choi YJ, Lu D, et al. 621 A phase I dose-escalation study of MFGR1877S, a human monoclonal anti-fibroblast growth factor receptor 3 (FGFR3) antibody, in patients (pts) with advanced solid tumors. Eur J Cancer. 2012;48:191–2.

[57] Tsang TY, Mohapatra G, Itamochi H, Mok SC, Birrer MJ. Abstract 1528: identification of FGFR3 as a potential therapeutic target gene for human clear cell ovarian cancer by global genomic analysis. Cancer Res 2014;74(19 Suppl.):1528.

[58] Zhang H, Lorianne M, Baker K, Sadra A, Bosch E, Brennan T, et al. FP-1039 (FGFR1:Fc), a soluble FGFR1 receptor antagonist, inhibits tumor growth and angiogenesis. Mol Cancer Ther 2007;6(11 Suppl.):B55.

[59] Garrido Lopez P, Felip E, Delord J-P, Paz-Ares L, Grilley-Olson JE, Gordon MS, et al. Multiarm, nonrandomized, open-label phase IB study to evaluate FP1039/GSK3052230 with chemotherapy in NSCLC and MPM with deregulated FGF pathway signaling. J Clin Oncol. 2014;32(15 Suppl.) [abstract TPS8120].

[60] Bjornsti MA, Houghton PJ. The TOR pathway: a target for cancer therapy. Nat Rev Cancer 2004;4(5):335–48.

[61] Philp AJ, Campbell IG, Leet C, Vincan E, Rockman SP, Whitehead RH, et al. The phosphatidylinositol 3′-kinase p85α gene is an oncogene in human ovarian and colon tumors. Cancer Res 2001;61(20):7426–9.

[62] Levine DA, Bogomolniy F, Yee CJ, Lash A, Barakat RR, Borgen PI, et al. Frequent mutation of the PIK3CA gene in ovarian and breast cancers. Clin Cancer Res 2005;11(8):2875–8.

[63] Altomare DA, Wang HQ, Skele KL, De Rienzo A, Klein-Szanto AJ, Godwin AK, et al. AKT and mTOR phosphorylation is frequently detected in ovarian cancer and can be targeted to disrupt ovarian tumor cell growth. Oncogene 2004;23(34):5853–7.

[64] Kolasa IK, Rembiszewska A, Felisiak A, Ziolkowska-Seta I, Murawska M, Moes J, et al. PIK3CA amplification associates with resistance to chemotherapy in ovarian cancer patients. Cancer Biol Ther 2009;8(1):21–6.

[65] Shapiro GI, Rodon J, Bedell C, Kwak EL, Baselga J, Braña I, et al. Phase I safety, pharmacokinetic, and pharmacodynamic study of SAR245408 (XL147), an oral pan-class I PI3K inhibitor, in patients with advanced solid tumors. Clin Cancer Res 2014;20(1):233–45.

[66] Sarker D, Ang JE, Baird R, Kristeleit R, Shah K, Moreno V, et al. First-in-human phase I study of pictilisib (GDC-0941), a potent pan-class I phosphatidylinositol-3-kinase (PI3K) inhibitor, in patients with advanced solid tumors. Clin Cancer Res 2015;21(1):77–86.

[67] Hong DS, Bowles DW, Falchook GS, Messersmith WA, George GC, O'Bryant CL, et al. A multicenter phase I trial of PX-866, an oral irreversible phosphatidylinositol 3-kinase inhibitor, in patients with advanced solid tumors. Clin Cancer Res 2012;18(15):4173–82.

[68] Bendell JC, Rodon J, Burris HA, de Jonge M, Verweij J, Birle D, et al. Phase I, dose–escalation study of BKM120, an oral pan-class I PI3K inhibitor, in patients with advanced solid tumors. J Clin Oncol 2012;30(3):282–90.

[69] Yap TA, Yan L, Patnaik A, Fearen I, Olmos D, Papadopoulos K, et al. First-in-man clinical trial of the oral pan-AKT inhibitor MK-2206 in patients with advanced solid tumors. J Clin Oncol 2011;29(35):4688–95.

[70] Behbakht K, Sill MW, Darcy KM, Rubin SC, Mannel RS, Waggoner S, et al. Phase II trial of the mTOR inhibitor, temsirolimus and evaluation of circulating tumor cells and tumor biomarkers in persistent and recurrent epithelial ovarian and primary peritoneal malignancies: a Gynecologic Oncology Group study. Gynecol Oncol 2011;123(1):19–26.

[71] O'Reilly KE, Rojo F, She Q-B, Solit D, Mills GB, Smith D, et al. mTOR inhibition induces upstream receptor tyrosine kinase signaling and activates Akt. Cancer Res 2006;66(3):1500–8.

[72] Papadopoulos KP, Tabernero J, Markman B, Patnaik A, Tolcher AW, Baselga J, et al. Phase I safety, pharmacokinetic, and pharmacodynamic study of SAR245409 (XL765), a novel, orally administered PI3K/mTOR inhibitor in patients with advanced solid tumors. Clin Cancer Res 2014;20(9):2445–56.

[73] Elser C, Hirte H, Kaizer L, Mackay H, Bindra S, Tinker L, et al. Phase II study of MKC-1 in patients with metastatic or resistant epithelial ovarian cancer or advanced endometrial cancer. J Clin Oncol. 2009;27(15 Suppl.) [abstract 5577].

[74] Hisamatsu T, Mabuchi S, Matsumoto Y, Kawano M, Sasano T, Takahashi R, et al. Potential role of mTORC2 as a therapeutic target in clear cell carcinoma of the ovary. Mol Cancer Ther 2013;12(7):1367–77.

[75] Campbell IG, Russell SE, Choong DY, Montgomery KG, Ciavarella ML, Hooi CS, et al. Mutation of the PIK3CA gene in ovarian and breast cancer. Cancer Res 2004;64(21):7678–81.

[76] Boutros T, Chevet E, Metrakos P. Mitogen-activated protein (MAP) kinase/MAP kinase phosphatase regulation: roles in cell growth, death, and cancer. Pharmacol Rev 2008;60(3):261–310.

[77] Schubbert S, Shannon K, Bollag G. Hyperactive Ras in developmental disorders and cancer. Nat Rev Cancer 2007;7(4):295–308.

[78] Chapman PB, Hauschild A, Robert C, Haanen JB, Ascierto P, Larkin J, et al. Improved survival with vemurafenib in melanoma with BRAF V600E mutation. New Engl J Med 2011;364(26):2507–16.

[79] Singer G, Oldt III R, Cohen Y, Wang BG, Sidransky D, Kurman RJ, et al. Mutations in BRAF and KRAS characterize the development of low-grade ovarian serous carcinoma. J Natl Cancer Inst 2003;95(6):484–6.

[80] Farley J, Brady WE, Vathipadiekal V, Lankes HA, Coleman R, Morgan MA, et al. Selumetinib in women with recurrent low-grade serous carcinoma of the ovary or peritoneum: an open-label, single-arm, phase 2 study. Lancet Oncol 2013;14(2):134–40.

[81] Monk BJ, Grisham RN, Marth C, Banerjee SN, Hilpert F, Coleman RL, et al. The MILO (MEK inhibitor in low-grade serous ovarian cancer)/ENGOT-ov11 study: a multinational, randomized, open-label phase 3 study of binimetinib (MEK162) versus physician's choice chemotherapy in patients with recurrent or persistent low-grade serous carcinomas of the ovary, fallopian tube, or primary peritoneum. J Clin Oncol. 2015;33(15 Suppl.) [abstract TPS5610].

[82] Robert C, Karaszewska B, Schachter J, Rutkowski P, Mackiewicz A, Stroiakovski D, et al. Improved overall survival in melanoma with combined dabrafenib and trametinib. New Engl J Med 2015;372(1):30–9.

[83] Kandoth C, McLellan MD, Vandin F, Ye K, Niu B, Lu C, et al. Mutational landscape and significance across 12 major cancer types. Nature 2013;502(7471):333–9.

[84] Ahmed AA, Etemadmoghadam D, Temple J, Lynch AG, Riad M, Sharma R, et al. Driver mutations in TP53 are ubiquitous in high grade serous carcinoma of the ovary. J Pathol 2010;221(1):49–56.

[85] Bieging KT, Mello SS, Attardi LD. Unravelling mechanisms of p53-mediated tumour suppression. Nat Rev Cancer 2014;14(5):359–70.

[86] Muller Patricia AJ, Vousden Karen H. Mutant p53 in cancer: new functions and therapeutic opportunities. Cancer Cell.2014;25(3):304–17.

[87] Brachova P, Thiel K, Leslie K. The consequence of oncomorphic TP53 mutations in ovarian cancer. Int J Mol Sci 2013;14(9):19257.

[88] Sepehrnia B, Paz IB, Dasgupta G, Momand J. Heat shock protein 84 forms a complex with mutant p53 protein predominantly within a cytoplasmic compartment of the cell. J Biol Chem 1996;271(25):15084–90.

[89] Peng Y, Chen L, Li C, Lu W, Chen J. Inhibition of MDM2 by hsp90 contributes to mutant p53 stabilization. J Biol Chem 2001;276(44):40583–90.

[90] Li D, Marchenko ND, Moll UM. SAHA shows preferential cytotoxicity in mutant p53 cancer cells by destabilizing mutant p53 through inhibition of the HDAC6-Hsp90 chaperone axis. Cell Death Differ 2011;18(12):1904–13.

[91] Liu H, Xiao F, Serebriiskii IG, O'Brien SW, Maglaty MA, Astsaturov I, et al. Network analysis identifies an HSP90-central hub susceptible in ovarian cancer. Clin Cancer Res 2013;19(18):5053–67.

[92] Lane DP. p53, guardian of the genome. Nature 1992;358(6381):15–16.

[93] Hirai H, Iwasawa Y, Okada M, Arai T, Nishibata T, Kobayashi M, et al. Small-molecule inhibition of Wee1 kinase by MK-1775 selectively sensitizes p53-deficient tumor cells to DNA-damaging agents. Mol Cancer Ther 2009;8(11):2992–3000.

[94] Wang Y, Li J, Booher RN, Kraker A, Lawrence T, Leopold WR, et al. Radiosensitization of p53 mutant cells by PD0166285, a novel G2 checkpoint abrogator. Cancer Res 2001;61(22):8211–17.

[95] Do KT, Wilsker D, Balasubramanian P, Zlott J, Jeong W, Lawrence SM, et al. Phase I trial of AZD1775 (MK1775), a wee1 kinase inhibitor, in patients with refractory solid tumors. J Clin Oncol. 2014;32(15 Suppl.) [abstract 2503].

[96] Leijen S, Beijnen JH, Schellens JH. Abrogation of the G2 checkpoint by inhibition of Wee-1 kinase results in sensitization of p53-deficient tumor cells to DNA-damaging agents. Curr Clin Pharmacol 2010;5(3):186–91.

[97] Leffers N, Lambeck AJA, Gooden MJM, Hoogeboom B-N, Wolf R, Hamming IE, et al. Immunization with a p53 synthetic long peptide vaccine induces p53-specific immune responses in ovarian cancer patients, a phase II trial. Int J Cancer 2009;125(9):2104–13.

[98] Leffers N, Vermeij R, Hoogeboom B-N, Schulze UR, Wolf R, Hamming IE, et al. Long-term clinical and immunological effects of p53-SLP® vaccine in patients with ovarian cancer. Int J Cancer 2012;130(1):105–12.

CHAPTER 9

Novel Chemotherapy Tools: Intraperitoneal Therapy, Dose-Dense Therapy

D.S. Dizon

Harvard Medical School, Boston, MA, United States

CONTENTS

INTRODUCTION

The role of cytoreductive surgery and chemotherapy in the treatment of advanced-stage epithelial ovarian cancer (EOC, defined as cancers of the ovary, fallopian tube, and peritoneum) is well established [1,2], with first-line standard chemotherapy generally consisting of both a platinum and taxane agent [3]. However, while conventional treatment has been administered on a three-weekly schedule, revisions to both the delivery and the timing of treatment have been shown to be more effective in several studies, both in terms of survival and on quality of life. This chapter will discuss alternative methods of chemotherapy administration in the first-line treatment of EOC, specifically discussing both intraperitoneal (IP) therapy and weekly (or dose-dense) administered treatment.

INTRAPERITONEAL THERAPY

EOC is a disease that spreads by a transcoelomic route to involve the peritoneal cavity at the time of initial diagnosis in most patients, felt to be the result of shedding of cancer cells from the primary tumor that are then disseminated via the peritoneal circulation [4]. Cancer cells appear to selectively invade the mesothelium of the peritoneal surface, forming micrometastases that are adequately supplied by diffusion alone until they attain sufficient size to recruit vascularization [5]. Because of the apparent localization of metastatic EOC, treatment in to this cavity via the direct infusion of anticancer drugs (IP therapy) has been long investigated as an ideal treatment alternative. Dedrick et al. proposed that the physiologic and anatomic characteristics

Translational Advances in Gynecologic Cancers. DOI: http://dx.doi.org/10.1016/B978-0-12-803741-6.00009-4

of the peritoneal lining formed a barrier from the blood compartment that would allow for significantly higher concentrations of chemotherapeutic agents to be achieved within the peritoneal cavity, while limiting systemic levels [6]. Subsequent pharmacokinetic studies demonstrated that IP chemotherapy results in high peritoneal to plasma peak concentration ratios, which varies due to drug-related factors including molecular size and lipid solubility [7]. For example, peak IP paclitaxel concentrations are 1000-fold greater than serum levels, compared with cisplatin where peak concentrations are 10- to 20-fold greater [8,9]. This means that higher drug concentrations and longer durations of tissue exposure can be achieved with IP rather than intravenous (IV) treatment, predominantly in the area most commonly involved by metastatic disease [10].

Early on it was recognized that tumor volume has an impact on treatment and currently, the role for IP treatment is limited to patients with little or no residual disease following cytoreductive surgery, since the direct penetration of chemotherapeutic agents into tumor tissue is likely only a few millimeters from the surface. Today, it is widely accepted that eligible patients for IP therapy include those with stage III optimally cytoreduced (to $\leq 1.0\,cm$) EOC and those who undergo a complete cytoreduction to no residual disease (an R0 resection). However, lessons from other trials have broadened the inclusion to those with stage II disease due to their higher risk of recurrence compared to women with stage I disease [11,12].

Randomized clinical trials have consistently demonstrated the benefits of IP therapy in the first-line treatment of women with EOC and at least nine randomized studies have been reported comparing IV and IP chemotherapy with the largest trials conducted by the Gynecologic Oncology Group (GOG) (Table 9.1) [13–15]. A 2011 meta-analysis included these trials and six others ($n = 2119$) and concluded that compared to IV treatment, IP therapy was associated with a significant reduction in the risk of death (eight studies, 2026 women; HR = 0.81; 95% confidence interval (CI): 0.72–0.90) and a significant improvement in progression-free survival (PFS) (five studies, 1311 women; HR = 0.78; 95% CI: 0.70–0.86) [16].

The current standard of care for most patients is based on GOG 172 trial, which assigned women with optimally cytoreduced stage III disease randomly assigned to standard treatment (IV paclitaxel $135\,mg/m^2$ over 24 h on day 1 plus IV cisplatin $75\,mg/m^2$ on day 2) or to an experimental regimen (IV paclitaxel on day 1, IP cisplatin $100\,mg/m^2$ on day 2, and IP paclitaxel $60\,mg/m^2$ on day 8) [15]. The results of this trial are summarized in Table 9.1 which reflects the significant survival advantages associated with this IP-containing regimen. However, multiple concerns have limited its widespread use,

Table 9.1 Phase III GOG Trials in IP Therapy

Trial	Arms	N	Main Outcomes	Toxicities
GOG 104/ SWOG 8501	IV CTX + IV CDDP vs IV CTX + IP CDDP	279	pCR (%) 36% vs 47%	Abdominal symptoms (2% vs 18%)
		287	Median OS: 41 vs 49 m	Two deaths in IP arm G3/4 Neutropenia (69% vs 56%)
				Tinnitus/Hearing loss (29% vs 12%)
GOG 114/ SWOG 9927/ ECOG GO114	IV PAC + IV CDDP vs IV Carbo + IV PAC + IP CDDP	227	Median PFS: 22 vs 28 m	G4 neutropenia (13% vs 28%)
				G4 thrombocytopenia (1% vs 24%)
		235	Median OS: 52 vs 63 m	G3/4 GI toxicity (17% vs 37%); Nephrotoxicity (<2% vs 5%)
GOG 172	IV CDDP + IV PAC vs IV PAC + IP CDDP + IP PAC	215	pCR (%): 41% vs 57%	G3/4 Leukopenia (63% vs 74%)
				G3/4 GI toxicity (24% vs 45%)
		214	Median PFS: 18 vs 28 m	G3/4 Neuropathy (9% vs 19%)
			Median OS: 50 vs 67 m	G3/4 Pain (1% vs 11%) G3/4 Renal (1% vs 6%)

From Alberts DS, Liu PY, Hannigan EV, O'Toole R, Williams SD, Young JA, et al. Intraperitoneal cisplatin plus intravenous cyclophosphamide versus intravenous cisplatin plus intravenous cyclophosphamide for stage III ovarian cancer. N Engl J Med. 1996;335(26):1950–5; Markman M, Bundy BN, Alberts DS, Fowler JM, Clark-Pearson DL, Carson LF, et al. Phase III trial of standard-dose intravenous cisplatin plus paclitaxel versus moderately high-dose carboplatin followed by intravenous paclitaxel and intraperitoneal cisplatin in small-volume stage III ovarian carcinoma: an intergroup study of the Gynecologic Oncology Group, Southwestern Oncology Group, and Eastern Cooperative Oncology Group. J Clin Oncol. 2001;19(4):1001–7; Armstrong DK, Bundy B, Wenzel L, Huang HQ, Baergen R, Lele S, et al. Intraperitoneal cisplatin and paclitaxel in ovarian cancer. N Engl J Med. 2006;354(1):34–43.

including significantly more toxicity (when compared to IV-only treatment) and questions regarding the benefit of IP in the context of additional doses used (e.g., day 8 treatment with IP paclitaxel), suggesting a dose-dense approach may be responsible for the benefit. With regard to toxicity, a modification of this regimen was shown to be more feasible by the GOG in GOG 9921 in which day 1 treatment consisted of both IV paclitaxel ($135\,mg/m^2$) and IP cisplatin ($75\,mg/m^2$) followed by IP paclitaxel on day 8 ($60\,mg/m^2$) [17]. Of 23 patients enrolled, 20 (95%) completed all six cycles, which is a huge improvement over GOG 172, where only 42% of patients were able to complete the specified number of IP treatments (six).

Given the toxicities of cisplatin, there has been a long interest in the role of carboplatin as IP treatment and several studies have been published, though no randomized trials have been performed. Nagao and colleagues reported a feasibility study that included 20 women who received a GOG 172-like regimen with carboplatin (AUC 6) administered on day 1 (following IV paclitaxel 135 mg/m^2) [18]. The completion rate was 60% and serious toxicities included allergy ($n = 4$) and ileus ($n = 1$). The GOG also conducted evaluation of IP carboplatin in a phase I study with an expanded cohort [19]. The maximum tolerated dose of carboplatin was established at an AUC 6, when administered with IV paclitaxel on day 1 and IP paclitaxel on day 8. At this dose level, 35% experienced a dose-limiting toxicity including grade 4 thrombocytopenia ($n = 1$), grade 3 neutropenic fever ($n = 3$), at least a two-week delay due to neutropenia ($n = 1$), grade 3 liver function abnormalities ($n = 1$), and grade 3 infection ($n = 1$).

Perhaps the more immediate question is whether IP treatment is more effective than other options for the first-line treatment of ovarian cancer, including dose-dense IV therapy with or without the incorporation of bevacizumab. To this end, the GOG has completed accrual to GOG-252 (NCT00951496), which enrolled patients with stage II, III, or IV EOC and randomly assigned them to one of three arms: IV carboplatin, IV paclitaxel (weekly), and IV bevacizumab (arm 1) versus IP carboplatin, IV paclitaxel, (weekly) and IV bevacizumab (arm 2) versus IV paclitaxel, IP cisplatin, IP paclitaxel, and IV bevacizumab (arm 3). This trial has also completed accrual and results are eagerly anticipated.

Beyond first-line administration, there has been some interest in utilizing IP treatment as consolidation therapy (Table 9.2) [20–26]. However, no

Table 9.2 Phase II Consolidation Trials in Ovarian Cancer

Regimen	Eligibility	n	PFS	OS
Cisplatin 200/m^2 × 3c[a]	St 2–4, cCR	31	35 m	60% 5 years
Mitoxantrone 20 mg × 6c[b]	St 2–4, pCR	50	NR	59.8% 5 years
Cisplatin 80/m^2 × 3c OR Mitoxantrone 10/m^2 × 3c[c]	St 3, pCR	41	18 m	NR
		10	18 m	
Mitoxantrone 10/m^2 q2w × 9c vs FUDR 3 g/day × 3d q3w × 6c[d]	≤1 cm RD after SLL	39	11 m	21 m
		28	25 m	38 m
Cisplatin 100/m^2 + Etoposide 200/m^{2e}	St 2–4, pCR	36	NR	NR
Cisplatin 100/m^2 × 3c[f]	St 3, pCR	30	50 m	69.1 m
Cisplatin 75/m^2 D1+ Gemcitabine 500/m^2 D1 and D8[g]	St 3, cCR	30	15.9 m	43.5 m

[a]Menczer et al. Gynecol Oncol. 1992; 222–5.
[b]Dufour et al. Cancer, 1994; 73:1865–9.
[c]Tarraza et al. Gynecol Oncol. 1993; 50:287–90.
[d]Muggia et al. Gynecol Oncol. 1996; 61:395–402.
[e]Barakat et al. Gynecol Oncol. 1998; 69:17–22.
[f]Topuz et al. Gynecol Oncol. 2004; 92:147–51.
[g]Sabbatini et al. Clin Cancer Res 2004; 10:2962–7.

phase III trials have been completed in this context. In addition, a series of smaller studies evaluating IP treatments for recurrent disease was carried out (Table 9.3) [27–32]. More recently, the GOG explored an exclusively administered IP regimen of carboplatin and bortezomib (administered on day 1 of 21 days) for women with relapsed ovarian cancer in the GOG 9921 trial [33]. Originally, carboplatin was fixed at an AUC 5, but due to dose-limiting toxicities, it was reduced to an AUC 4 with successful escalation of bortezomib dosing. A total of 33 patients were enrolled and while no maximum tolerated dose of bortezomib was reached, the overall response rate among patients with measurable disease ($n = 21$) was 19%; an additional 67% had stable disease. The highest dose of bortezomib tested was $2.5 \, mg/m^2$ and when combined with carboplatin AUC 4 there were no Dose limiting toxicities (DLTs) seen (0/6 patients).

Given the increased use of neoadjuvant chemotherapy for women with advanced ovarian cancer, there is also interest in the administration of IP therapy following interval cytoreduction. However, there are no data to inform this indication and it is not clear whether the results of adjuvant IV therapeutic trials will be applicable to these patients. The National Cancer Institute of Canada has completed enrollment in a randomized trial that tests this strategy (OV.21) and results are eagerly awaited.

Table 9.3 Phase II IP Trials in Recurrent Ovarian Cancer

Regimen (mg/m²)	Cycles	N	RR (%)	PFS	OS	Toxicity
Cisplatin 60–90[a]	4	23	56.5	NR	30 m	8/23 catheter complications
Cisplatin 100–105 + Cytarabine 600–900[b]	5	39	39	NR	NR	13% GI 8% renal 10% infectious
Carboplatin 200–300 + Etoposide 100[c]	6	46	38	NR	NR	17% catheter 23% emesis
Cisplatin 100–200+ Etoposide 350[d]	6	35	NR	13.7 m	21.8 m	Cumul. renal toxicity
Mitoxantrone 20–30[e]	NR	19	79	NR	NR	
Cisplatin 90/m² + 5-FU 1040 mg[f]	8	24	27	7 m	15.5 m	1 G4 neutrop. 11 G3–4 GI

[a]ten Tije et al. Oncology 1992;49:442–4.
[b]Markman, JCO 1991; 9:204–9.
[c]Markman, Gynecol Oncol. 1992; 47:353–7.
[d]Malmström, Gynecol Oncol. 1993; 49:166–71.
[e]Lorusso, Eur J Gynecol Oncol. 1994;15:75–80.
[f]Morgan et al. Gynecol Oncol. 2000; 77:433–8.

DOSE-DENSE THERAPY

Two hypotheses provide the foundation for the role of dose intensification of chemotherapy. The first is the Goldie–Coldman hypothesis, which proposes that drug-resistance is attributable to genetic mutations that arise at a measurable level that is proportional to tumor size [34]. In order to prevent the generation of resistant clones, they proposed that the use of dose-intense regimens comprised of alternating agents is likely to be effective. This is complemented by the Norton–Simon hypothesis, which proposed that the growth rate of a tumor is a function of its size such that smaller tumors will grow at a faster rate than larger ones. For cancers in which surgical cytoreduction is an important part of treatment (as in ovarian cancer), it posits that tumor regrowth is likely to increase following surgery, resulting in a greater chance of resistant clones to emerge [35]. Hence, it proposes that optimal treatment be administered at the shortest intervals possible to have maximal effectiveness. The concept of dose-dense therapy has been widely established in other areas of oncology, most notably in the treatment of breast cancer, where dose-dense therapy is a preferred option in the adjuvant setting and in the use of weekly paclitaxel as a standard option for women with metastatic disease.

Its role in ovarian cancer was established recently with the seminal Japanese Gynecologic Oncology Group (JGOG) 3016 trial that evaluated standard therapy (paclitaxel 180 mg/m^2 plus carboplatin AUC 6 every 21 days) versus dose-dense therapy (carboplatin AUC 6 every 21 days plus weekly paclitaxel 80 mg/m^2) in over 600 women with stage II–IV EOC. With a median of 6 years of follow-up, the rate of overall survival (OS) at 5 years significantly favored dose-dense over standard therapy (58.7% vs 51.1%, respectively; HR = 0.79; 95% CI: 0.63–0.99) [36].

The US GOG launched GOG 262 to further evaluate dose-dense therapy in a non-Asian population [37]. In this trial, over 680 women with stage II–IV EOC, optimal or suboptimally cytoreduced, were randomly assigned to standard every 3 week treatment versus dose-dense therapy (similar to the JGOG trial). In both arms, the administration of bevacizumab was optional; if administered it was given both during chemotherapy and as maintenance treatment until disease progression. Compared to standard therapy, dose-dense treatment was associated with a significantly higher rate of serious neuropathy (26% vs 18%, $p < .001$) but a significantly lower rate of grade 3 or 4 neutropenia (72% vs 83%, $p = .012$). For the overall cohort, there was no difference in PFS or OS seen among the two arms. However, among patients not treated with bevacizumab ($n = 112$), there was a significant improvement in PFS with dose-dense compared to standard treatment (14 vs 10 months, $p = .033$). OS outcomes were not yet mature enough to warrant conclusions.

Another example of dose-dense therapy comes from the Multicenter Italian Trial in Ovarian Cancer 7 (MITO-7) trial which enrolled over 800 women with stage IC–IV ovarian cancer to treatment using standard therapy (carboplatin plus paclitaxel every 3 weeks) versus a dose-dense regimen consisting of carboplatin (AUC 2) plus paclitaxel ($80 \, mg/m^2$) weekly every 3 weeks for six complete cycles [38]. Dose-dense treatment resulted in similar PFS (HR = 0.96; 95% CI: 0.80–1.16) with no difference in the estimated survival at 2 years (HR = 1.20; 95% CI: 0.90–1.61). However, dose-dense therapy was better tolerated with preserved quality of life scores for longer as treatment progressed. In addition, it was associated with lower grade 3/4 toxicities, including neutropenia (42% vs 50% with standard treatment), thrombocytopenia (1% vs 7%), and neuropathy (6% vs 17%). While no comparative data exists between this regimen and a regimen as employed by the JGOG study, which as above, employed carboplatin every 3 weeks with dose-dense paclitaxel, it is our preference to utilize this regimen whenever concerns exist over one's ability to tolerate standard therapy otherwise.

CONCLUSIONS

For women with ovarian cancer, there is no longer a singular standard treatment that should be applied to all patients. Randomized trial data support the role of IP therapy for women with optimally cytoreduced ovarian cancer, which has produced the longest survival advantages of any adjuvant treatment. However, the role of dose-dense therapy has emerged as a potential alternative option, especially for patients in whom IP therapy is not an option. Future work will improve upon these options, including an exploration on the incorporation of novel agents, both during and as extended (or maintenance) therapy.

LIST OF ACRONYMS AND ABBREVIATIONS

EOC   Epithelial ovarian cancer
IP   Intraperitoneal
IV   Intravenous
OS   Overall survival
PFS   Progression-free survival
SWOG   Southwestern Oncology Group

References

[1] Bristow RE, Tomacruz RS, Armstrong DK, Trimble EL, Montz FJ. Survival effect of maximal cytoreductive surgery for advanced ovarian carcinoma during the platinum era: a meta-analysis. J Clin Oncol 2002;20(5):1248–59.

[2] Högberg T, Glimelius B, Nygren P, SBU-group Swedish Council of Technology Assessment in Health Care. A systematic overview of chemotherapy effects in ovarian cancer. Acta Oncol Stockh Swed 2001;40(2–3):340–60.

[3] Kyrgiou M, Salanti G, Pavlidis N, Paraskevaidis E, Ioannidis JPA. Survival benefits with diverse chemotherapy regimens for ovarian cancer: meta-analysis of multiple treatments. J Natl Cancer Inst 2006;98(22):1655–63.

[4] Lengyel E. Ovarian cancer development and metastasis. Am J Pathol 2010;177(3):1053–64.

[5] Bamberger ES, Perrett CW. Angiogenesis in epithelian ovarian cancer. Mol Pathol MP 2002;55(6):348–59.

[6] Dedrick RL, Myers CE, Bungay PM, DeVita VT. Pharmacokinetic rationale for peritoneal drug administration in the treatment of ovarian cancer. Cancer Treat Rep 1978;62(1):1–11.

[7] Markman M. Intraperitoneal chemotherapy in the management of malignant disease. Expert Rev Anticancer Ther 2001;1(1):142–8.

[8] Markman M. Intraperitoneal chemotherapy. Semin Oncol 1991;18(3):248–54.

[9] Markman M, Rowinsky E, Hakes T, Reichman B, Jones W, Lewis JL, et al. Intraperitoneal administration of Taxol in the management of ovarian cancer. J Natl Cancer Inst Monogr 1993;15:103–6.

[10] Markman M. Intraperitoneal drug delivery of antineoplastics. Drugs 2001;61(8):1057–65.

[11] Young RC, Brady MF, Nieberg RK, Long HJ, Mayer AR, Lentz SS, et al. Adjuvant treatment for early ovarian cancer: a randomized phase III trial of intraperitoneal 32P or intravenous cyclophosphamide and cisplatin—a gynecologic oncology group study. J Clin Oncol 2003;21(23):4350–5.

[12] Bell J, Brady MF, Young RC, Lage J, Walker JL, Look KY, et al. Randomized phase III trial of three versus six cycles of adjuvant carboplatin and paclitaxel in early stage epithelial ovarian carcinoma: a Gynecologic Oncology Group study. Gynecol Oncol 2006;102(3):432–9.

[13] Alberts DS, Liu PY, Hannigan EV, O'Toole R, Williams SD, Young JA, et al. Intraperitoneal cisplatin plus intravenous cyclophosphamide versus intravenous cisplatin plus intravenous cyclophosphamide for stage III ovarian cancer. N Engl J Med 1996;335(26):1950–5.

[14] Markman M, Bundy BN, Alberts DS, Fowler JM, Clark-Pearson DL, Carson LF, et al. Phase III trial of standard-dose intravenous cisplatin plus paclitaxel versus moderately high-dose carboplatin followed by intravenous paclitaxel and intraperitoneal cisplatin in small-volume stage III ovarian carcinoma: an intergroup study of the Gynecologic Oncology Group, Southwestern Oncology Group, and Eastern Cooperative Oncology Group. J Clin Oncol 2001;19(4):1001–7.

[15] Armstrong DK, Bundy B, Wenzel L, Huang HQ, Baergen R, Lele S, et al. Intraperitoneal cisplatin and paclitaxel in ovarian cancer. N Engl J Med 2006;354(1):34–43.

[16] Jaaback K, Johnson N, Lawrie TA. Intraperitoneal chemotherapy for the initial management of primary epithelial ovarian cancer. Cochrane Database Syst Rev 2011(11):CD005340.

[17] Dizon DS, Sill MW, Gould N, Rubin SC, Yamada SD, Debernardo RL, et al. Phase I feasibility study of intraperitoneal cisplatin and intravenous paclitaxel followed by intraperitoneal paclitaxel in untreated ovarian, fallopian tube, and primary peritoneal carcinoma: a gynecologic oncology group study. Gynecol Oncol 2011;123(2):182–6.

[18] Nagao S, Iwasa N, Kurosaki A, Nishikawa T, Ohishi R, Hasegawa K, et al. Intravenous/intraperitoneal paclitaxel and intraperitoneal carboplatin in patients with epithelial ovarian, fallopian tube, or peritoneal carcinoma: a feasibility study. Int J Gynecol Cancer 2012;22(1):70–5.

[19] Gould N, Sill MW, Mannel RS, Thaker PH, Disilvestro P, Waggoner S, et al. A phase I study with an expanded cohort to assess the feasibility of intravenous paclitaxel, intraperitoneal

carboplatin and intraperitoneal paclitaxel in patients with untreated ovarian, fallopian tube or primary peritoneal carcinoma: a Gynecologic Oncology Group study. Gynecol Oncol 2012;125(1):54–8.

[20] Menczer J, Ben-Baruch G, Rizel S, Brenner H. Intraperitoneal chemotherapy versus no treatment in patients with ovarian carcinoma who are in complete clinical remission. Cancer 1992;70(7):1956–9.

[21] Dufour P, Bergerat JP, Barats JC, Giron C, Duclos B, Dellenbach P, et al. Intraperitoneal mitoxantrone as consolidation treatment for patients with ovarian carcinoma in pathologic complete remission. Cancer 1994;73(7):1865–9.

[22] Tarraza HM, Boyce CR, Smith WG, Jones MA. Consolidation intraperitoneal chemotherapy in epithelial ovarian cancer patients following negative second-look laparotomy. Gynecol Oncol 1993;50(3):287–90.

[23] Muggia FM, Liu PY, Alberts DS, Wallace DL, O'Toole RV, Terada KY, et al. Intraperitoneal mitoxantrone or floxuridine: effects on time-to-failure and survival in patients with minimal residual ovarian cancer after second-look laparotomy—a randomized phase II study by the Southwest Oncology Group. Gynecol Oncol 1996;61(3):395–402.

[24] Barakat RR, Almadrones L, Venkatraman ES, Aghajanian C, Brown C, Shapiro F, et al. A phase II trial of intraperitoneal cisplatin and etoposide as consolidation therapy in patients with Stage II-IV epithelial ovarian cancer following negative surgical assessment. Gynecol Oncol 1998;69(1):17–22.

[25] Topuz E, Eralp Y, Saglam S, Saip P, Aydiner A, Berkman S, et al. Efficacy of intraperitoneal cisplatin as consolidation therapy in patients with pathologic complete remission following front-line therapy for epithelial ovarian cancer. Consolidative intraperitoneal cisplatin in ovarian cancer. Gynecol Oncol 2004;92(1):147–51.

[26] Sabbatini P, Aghajanian C, Leitao M, Venkatraman E, Anderson S, Dupont J, et al. Intraperitoneal cisplatin with intraperitoneal gemcitabine in patients with epithelial ovarian cancer: results of a phase I/II Trial. Clin Cancer Res 2004;10(9):2962–7.

[27] ten Tije BJ, Wils J. Intraperitoneal cisplatin in the treatment of refractory or recurrent advanced ovarian carcinoma. Oncology 1992;49(6):442–4.

[28] Markman M, Reichman B, Hakes T, Jones W, Lewis JL, Rubin S, et al. Responses to second-line cisplatin-based intraperitoneal therapy in ovarian cancer: influence of a prior response to intravenous cisplatin. J Clin Oncol 1991;9(10):1801–5.

[29] Markman M, Reichman B, Hakes T, Rubin S, Jones W, Lewis JL, et al. Phase 2 trial of intraperitoneal carboplatin and etoposide as salvage treatment of advanced epithelial ovarian cancer. Gynecol Oncol 1992;47(3):353–7.

[30] Malmström H, Rasmussen S, Simonsen E. Intraperitoneal high-dose cisplatin and etoposide with systemic thiosulfate protection in second-line treatment of advanced ovarian cancer. Gynecol Oncol 1993;49(2):166–71.

[31] Lorusso V, Catino A, Gargano G, Fioretto A, Berardi F, de Lena M. Mitoxantrone in the treatment of recurrent ascites of pretreated ovarian carcinoma. Eur J Gynaecol Oncol 1994;15(1):75–80.

[32] Morgan RJ, Braly P, Leong L, Shibata S, Margolin K, Somlo G, et al. Phase II trial of combination intraperitoneal cisplatin and 5-fluorouracil in previously treated patients with advanced ovarian cancer: long-term follow-up. Gynecol Oncol 2000;77(3):433–8.

[33] Dizon DS, Brady WE, Lankes HA, Jandial DD, Howell, S. B., Schilder RJ. Results of a phase I pharmacokinetic study of intraperitoneal bortezomib (B) and carboplatin (C) in patients with persistent or recurrent ovarian cancer (OC): An NRG/Gynecologic Oncology Group study. Available from: http://meetinglibrary.asco.org/content/145407-156; [accessed 29.11.15].

[34] Coldman AJ, Goldie JH. Impact of dose-intense chemotherapy on the development of permanent drug resistance. Semin Oncol 1987;14(Suppl. 4):29–33.

[35] Simon R, Norton L. The Norton-Simon hypothesis: designing more effective and less toxic chemotherapeutic regimens. Nat Clin Pract Oncol 2006;3(8):406–7.

[36] Katsumata N, Yasuda M, Isonishi S, Takahashi F, Michimae H, Kimura E, et al. Long-term results of dose-dense paclitaxel and carboplatin versus conventional paclitaxel and carboplatin for treatment of advanced epithelial ovarian, fallopian tube, or primary peritoneal cancer (JGOG 3016): a randomised, controlled, open-label trial. Lancet Oncol 2013;14(10):1020–6.

[37] Chan JK, Brady MF, Penson RT, Huang H, Birrer MJ, Walker JL, et al. Weekly vs. every-3-week paclitaxel and carboplatin for ovarian cancer. N Engl J Med 2016;374(8):738–48.

[38] Pignata S, Scambia G, Katsaros D, Gallo C, Pujade-Lauraine E, De Placido S, et al. Carboplatin plus paclitaxel once a week versus every 3 weeks in patients with advanced ovarian cancer (MITO-7): a randomised, multicentre, open-label, phase 3 trial. Lancet Oncol 2014;15(4):396–405.

CHAPTER 10

Updates on Rare Epithelial Ovarian Carcinoma

J. Bergstrom, I.-M. Shih and A.N. Fader
Johns Hopkins Medicine, Baltimore, MD, United States

INTRODUCTION: EPITHELIAL OVARIAN CANCER SUBTYPES

In 2015, approximately 21,290 new cases of ovarian cancer were diagnosed in the United States and more than 14,000 women died of their disease [1]. Ovarian cancer is the fifth leading cause of cancer death among women and the most common cause of cancer death from a gynecologic malignancy [2]. Approximately 90% of malignant ovarian tumors are epithelial in origin, with the five most common epithelial ovarian cancer (EOC) subtypes reviewed in this chapter. These include high-grade serous carcinoma (HGSC), low-grade serous carcinoma (LGSC), endometrioid carcinoma, mucinous carcinoma, and clear cell carcinoma (CCC). Each of these subtypes differs in their molecular genetics, pathogenesis, clinical behavior, and overall prognosis [3]. HGSC is by far the most common subtype of EOC, accounting for nearly 70% of cases followed by clear cell (12%), endometrioid (11%), mucinous (3%), and LGSC (3%). Since CCC, endometrioid carcinoma, LGSC, and mucinous carcinoma are relatively uncommon as compared to HGSC, they are generally referred to rare EOC. Historically, women with ovarian cancer were treated in a similar fashion, despite the significant heterogeneity of clinicopathological characteristics of these different subtypes. Recently, global genome-wide analysis and molecular genetic studies have helped elucidate the molecular landscapes of different subtypes of ovarian cancer, allowing for a more sophisticated and tailored approach to ovarian cancer treatment [4].

Kurman and Shih [5] suggested dividing EOC into two distinct groups—type I and type II tumors. Type I tumors include three subgroups: (1) endometriosis-related tumors which include endometrioid, CCC, and seromucinous

CONTENTS

181

Translational Advances in Gynecologic Cancers. DOI: http://dx.doi.org/10.1016/B978-0-12-803741-6.00010-0

carcinomas, (2) LGSC, and (3) mucinous carcinomas and malignant Brenner tumors. Type II tumors include HGSC, carcinosarcoma, and undifferentiated carcinoma. Type I tumors are more indolent and genetically stable than their type II counterparts. Additionally, type I ovarian carcinomas often exhibit somatic mutations in *CTNNB1*, *ARID1A*, *KRAS*, *BRAF*, *PIK3CA*, *PTEN*, *ERBB2* and mismatch repair genes, and are diagnosed at earlier stages. In contrast, HGSC, the prototype of type II carcinoma, is highly aggressive, is most often diagnosed at an advanced stage, and harbors somatic alterations in *TP53* in virtually all of the cases.

As our understanding of rare EOCs grows, our treatment strategies will also evolve. Research pertaining to these uncommon malignancies remains challenging, given their low incidence. However, the initiation of cooperative group clinical trials focused on LGSC, CCC, and mucinous carcinoma within the Gynecologic Oncology Group (GOG) Rare Tumor Committee is an important step toward developing novel treatment strategies [3]. In this chapter, we review the unique molecular, genetic and clinical characteristics of the rare EOC tumor types and critically appraise the literature regarding the best treatment options for these tumors.

LOW-GRADE SEROUS CARCINOMA

Epidemiology

Serous carcinoma comprises almost 75% of all EOCs, with approximately 3% characterized by low-grade disease [3]. Women with LGSC are often diagnosed at a younger age, compared to those with high-grade disease. Additionally, although those with LGSC often experience longer survival than women with HGSC, overall survival outcomes remain poor for this cohort, given that most will be diagnosed at an advanced stage and because of the poor response of LGSC tumors to conventional chemotherapy agents [6].

Histology/Pathogenesis

A two-tiered grading system has been utilized to describe serous ovarian carcinoma (low vs high grade). This system is based primarily on the assessment of nuclear atypia with the mitotic rate as a secondary feature. LGSC has a mitotic rate of less than 12 mitoses per 10 high power field (HPF) where HGSC has a rate of greater than 12 mitoses per 10 HPF. Mild-to-moderate cytologic atypia is seen with low-grade disease and marked nuclear atypia is seen in high-grade carcinoma (Table 10.1) [7,8].

Most HGSCs are thought to originate from a precursor lesion, the serous tubal intraepithelial carcinoma (STIC). This has been identified most often

Table 10.1 Distinctions Among the Type I and Type II Epithelial Ovarian Carcinoma Subtypes

Histology	HGSC	LGSC	CCC	Mucinous Carcinoma	Endometrioid Carcinoma
Proportion of cases	70%	3%	12%	3%	11%
Nuclear atypia	+++	+			
Precursor lesions	STIC	Serous borderline tumor	Endometriosis		Endometriosis Oxidative stress from Iron overload
Mitosis/HPF	>12	≤12	≤12		
Molecular and protein expression					
Estrogen (ER)/progesterone (PR)	–	+	–		+
KRAS	–	+		+	+
BRAF	–	+		+	+
NRAS		+			
Ki67	+	–		–	–
P16	+				–
P53 mutations	+	–	–	–	–
MAP kinase	–	+		+	+
Wilms' tumor 1 (WT1)			–		–
PI3K pathway	–	+(40%)	+		+
Inactivation of PTEN mutations		+(3–8%)			+
IFG receptor expression (binding protein-1)		+	+	+	
PAX-2 expression		+			
PAX-8					+
Her 2				+	
Beta-catenin					+
PID3CA					+
ARID1A			+		+
MSI			+		+
Myeloperoxidase				+	
Tissue plasminogen activator				+	
Matrix metalloproteinase-9				+	
Myeloperoxidase				+	
Vimentin					+

in the distal, fimbriated portion of the fallopian tube. It is now believed that most HGSC "ovarian" cancers are fallopian tube in origin, with implantation of malignant cells from tubal carcinoma migrating on to the adjacent ovary [5,9]. Conversely, the precursor lesion to low-grade serous ovarian carcinomas is believed to be the serous borderline tumor [5,10,11].

Molecular Pathways

There is growing evidence of dual pathways for LGSC vs HGSC based upon distinct molecular and genetic differences in these tumors [11,12]. There is marked differential tumoral protein expression with immunohistochemical comparisons. Important differences include significantly higher expression of MIB1, C-KIT, HER-2/neu, and p53 mutations among HGSC as compared to LGSC. Conversely, KRAS and BRAF are more frequently seen in LGSC [11]. Estrogen and progesterone receptors are also more likely to be expressed in LGSC (Table 10.1) [10]. Interestingly, BRCA1 and BRCA2 mutations are infrequently observed with LGSC but are more commonly seen in women diagnosed with HGSC of the ovary [13,14]. Phosphatidylinositol 3-kinase (PI3K) and AKT, the downstream effectors of the insulin-like growth factor pathway, are also thought to play an important role in the pathogenesis of low-grade serous ovarian carcinoma [10].

Clinical Considerations

Women with LGSC tend to be younger at the time of diagnosis as well as have an improved overall survival as compared to women with HGSC [7,15]. In a retrospective review, it was found that the 3-year progression-free survival for LGSC was 55.9% vs 27.7% for HGSC. The overall survival at 3 years was 90.5% for LGSC as compared to 67.6% for HGSC [7]. However, these superior outcomes are not observed in all women with LGSC, as many are diagnosed at an advanced stage and will experience at least one more cancer recurrences. Further, the relatively slow proliferative activity of LGSC renders them less responsive to conventional cytotoxic chemotherapy than HGSC.

In 75% of cases, women with newly diagnosed EOC will present with stage III or IV disease. It is well established in this setting that survival improves incrementally with decreasing residual disease volume after cytoreductive surgery. This has led to a surgical paradigm advocating for radical techniques aimed at achieving microscopic residual disease at primary or interval cytoreductive surgery. For women with LGSC, cytoreduction to microscopic disease is of the utmost importance. Fader et al. [6] underscored the importance of cytoreductive surgery, in this population. In an ancillary analysis of randomized, cooperative group ovarian cancer trial, women with LGSC who were cytoreduced to no gross residual disease experienced a doubling in their median progression-free survival (33.2 months vs 14.1 months) and overall survival (96 months vs 42 months) as compared to women with any residual disease. In fact, the only factor related to survival in this ancillary analysis of a GOG phase III trial was residual disease after primary cytoreductive surgery for women with LGSC [6].

Treatment

The mainstay of treatment for women with LGSC consists of primary radical cytoreductive surgery. There is compelling evidence to suggest that LGSC is relatively resistant to cytotoxic chemotherapy, such as commonly used agents carboplatin and paclitaxel as well as cisplatin, gemcitabine, and cyclophosphamide [16].

A phase II study that demonstrated an objective tumor response using RECIST criteria for women with recurrent LGSC using selumetinib, an inhibitor of MEK1/2, showed promising results. Of the 51 patients enrolled, 15% had a response to treatment and 65% had stable disease, for an overall clinical benefit of 80% [17]. Gershenson et al. [18] investigated the efficacy of hormonal therapy for LGSCs. Although the overall response rate was 9%, patients experienced a progression-free survival of at least 6 months in 61% of the entire population.

The National Comprehensive Cancer Network (NCCN) guidelines suggest the preferred treatment for all EOCs is radical cytoreductive surgery followed by platinum/taxane chemotherapy [19]. However, given the relative chemoresistance of LGSC, alternative treatment options may be considered [16]. Given the robust ER/PR expression in LGSC and the moderate antitumor activity that hormonal therapy has demonstrated, this could be an alternative [18]. The optimal duration of hormonal therapy is not known but could be considered for use indefinitely until disease recurrence, progression, or adverse reactions.

Recurrences should also be managed with radical cytoreductive surgery if the patient is eligible to undergo surgery and otherwise an appropriate candidate for a cytoreductive procedure, with the goal of no gross residual disease, if possible. Treatment options after surgery, or in the event surgery is not feasible, would include hormonal therapy, bevacizumab, clinical trials, or cytotoxic chemotherapy [10].

CLEAR CELL CARCINOMA

Epidemiology

CCC of the ovary comprises approximately 12% of all ovarian epithelial cancers. CCC is the second most common subtype, in the United States, after HGSC, and is even more common in Japan, accounting for up to 20% of epithelial ovarian carcinoma in Asian women [20]. Data from the Surveillance, Epidemiology, and End Results Program between 1988 and 2001 were evaluated for information regarding CCC. Of the 28,082 women with epithelial ovarian carcinomas, 1411 (5%) had CCC. Women diagnosed with CCC tend to be younger, more likely to be Asian [21]. Studies have shown that CCC is more likely to be diagnosed at an earlier stage and therefore have an overall improved survival but when adjusted for stage, women with advanced stage disease do worse than those with serous ovarian carcinoma [22,23].

Histology/Pathogenesis

Histologically, CCC can display papillary, tubulocystic, and solid patterns. Microscopically CCC is characterized by clear cells and hobnail cells. They tend to be genetically more stable than HGSC with a low mitotic rate, exceeding 10 per 10 HPF in only one-fourth of cases. CCC is not graded and is considered to be high grade in all cases [10,24]. Cells in CCC share some phenotypic and morphologic characteristics with gestational endometrium that is undergoing hormone-induced hyperplastic change, otherwise known as the Arias-Stella reaction. There is substantial data that suggests that CCC of the ovary is the most common subtype of ovarian carcinoma in women who have associated endometriosis [25]. Repeated hemorrhage from endometriosis causes iron-mediated oxidative stress which accelerates carcinogenesis in this population by modifying the host genomic DNA stability [25]. CCC tends to be hormone independent, therefore rarely expresses ER/PR but often overexpresses hepatocyte nuclear factor (HNF-1B). HNF-1B influences several significant genetic alterations, including endometrial differentiation and regeneration, glycogen synthesis, detoxification, ion exchange, decidualization, and cell cycle regulation [10,25].

Molecular Pathways

The most common molecular genetic alteration in CCCs of the ovary is the somatic inactivating mutations in the gene encoding the AT-rich interacting domain containing protein 1A (ARID1A). These mutations lead to a loss of expression of the protein ARID1A which is a subunit of the tumor suppressor SWI/SNF chromatin remodeling complex [26]. PI3K/AKT is often activated in CCC as well. Activation of PI3K promotes carcinogenesis by multiple mechanisms, including proliferation, inhibition of apoptosis, cell adhesion, and transformation [5,26]. Activation of the PI3K/AKT pathway due to loss of PTEN expression is rare in high-grade serous ovarian carcinomas (<5%) but has been commonly found in CCC of the ovary (up to 40%) [26]. MET gene amplification was demonstrated by PCR in 37% of CCCs and those exhibiting this amplification were noted to have a poorer prognosis [27].

Clinical Considerations

CCC of the ovary is often diagnosed at an earlier stage. However, when it is diagnosed at an advanced stage, women have a worse than those diagnosed with serous ovarian carcinoma [22,23]. This may be in part due to the relative chemoresistance exhibited by CCC [28]. In a review of patients with CCC diagnosed between 1944 and 1981, women with CCC were more likely to be diagnosed as stage I (50% vs 31%), presented with large pelvic primary tumors greater than 10 cm (73% vs 29%) as compared to those with serous carcinomas. When comparing recurrent disease, lymph node involvement was noted in 40% of CCC as compared to only 7% of serous ovarian cancers. Parenchymal organ involvement was also far more common in the CCC group (40% vs 13%) [29]. Women with CCC also have been found to have a 2.5-fold increase in venous thromboembolism [26].

Treatment

There is a significant survival benefit if patients can be surgically cytoreduced to microscopic disease, and therefore, radical surgery is the mainstay of primary treatment [28]. As discussed previously, CCC is relatively chemoresistant. Theories related to this chemoresistance include decreased drug accumulation, low cell proliferation, and increased drug detoxification [10]. Rose et al. [30] investigated the combination of platinum and gemcitabine to potentiate the effects of the platinum drug and overcome platinum resistance. In-field radiation therapy was used in a study by MD Anderson that revealed improved outcomes for local recurrence control, especially in those with CCC, suggesting radiation therapy could play a role in recurrent and primary treatment of disease [31]. Vascular endothelial growth factor (VEGF) is frequently expressed in CCC and those with greater expression of VEGF tend to have shorter survival. Bevacizumab (VEGF inhibitor) is a promising therapeutic alternative for both primary and recurrent treatments [32]. Per NCCN guidelines, the current recommendations are to perform surgical staging for early-stage disease and cytoreductive surgery for advanced stage disease followed by a platinum/taxane-based chemotherapy [19].

ENDOMETRIOID OVARIAN CARCINOMA

Epidemiology

Endometrioid carcinoma of the ovary accounts for 11% of all epithelial ovarian carcinomas [3]. They are commonly diagnosed at an earlier stage as compared to other epithelial ovarian carcinomas, are typically well differentiated, and can be associated with endometriosis [33,34]. The average age of diagnosis is 56 years of age [30].

Histology/Pathogenesis

Histologically, endometrioid carcinoma of the ovary is often glandular with a confluent or cribriform pattern that resembles uterine carcinoma. Cytologic atypia is typically mild to moderate [33]. Endometrioid carcinoma of the ovary must have at least one of the following characteristics: glands typical of endometrioid adenocarcinoma, foci of squamous differentiation, and/or an adenofibromatous component [35]. Endometriosis has been noted to be a precursor lesion for endometrioid ovarian carcinoma as well as clear cell ovarian carcinoma. It is thought that DNA mutations that can ultimately lead to cancer are caused by oxidative stress of the iron overload associated with endometriosis [36].

Molecular Pathways

Multiple genetic mutations have been found in ovarian endometrioid carcinomas. Somatic mutations in the tumor suppressor, ARID1A, can be found in one-third of cases.

Genetic alterations in PTEN and beta-catenin have been frequently found in low-stage endometrioid ovarian carcinomas, which is similar to endometrioid carcinomas within the uterine corpus [37,38]. PTEN is located on chromosome 10q23 and is a tumor suppressor gene frequently altered in endometrioid uterine cancers but is not mutated in serous uterine carcinomas. PTEN mutations are more frequent in endometrioid ovarian carcinoma subtypes which raises the hypothesis that endometrioid carcinomas arise from a distinct molecular pathway as compared to other epithelial ovarian carcinoma subtypes, potentially related to ectopic endometrial tissue or endometriosis [39]. Microsatellite instability has also been noted in 12.5–50% of endometrioid ovarian carcinomas and is even more common in cases of synchronous primary carcinomas of the ovary and uterus [40].

Clinical Considerations

Women diagnosed with endometrioid ovarian carcinoma are often diagnosed at an earlier age, with median age of 56, compared to women with other ovarian carcinoma subtypes, with median age of 64 years. The majority are diagnosed at an early stage, with 47% diagnosed as stage I and approximately 15% diagnosed as stage II [30]. As discussed previously, endometrioid ovarian carcinomas are associated with endometriosis. In endometriosis-associated ovarian carcinoma, women tend to be premenopausal and present at an early stage, leading to improved outcomes for this population compared to those diagnosed with non-endometriosis-associated ovarian cancers [41]. Endometrioid histology has been associated with improved overall survival compared to serous ovarian carcinomas, despite similar response rates to platinum chemotherapy [42].

Treatment

Currently, the standard of care for women with endometrioid ovarian carcinoma is to undergo surgical staging and debulking if indicated. For those found to have stage IA grade 3 or stages IB–IC grades 2–3 IV, carboplatin/paclitaxel therapy should be considered. Intraperitoneal chemotherapy should be considered for those women with stage II–IV disease [19]. A post hoc analysis of GOG 157 investigated the benefit of 6 vs 3 cycles of chemotherapy in early-stage nonserous histologies. This study suggested early-stage, nonserous histologies had a similar recurrence-free survival with 3 cycles of chemotherapy as compared to 6 cycles [43]. Hormonal therapy may be considered, specifically in the recurrent setting, given that endometrioid ovarian carcinoma is often ER/PR positive [10]. Secondary cytoreductive surgery may also be considered for recurrence if disease is limited.

MUCINOUS OVARIAN CARCINOMA

Epidemiology

Mucinous ovarian carcinoma accounts for approximately 3% of all epithelial ovarian carcinomas [3]. Historical literature describes the incidence of mucinous ovarian carcinoma to be as high as 11–16%, but in actuality, the incidence is closer to 3–4%. This discrepancy is thought to be due to misclassification of gastrointestinal primary tumors, misclassification of mucinous borderline tumors as invasive carcinoma, and selection biases at large referral hospitals [44–45]. Up to 80% of mucinous epithelial tumors that are found in the ovary are from an extra ovarian primary origin [45]. Sites of primary tumors resulting in metastatic mucinous ovarian carcinomas are gastrointestinal (45%), pancreas (20%), cervix (13%), breast (8%), uterus (5%), and unknown primary (10%). The majority of primary mucinous ovarian carcinoma is diagnosed as early-stage disease [45,46].

Histology/Pathogenesis

Mucinous carcinoma of the ovary is classified as either intraepithelial (noninvasive) carcinoma or invasive carcinoma. Invasive carcinoma is suggested when stromal invasion exceeds 5 mm. Intraepithelial or noninvasive mucinous carcinomas are characterized by marked epithelial atypia but no stromal invasion is present. Invasive mucinous carcinomas are further divided into expansile and infiltrative types. The infiltrative type is noted to be more aggressive as compared to the expansile type [45,47–48]. It is common to find an invasive mucinous carcinoma of the ovary adjacent to a mucinous borderline tumor or benign mucinous cystadenoma, implying that these latter entities could be precursor lesions [45,50].

Molecular Pathway

The RAS family of G proteins is part of a pathway that potentiates growth signals from the cell surface to the nucleus. Activating RAS mutations render

GTPase proteins to be constitutively activated and thus have uninhibited growth patterns. It has been found that RAS mutations, specifically KRAS, are found commonly in mucinous ovarian carcinomas but not in other subtypes of ovarian carcinomas. These mutations are found in up to 50% of mucinous ovarian carcinomas, as opposed to only 5% of serous and 10% of endometrioid ovarian carcinomas [45,49].

BRCA mutations, which are common in serous ovarian carcinoma, are rarely found in mucinous ovarian carcinomas [51,52]. Tumor suppressor gene p53 also plays an important role in the pathogenesis of serous ovarian carcinomas but is not thought to play a prominent role in mucinous carcinoma. P53 mutations are observed in up to 58% of serous carcinoma but only 17% of mucinous ovarian carcinoma of the ovary [45,53]. Finally, HER2 amplification is seen in close to 19% of mucinous ovarian carcinomas [54].

Clinical Considerations

As discussed earlier, the majority of mucinous carcinomas involving the ovary represent metastatic disease from other organs as opposed to primary ovarian carcinomas. Bilateral mucinous tumors are metastatic in approximately 94% of cases. Size of unilateral tumors can be helpful in predicting if a primary ovarian carcinoma exists. For unilateral tumors that are greater than 10 cm upward of 80% are a primary ovarian malignancy as opposed tumors less than 10 cm are metastasis in 88% of the cases [45,55].

Eighty-three percent of women with mucinous ovarian carcinomas are diagnosed at stage I disease as compared to only 4% of serous ovarian carcinomas [44]. In a retrospective study by Schmeler et al. [56], there were no cases of isolated lymphatic metastases in women with primary mucinous carcinoma of the ovary with apparent stage IA disease, implying that routine lymphadenectomy may not be required in this population.

Treatment

Unlike other epithelial ovarian carcinomas, during primary surgery, routine lymphadenectomy is not recommended in apparent early-stage disease given the low prevalence of nodal metastases (<1%) [56]. Mucinous ovarian carcinoma is rarely diagnosed in advanced stage disease, but when it is, women fair significantly worse as compared to those with serous ovarian carcinoma.

Mucinous ovarian carcinoma is considered to be platinum resistant [57]. Women with advanced stage mucinous ovarian carcinoma undergoing standard platinum-based chemotherapy were compared to women with other histologic subtypes of epithelial ovarian carcinoma by Hess. This study demonstrated a significantly reduced overall survival of 12 months vs 36.7

months for mucinous ovarian carcinomas vs other subtypes, suggesting that other therapeutic approaches should be considered [58]. Winter et al. [59] reviewed data from six GOG phase III trials using adjuvant chemotherapy with platinum/taxane-based chemotherapy. Only 2% of these patients had mucinous ovarian carcinoma. The women with mucinous ovarian carcinoma had significantly worse progression-free survival and overall survival compared to serous ovarian carcinomas, 10.5 vs 16.9 and 14.8 and 45.2 months, respectively.

A phase II study using irinotecan and mitomycin-C for patients with clear cell and mucinous ovarian carcinoma that was platinum refractory showed promising results. The overall survival was 15.3 months for these platinum refractory patients [60]. There is hope that the combination of oxaliplatin and 5-fluorouracil (5-FU) may be effective for mucinous adenocarcinomas of the ovary after promising results in a murine model [61]. It is reasonable to consider 6 cycles of intravenous oxaliplatin and 5-FU with or without bevacizumab in women with stage IC–IV disease given the efficacy in other mucinous cancer subtypes [10].

FUTURE DIRECTIONS

Currently, there are multiple exciting trials under way to further our understanding of rare type I epithelial ovarian carcinomas and new treatment strategies. For LGSC of the ovary or peritoneum, cooperative group and industry-sponsored studies investigating the role of MEK inhibition or inhibitors of PI3K/AKT are under way. The MEK inhibitor in low-grade serous ovarian cancer (MILO) is a multinational, randomized trial investigating the use of binimetinib (a MEK1/2 inhibitor) vs physician's choice of chemotherapy in patients with recurrent or persistent LGSCs [62]. Another phase II/III trial, GOG/NRG 281, a cooperative group study investigating the MEK1/2 inhibitor trametinib, aims to treat patients with recurrent or progressive low-grade serous ovarian carcinoma comparing the treatment group with aromatase inhibition or cytotoxic chemotherapy [63,64].

For CCC, current trials are focused on investigating the efficacy of inhibitors of VEGFs and platelet-derived growth factors (PDGFs). GOG 254, a phase II study of sunitinib, an inhibitor of multiple tyrosine kinases (PDGF, VEGF), in the treatment of women with persistent or recurrent CCC is currently undergoing analysis [65,66].

There are few ongoing studies for endometrioid carcinomas of the ovary. Potential future directions include study of hormonal therapy for both primary and recurrent diseases and radiotherapy. Given the rarity of endometrioid ovarian carcinomas, studies will need to be collaborative.

Finally, in a phase II Japanese study, women with advanced or recurrent mucinous ovarian carcinoma are receiving treatment with oxaliplatin and S1, an orally active drug combining tegafur (converted to fluorouracil), gimeracil (dihydropyrimidine dehydrogenase inhibitor), and oteracil (thought to improve treatment-related gastrointestinal side effects). Unfortunately a four-arm international GOG study of carboplatin and paclitaxel with or without bevacizumab compared with capecitabine and oxaliplatin with or without bevacizumab used in women with stage II-IV mucinous ovarian carcinoma was closed due to low accrual.

CONCLUSIONS

Type I ovarian carcinoma, including low-grade serous, clear cell, endometrioid, and mucinous carcinomas, are a group of rare, but clinically important, EOC subtypes. As demonstrated earlier, each subtype is distinctly different in regard to histology, molecular pathways, genetics, pathogenesis, and clinical behavior. To treat each of these distinct entities in an identical manner is antiquated, given our enhanced understanding of the differences in these subtypes. It will be critical for institutions and cooperative groups to continue collaborating to advance our understanding of best practices in treatment of rare gynecologic malignancies and to develop innovative, personalized treatment plans that improve outcome and survival for women with uncommon gynecologic cancers.

References

[1] http://seer.cancer.gov/statfacts/html/ovary.html [accessed October 2015].

[2] http://www.cancer.org/cancer/ovariancancer/index [accessed October 2015].

[3] http://www.gog.org/Spring2013newsletter.pdf [accessed October 2015].

[4] Farley J, Ozbun L, Birrer M. Genomic analysis of epithelial ovarian cancer. Cell Res 2008;18(5):538–48.

[5] Kurman R, Shih I. The origin and pathogenesis of epithelial ovarian cancer: a proposed unifying theory. Am J Surg Pathol 2010;34(3):433–43.

[6] Fader AN, Java J, Ueda S, Bristow R, Armstrong D, Bookman M, et al. Survival in women with grade 1 serous ovarian carcinoma. Obstet Gynecol 2013;122(2, Part 1):225–32.

[7] Bodurka D, Deavers M, Tian C, Sun C, Malpica A, Coleman R, et al. Reclassification of serous ovarian carcinoma by a 2-tier system. Cancer 2011;118(12):3087–94.

[8] Malpica A, Deavers M, Lu K, Bodurka D, Atkinson E, Gershenson D, et al. Grading ovarian serous carcinoma using a two-tier system. Am J Surg Pathol 2004;28(4):496–504.

[9] Medeiros F, Muto M, Lee Y, Elvin J, Callahan M, Feltmate C, et al. The tubal fimbria is a preferred site for early adenocarcinoma in women with familial ovarian cancer syndrome. Am J Surg Pathol 2006;30(2):230–6.

[10] Groen R, Gershenson D, Fader A. Updates and emerging therapies for rare epithelial ovarian cancers: one size no longer fits all. Gynecol Oncol 2015;136(2):373–83.

[11] Mishra S, Crasta J. An immunohistochemical comparison of P53 and Bcl-2 as apoptotic and MIB1 as proliferative markers in low-grade and high-grade ovarian serous carcinomas. Int J Gynecol Cancer 2010;20(4):537–41.

[12] Singer G, Cope L, Dehari R, Hartmann A, Cao D, et al. Patterns of p53 mutations separate ovarian serous borderline tumors and low- and high-grade carcinomas and provide support for a new model of ovarian carcinogenesis. Am J Surg Pathol 2005;29(2):218–24.

[13] Risch H, McLaughlin J, Cole D, Rosen B, Bradley L, Kwan E, et al. Prevalence and penetrance of germline BRCA1 and BRCA2 mutations in a population series of 649 women with ovarian cancer. Am J Hum Genet 2001;68(3):700–10.

[14] Press J, De Luca A, Boyd N, Young S, Troussard A, Ridge Y, et al. Ovarian carcinomas with genetic and epigenetic BRCA1 loss have distinct molecular abnormalities.. BMC Cancer 2008;8(1):17.

[15] Gershenson D, Sun C, Lu K, Coleman R, Sood A, Malpica A, et al. Clinical behavior of stage II–IV low-grade serous carcinoma of the ovary. Obstetr Gynecol 2006;108(2):361–8.

[16] Santillan A, Kim Y, Zahurak M, Gardner G, Giuntoli R, Shih I, et al. Differences of chemoresistance assay between invasive micropapillary/low-grade serous ovarian carcinoma and high-grade serous ovarian carcinoma. Int J Gynecol Cancer 2007;17(3):601–6.

[17] Farley J, Brady W, Vathipadiekal V, Lankes H, Coleman R, Morgan M, et al. Selumetinib in women with recurrent low-grade serous carcinoma of the ovary or peritoneum: an open-label, single-arm, phase 2 study. Lancet Oncol 2013;14(2):134–40.

[18] Gershenson D, Sun C, Iyer R, Wong K, Kavanagh J, Malpica A, et al. Hormonal therapy for recurrent low-grade serous carcinoma of the ovary or peritoneum. Gynecol Oncol 2012;125:S35.

[19] http://www.nccn.org/professionals/physician_gls/pdf/ovarian.pdf [accessed October 2015].

[20] Itamochi H, Kigawa J, Terakawa N. Mechanisms of chemoresistance and poor prognosis in ovarian clear cell carcinoma. Cancer Sci 2008;99(4):653–8.

[21] Chan J, Teoh D, Hu J, Shin J, Osann K, Kapp D. Do clear cell ovarian carcinomas have poorer prognosis compared to other epithelial cell types? A study of 1411 clear cell ovarian cancers. Gynecol Oncol 2008;109(3):370–6.

[22] Kennedy A, Biscotti C, Hart W, Webster K. Ovarian clear cell adenocarcinoma. Gynecol Oncol 1989;32(3):342–9.

[23] Mizuno M, Kikkawa F, Shibata K, Kajiyama H, Ino K, Kawai M, et al. Long-term follow-up and prognostic factor analysis in clear cell adenocarcinoma of the ovary. J Surg Oncol 2006;94(2):138–43.

[24] Barakat R. Principles and practice of gynecologic oncology. Philadelphia, PA: Wolters Kluwer Health/Lippincott Williams & Wilkins; 2013.

[25] Kobayashi H. Toward an understanding of the pathophysiology of clear cell carcinoma of the ovary (Review). Oncol Lett 2013.

[26] Samartzis E, Noske A, Dedes K, Fink D, Imesch P. ARID1A mutations and PI3K/AKT pathway alterations in endometriosis and endometriosis-associated ovarian carcinomas. IJMS 2013;14(9):18824–49.

[27] Yamashita Y, Akatsuka S, Shinjo K, Yatabe Y, Kobayashi H, Seko H, et al. Met is the most frequently amplified gene in endometriosis-associated ovarian clear cell adenocarcinoma and correlates with worsened prognosis. PLoS ONE 2013;8(3):e57724.

[28] Takano M, Tsuda H, Sugiyama T. Clear cell carcinoma of the ovary: Is there a role of histology-specific treatment? J Exp Clin Cancer Res 2012;31(1):53.

[29] Jenison E, Montag A, Griffiths C, Welch W, Lavin P, Greer J, et al. Clear cell adenocarcinoma of the ovary: a clinical analysis and comparison with serous carcinoma. Gynecol Oncol 1989;32(1):65–71.

[30] Rose P. Gemcitabine reverses platinum resistance in platinum-resistant ovarian and peritoneal carcinoma. Int J Gynecol Cancer 2005;15(s1):18–22.

[31] Brown A, Jhingran A, Klopp A, Schmeler K, Ramirez P, Eifel P. Involved-field radiation therapy for locoregionally recurrent ovarian cancer. Gynecol Oncol 2013;130(2):300–5.

[32] Mabuchi S, Kawase C, Altomare D, Morishige K, Hayashi M, Sawada K, et al. Vascular endothelial growth factor is a promising therapeutic target for the treatment of clear cell carcinoma of the ovary. Mol Cancer Ther 2010;9(8):2411–22.

[33] Seidman J, Horkayne-Szakaly I, Haiba M, Boice C, Kurman R, Ronnett B. The histologic type and stage distribution of ovarian carcinomas of surface epithelial origin. Int J Gynecol Pathol 2004;23(1):41–4.

[34] Kline R, Wharton J, Atkinson E, Burke T, Gershenson D, Edwards C. Endometrioid carcinoma of the ovary: retrospective review of 145 cases. Int J Gynecol Obstetr 1991;36(2):169–70.

[35] Tornos C, Silva E, Ordonez N, Gershenson D, Young R, Scully R. Endometrioid carcinoma of the ovary with a prominent spindle-cell component, a source of diagnostic confusion a report of 14 cases. Am J Surg Pathol 1995;19(12):1343–53.

[36] Yamada Y, Shigetomi H, Onogi A, Haruta S, Kawaguchi R, Yoshida S, et al. Redox-active iron-induced oxidative stress in the pathogenesis of clear cell carcinoma of the ovary. Int J Gynecol Cancer 2011;14(1):32–40.

[37] Catasús L, Bussaglia E, Rodríguez I, Gallardo A, Pons C, Irving J, et al. Molecular genetic alterations in endometrioid carcinomas of the ovary: similar frequency of beta-catenin abnormalities but lower rate of microsatellite instability and PTEN alterations than in uterine endometrioid carcinomas. Hum Pathol 2004;35(11):1360–8.

[38] Palacios J, Gamallo C. Mutations in beta-catenin gene (CTNNB1) in endometrioid ovarian carcinomas. Cancer Res 1998;58:2095.

[39] Obata K, Morlan S, Watson R, Hitchcock A, Chenevix-Trech G, Thomas E. Frequent PTEN/MMAC mutations in endometrioid but not serous or mucinous epithelial ovarian tumors. Cancer Res 1998;58:1334.

[40] Irving J, Catasús L, Gallardo A, Bussaglia E, Romero M, Matias-Guiu X, et al. Synchronous endometrioid carcinomas of the uterine corpus and ovary: alterations in the β-catenin (CTNNB1) pathway are associated with independent primary tumors and favorable prognosis. Hum Pathol 2005;36(6):605–19.

[41] Modesitt S. Ovarian and extraovarian endometriosis-associated cancer. Obstetr Gynecol 2002;100(4):788–95.

[42] Storey D, Rush R, Stewart M, Rye T, Al-Nafussi A, Williams A, et al. Endometrioid epithelial ovarian cancer. Cancer 2008;112(10):2211–20.

[43] Chan J, Tian C, Fleming G, Monk B, Herzog T, Kapp D, et al. The potential benefit of 6 vs. 3 cycles of chemotherapy in subsets of women with early-stage high-risk epithelial ovarian cancer: an exploratory analysis of a Gynecologic Oncology Group study. Gynecol Oncol 2010;116(3):301–6.

[44] Seidman J, Horkayne-Szakaly I, Haiba M, Boice C, Kurman R, Ronnett B. The histologic type and stage distribution of ovarian carcinomas of surface epithelial origin. Int J Gynecol Pathol 2004;23(1):41–4.

[45] Frumovitz M, Schmeler K, Malpica A, Sood A, Gershenson D. Unmasking the complexities of mucinous ovarian carcinoma. Gynecol Oncol 2010;117(3):491–6.

[46] Shimada M, Kigawa J, Ohishi Y, Yasuda M, Suzuki M, Hiura M, et al. Clinicopathological characteristics of mucinous adenocarcinoma of the ovary. Gynecol Oncol 2009;113(3):331–4.

[47] Lee K, Scully R. Mucinous tumors of the ovary. Am J Surg Pathol 2000;24(11):1447–64.

[48] Riopel M, Ronnett B, Kurman R. Evaluation of diagnostic criteria and behavior of ovarian intestinal-type mucinous tumors. Am J Surg Pathol 1999;23(6):617–35.

[49] Gemignani M, Schlaerth A, Bogomolniy F, Barakat R, Lin O, Soslow R, et al. Role of KRAS and BRAF gene mutations in mucinous ovarian carcinoma. Gynecol Oncol 2003;90(2):378–81.

[50] Mok S, Bell D, Knapp R, Fishbaugh P, Welch W, Muto M. Mutation of K-ras protoon-cogene in human ovarian epithelial tumors of borderline malignancy. Cancer Res 1993;53:1489–92.

[51] Tonin P, Maugard C, Perret C, Mes-Masson A, Provencher D. A review of histopathological subtypes of ovarian cancer in BRCA-related French Canadian cancer families. Fam Cancer 2007;6(4):491–7.

[52] Evans D, Young K, Bulman M, Shenton A, Wallace A, Lalloo F. Probability of BRCA1/2 mutation varies with ovarian histology: results from screening 442 ovarian cancer families. Clin Genet 2008;73(4):338–45.

[53] Schuijer M, Berns E. TP53 and ovarian cancer. Hum Mut 2003;21(3):285–91.

[54] McAlpine J, Wiegand K, Miller M, Adamiak A, Koebel M, Vang R, et al. HER2 overexpression and amplification is present in a subset of ovarian mucinous carcinomas and can be targeted with trastuzumab therapy. Gynecol Oncol 2010;116(3):593–4.

[55] Seidman J, Kurman R, Ronnett B. Primary and metastatic mucinous adenocarcinomas in the ovaries. Am J Surg Pathol 2003;27(7):985–93.

[56] Schmeler K, Tao X, Frumovitz M, Deavers M, Sun C, Sood A, et al. Prevalence of lymph node metastasis in primary mucinous carcinoma of the ovary. Obstetr Gynecol 2010;116(2, Part 1):269–73.

[57] Zaino R, Brady M, Lele S, Michael H, Greer B, Bookman M. Advanced stage mucinous adenocarcinoma of the ovary is both rare and highly lethal. Cancer 2010;117(3):554–62.

[58] Hess V. Mucinous epithelial ovarian cancer: a separate entity requiring specific treatment. Journal of Clinical Oncology 2004;22(6):1040–4.

[59] Winter W, Maxwell G, Tian C, Carlson J, Ozols R, Rose P, et al. Prognostic factors for stage III epithelial ovarian cancer: a Gynecologic Oncology Group study. J Clin Oncol 2007;25(24):3621–7.

[60] Shimizu Y, Umezawa S, Hasumi K, Yamauchi K, Silverberg S. Efficacy of a combination of irinotecan (CPT-11) with mitomycin-C (MMC) for clear cell carcinoma of the ovary (OCCA) which is intrinsically platinum-resistant. Eur J Cancer 1997;33:S119.

[61] Sato S, Itamochi H, Kigawa J, Oishi T, Shimada M, Sato S, et al. Combination chemotherapy of oxaliplatin and 5-fluorouracil may be an effective regimen for mucinous adenocarcinoma of the ovary: a potential treatment strategy. Cancer Sci 2009;100(3):546–51.

[62] http://meetinglibrary.asco.org/content/98477 [accessed October 2015].

[63] https://www.clinicaltrials.gov/ct2/show/NCT02101788 [accessed October 2015].

[64] Miller C, Oliver K, Farley J. MEK1/2 inhibitors in the treatment of gynecologic malignancies. Gynecol Oncol 2014;133(1):128–37.

[65] Wei W, Dizon D, Vathipadiekal V, Birrer M. Ovarian cancer: genomic analysis. Ann Oncol 2013;24(Suppl. 10):x7–x15.

[66] https://www.clinicaltrial.gov [accessed October 2015].

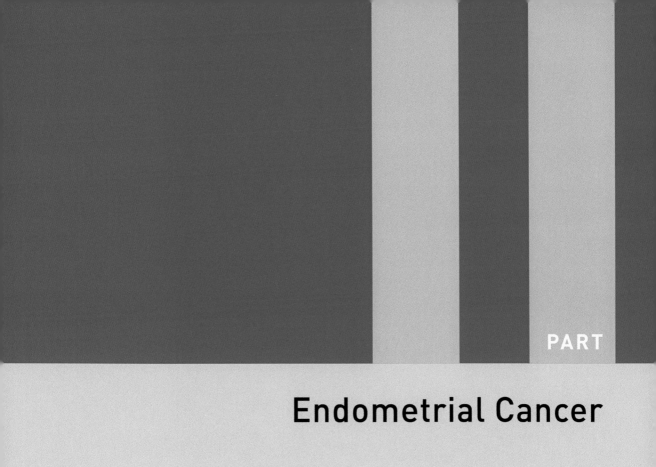

Endometrial Cancer

CHAPTER 11

Endometrial Cancer Genomics

M.E. Urick and D.W. Bell

National Institutes of Health, Bethesda, MD, United States

CONTENTS

INTRODUCTION

Uterine cancers are one of the most commonly diagnosed gynecologic malignancies, with more than 300,000 new cases among women worldwide each year [1,2]. Almost all uterine cancers are endometrial carcinomas (ECs) with the remaining uterine neoplasms including mesenchymal and trophoblastic tumors [3]. ECs are further classified into several distinct histopathological subtypes, including endometrioid ECs (EECs), serous ECs (SECs), clear cell ECs (CCECs), and uterine carcinosarcomas (UCSs) [4]. Adding to the complexity of EC, some tumors consist of an admixture of two or more histologies [5–11]. Difficulty in reproducibly classifying some ECs based on histopathology alone is a well-documented phenomenon [12–15], particularly for high-grade tumors, thus highlighting the need for molecular adjuncts to improve on current diagnostic approaches.

EECs are the most common histological subtype of EC, comprising 85–90% of diagnosed cases [3,16,17]. EECs are often associated with unopposed estrogen exposure and endometrial hyperplasia, and occur at a younger age (mean ages 56–64 years) relative to non-EECs [18–20]. The five-year recurrence-free survival rates for women with EEC are favorable, ranging from 68% to 83.1% [19,21–23], reflecting the fact that EEC can frequently be cured by surgical intervention [17,24]. Nonetheless, high-grade (G3) EECs have significantly poorer prognosis than lower grade EECs [25,26]. In one study, G3 EECs represented 15% of diagnosed ECs but accounted for 27% of EC-related deaths whereas lower grade (G1 and G2) EECs represented 72% of diagnoses but only 26% of deaths [27]. In terms of overall survival and disease-specific survival, it remains unclear whether high-grade EECs and non-EECs differ [27–33].

Translational Advances in Gynecologic Cancers. DOI: http://dx.doi.org/10.1016/B978-0-12-803741-6.00011-2

In contrast to EECs, SECs and CCECs tend to occur in older, postmenopausal women (mean ages 53–75 years) [5–8,27,34–40], are generally not associated with unopposed estrogen exposure or hyperplasia, are frequently associated with extrauterine disease, and have a high propensity to recur [6,41]. SECs, which are the most common of the non-EECs, comprise only 2–12% of all diagnosed ECs [16,22,27,35,37,42–45]. Despite their rarity at diagnosis, SECs contribute to a disproportionately high percentage (39% in one study) of EC-related deaths [27]. Five-year disease-free survival rates for women with SEC are 18–55% [22,23,27,37] and recurrence rates are 38–52% [46–48]. Together with CCEC, SECs account for 50% of EC relapses [49]. CCECs represent 2–7% of diagnosed ECs [16,22,44,50–54] and account for ~8% of EC deaths [27]. In terms of their clinical behavior, CCECs fall between EECs and SECs, with five-year disease-free survival rates of 38–50% [23,53,55] and relapse rates of 5–42% [47,48,51,53,55].

UCSs, previously known as malignant mixed mesodermal (or Müllerian) tumors (MMMTs) of the uterus, are rare metaplastic carcinomas that represent 2–4% of diagnosed EC cases [56–58]. UCSs are biphasic tumors with both carcinoma (epithelial) and sarcoma (mesenchymal) components that are generally believed to be monoclonal in origin [59–68]. Histologically, the carcinoma component typically is represented by intermediate-grade (G2) EEC, G3 EEC, SEC, or CCEC [69]. The sarcoma component can be homologous or heterologous [61,67] comprised of spindle cell, rhabdomyosarcoma, chondrosarcoma, osteosarcoma, leiomyosarcoma, and/or liposarcoma [69,70]. It is unclear whether homology of the sarcomatous component is prognostic for outcome [58,69–76].

UCS is one of the most clinically aggressive forms of EC. Patients with this tumor type have significantly poorer outcomes compared to patients with SEC, CCEC, or G3 EEC [48,77,78]. Overall five-year survival rates for patients with UCS range from 6% to 38% [58,79,80]. Almost half (41–45%) of UCS patients are diagnosed with advanced stage disease, as compared to 38–52% of SEC, 36% of CCEC, and 29–31% of high-grade EEC patients [27,77,81]. Moreover, 41–59% of UCS patients relapse [48,58,73,77,82–84].

With the development of massively parallel sequencing, also referred to as next generation sequencing (NGS), there has been a paradigm shift in the search for pathogenic driver genes of human cancer [85]. Using this approach, whole exome sequencing (WES) has enabled systematic searches for somatic mutations among all protein-encoding genes in the human genome. The major genomic alterations that drive EECs, SECs, and UCSs have been catalogued and have revealed important distinctions, as well as commonalities, both within and between histological subtypes (Fig. 11.1) [33,68,86–91]. Moreover, an integrated genomic analysis of EECs and SECs

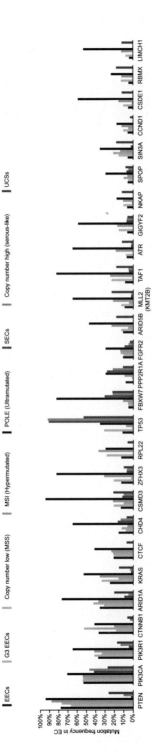

FIGURE 11.1

Mutation frequencies of significantly mutated genes (SMGs) (TCGA FDR CT < 0.02) in the TCGA study of endometrioid and serous ECs. The graph displays only SMGs mutated at a frequency >10% in at least one EC subtype. SMGs mutated at a frequency of <10% can be found in Tables 11.2–11.5. UCS values are combined frequencies from Table 11.5.

Table 11.1 Distribution of EC Histological Subtypes Among Molecular Subgroups [86]

	POLE (Ultramutated) (*n* = 17)	MSI (Hypermutated) (*n* = 65)	Copy Number Low (MSS) (*n* = 90)	Copy Number High (Serous-like) (*n* = 60)
EEC (non-G3)	9 (53%)	41 (63%)	82 (91%)	7 (12%)
G3 EEC	8 (47%)	24 (37%)	6 (7%)	9 (15%)
SEC	0	0	1 (1%)	41 (68%)
Mixed	0	0	1 (1%)	3 (5%)

by The Cancer Genome Atlas (TCGA) has shown that these two histological subtypes can be classified into four distinct molecular subgroups: *POLE* (ultramutated), microsatellite instability (MSI) (hypermutated), copy number low (MSS), and copy number high (serous-like) [86]. Within this classification, EECs populate each of the four subgroups whereas almost all (98%) SECs fall within the serous-like molecular subgroup (Table 11.1) [86].

Here, we discuss the major features of the genomic landscapes of EECs, SECs, and UCSs; we refer the reader to other recent literature on the current state of knowledge of the molecular pathogenesis of CCECs, which thus far have been genomically characterized in much less detail [13,23,66,89,92–100]. We have organized this chapter by histological subtypes and we review the genomics of *POLE* (ultramutated), MSI (hypermutated), and copy number low (microsatellite stable) (MSS) tumors within the EEC subsection, followed by a review of the genomics of the serous-like molecular subgroup within the SEC subsection.

THE GENOMICS OF EECs

The overall genomic landscape of EECs is comprised of a predominance of somatic mutations, frequent MSI, few copy number aberrations (Table 11.2), and increased methylation levels compared to SECs and normal endometrium [86]. Several of the major features of this landscape were initially uncovered in candidate gene studies and subsequently validated in comprehensive genomic studies. These features include frequent MSI (17–54%) and somatic mutations in *PTEN* (15–78%), *PIK3CA* (16–53%), *PIK3R1* (33–43%), *ARID1A* (35–58%), *KRAS* (7–26%), *FGFR2* (3–12%), and *CTNNB1* (8–37%) (Table 11.2). These alterations are found significantly more often among EECs than SECs [86] (Tables 11.2 and 11.4).

MSI reflects a mutator phenotype caused by underlying defects in DNA mismatch repair (MMR). In the context of sporadic EEC, most instances of MSI result from hypermethylation of the *MLH1* promoter [107,108], which is an

Table 11.2 Molecular Aberrations Found in EECs

Gene	Aberration	TCGA Aberration Frequency [86]	Range of Non-TCGA Reported Aberration Frequencies	Reference(s)
Mutations				
PTEN	Mutation	78% (136/175)	15–69%	[23,33] [a–f]
PIK3CA	Mutation	53% (93/175)	16–52%	[23,33,99,101] [a,c,d,e,g]
PIK3R1	Mutation	37% (65/175)	33–43%	[100] [a]
CTNNB1	Mutation	37% (64/175)	8–28%	[23,33,101] [a,e,h,i]
ARID1A	Mutation	35% (62/175)	48–58%	[33] [i]
KRAS	Mutation	25% (43/175)	7–26%	[23,33,101,102] [a,c,e,i]
CTCF	Mutation	21% (36/175)	25–85%	[103] [j]
CSMD3	Mutation	13% (23/175)	–	–
RPL22	Mutation	13% (22/175)	50–52%	[104] [k]
ZFHX3	Mutation	13% (22/175)	20%	[103]
TP53	Mutation	11% (20/175)	3–32%	[33,102] [a,c,i]
FGFR2	Mutation	11% (19/175)	3–12%	[23,101] [a]
ARID5B	Mutation	11% (19/175)	–	–
MLL4	Mutation	9% (16/175)	–	–
BCOR	Mutation	8% (14/175)	–	–
FAM135B	Mutation	7% (12/175)	–	–
GIGYF2	Mutation	6% (11/175)	–	–
NFE2L2	Mutation	6% (11/175)	–	–
SPOP	Mutation	6% (10/175)	0%	[89]
SIN3A	Mutation	6% (10/175)	–	–
CCND1	Mutation	6% (10/175)	–	–
FBXW7	Mutation	5% (9/175)	3–10%	[23,89] [c]
FOXA2	Mutation	5% (9/175)	–	–
INPP4A	Mutation	5% (9/175)	–	–
CSDE1	Mutation	5% (9/175)	–	–
RBMX	Mutation	5% (8/175)	–	–
SGK1	Mutation	5% (8/175)	–	–
LIMCH1	Mutation	4% (7/175)	–	–
RRN3P2	Mutation	4% (7/175)	–	–
SOX17	Mutation	4% (7/175)	–	–
MECOM	Mutation	3% (6/175	–	–
HIST1H2BD	Mutation	3% (5/175)	–	–
SMTNL2	Mutation	3% (5/175)	–	–
METL14	Mutation	2% (4/175)	–	–
RRAS2	Mutation	2% (4/175)	–	–
MIR543	Mutation	2% (3/175)	–	–
PNN	Mutation	2% (3/175)	–	–

(Continued)

Table 11.2 Molecular Aberrations Found in EECs (Continued)

Gene	Aberration	TCGA Aberration Frequency [86]	Range of Non-TCGA Reported Aberration Frequencies	Reference(s)
Copy Number Aberrations				
CCND1	Amp	3% (5/186)	3%	l
IGFR1	Amp	1% (2/186)	–	–
WWOX	Loss	3% (5/186)	38%	m
PTEN	Loss	2% (4/186)	–	–
MSI Status				
MSI	MSI	40%	17–54%	[89,101,102,105,106] e,f,n

This table is filtered to significantly mutated (FDR CT < 0.02) and copy number aberrant (FDR CT < 0.15) genes identified by TCGA in non-ultramutated EECs and copy number clusters 2 and 3, respectively.
aCheung LW, et al., High frequency of PIK3R1 and PIK3R2 mutations in endometrial cancer elucidates a novel mechanism for regulation of PTEN protein stability. Cancer Discov 2011;2:170–85.
bDjordjevic B, et al., Clinical assessment of PTEN loss in endometrial carcinoma: immunohistochemistry outperforms gene sequencing. Mod Pathol 2012;25(5):699–708.
cGarcia-Dios DA, et al., High-throughput interrogation of PIK3CA, PTEN, KRAS, FBXW7 and TP53 mutations in primary endometrial carcinoma. Gynecol Oncol 2013;128(2):327–34.
dHayes MP, et al., PIK3CA and PTEN mutations in uterine endometrioid carcinoma and complex atypical hyperplasia. Clin Cancer Res 2006;12(20 Pt 1):5932–5.
eKonopka B, et al., PIK3CA mutations and amplification in endometrioid endometrial carcinomas: relation to other genetic defects and clinicopathologic status of the tumors. Hum Pathol 2011;42(11):1710–19.
fTashiro H, et al., Mutations in PTEN are frequent in endometrial carcinoma but rare in other common gynecological malignancies. Cancer Res 1997;57(18):3935–40.
gVelasco A, et al., PIK3CA gene mutations in endometrial carcinoma: correlation with PTEN and K-RAS alterations. Hum Pathol 2006;37(11):1465–72.
hFukuchi T, et al., Beta-catenin mutation in carcinoma of the uterine endometrium. Cancer Res 1998;58(16):3526–8.
iMcConechy MK, et al., Ovarian and endometrial endometrioid carcinomas have distinct CTNNB1 and PTEN mutation profiles. Mod Pathol 2014;27(1):128–34.
jHoivik EA, et al., Hypomethylation of the CTCFL/BORIS promoter and aberrant expression during endometrial cancer progression suggests a role as an Epi-driver gene. Oncotarget 2014;5(4):1052–61.
kFerreira AM, et al., High frequency of RPL22 mutations in microsatellite-unstable colorectal and endometrial tumors. Hum Mutat 2014;35(12):1442–5.
lMoreno-Bueno G, et al., Cyclin D1 gene (CCND1) mutations in endometrial cancer. Oncogene 2003;22(38):6115–18.
mPluciennik E, et al., The WWOX tumor suppressor gene in endometrial adenocarcinoma. Int J Mol Med 2013;32(6):1458–64.
nKanopiene D, et al., Impact of microsatellite instability on survival of endometrial cancer patients. Medicina 2014;50(4):216–21.

early event [109]. MSI in EEC is significantly associated with higher tumor grade [105]. However, there are conflicting data as to whether MSI status is associated with clinical outcomes among EC patients [110].

PIK3CA, *PIK3R1*, and *PTEN* encode critical components of the PI3K pathway, which regulates a number of cellular processes including proliferation/growth, survival, metabolism, glucose homeostasis, transcription, and protein synthesis. Overall, 68–93% of EECs have at least one mutation in these genes

[86,99,100,111]. It is noteworthy that *PIK3CA* and *PTEN* exhibit unique spectrums of mutation in EECs compared to cancers originating in other tissues. Whereas mutations in exons 1–7 of *PIK3CA* are rare in most human cancers, EECs exhibit frequent *PIK3CA* mutations in these exons, with a particular aggregation in the first coding exon, and a high frequency of *PTEN* alterations at amino acid R130 [86,99]. EECs also exhibit a higher frequency of *PIK3R1* mutation than tumor types [100] and the mutually exclusive nature of most *PIK3CA* and *PIK3R1* mutations implies functional redundancy in activating the PI3K pathway [86,100]. The frequent co-occurrences of *PTEN* and *PIK3CA* mutations [99,100,112] are other unique genomic attributes of EECs compared to tumors from other lineages. Taken as a whole, the high frequency and recurrent nature of aberrations within the PI3K pathway indicates the importance of this pathway in EECs and is of potential therapeutic relevance; the unique spectrum of PI3K pathway aberrations in EC emphasizes the need to tailor the molecular characterization of multiple PI3K pathway members in clinical studies of EC patients.

ARID1A encodes the BAF250A subunit of the SWI/SNF chromatin remodeling complex and is believed to be a tumor suppressor gene [113]. Loss of BAF250A expression occurs in 26–40% of EECs [114–116] and correlates with *ARID1A* mutation [116]. BAF250A loss is also present in 16% of endometrial hyperplasias with atypia, and therefore appears to be an early event in EEC development [117]. The incidence of BAF250A loss is higher among MSI EECs and MMR-deficient EECs than among MSS or MMR-proficient EECs [118–120]. It has been speculated that BAF250A loss might result in *MLH1* promoter methylation [119]. *ARID1A* mutations frequently co-occur with *PIK3CA* and *PTEN* mutations in EECs [33,90]. Whether *ARID1A* alterations are clinically relevant in EC remains to be determined. However, it is noteworthy that preclinical studies in colon and breast cancer cell lines and xenografts indicate that *ARID1A* depletion results in sensitivity to PARP inhibition [121].

CTNNB1 encodes β-catenin, a key component of the canonical WNT signaling pathway. *CTNNB1* mutations are more frequent in MSS EECs compared with MSI EECs, and in low-grade versus high-grade EECs [86,101]. Unsupervised clustering of RNA expression data for EECs within the TCGA study identified four transcriptome clusters, one of which was characterized by a high frequency (87%) of *CTNNB1* mutations, low-grade and low-stage tumors, and younger, obese patients [122]. Patients within this cluster exhibited decreased overall survival compared to patients in a cluster that was predominated by low-grade and low-stage tumors but which had relatively few *CTNNB1* mutations [86]. The KRAS GTPase, a critical component of the MAP-kinase signaling pathway, has also been implicated in the regulation of the WNT pathway in certain cellular contexts [123]. Mutual exclusivity between *CTNNB1* and *KRAS* mutations is observed in EECs, and in molecular subtypes comprised predominantly of EECs, suggesting functional redundancy [86,101].

Genomic Subgroups Predominated by EECs

As noted previously, the integrated genomic analysis of EECs and SECs by TCGA characterized the overall genomic landscape of these histological subtypes and defined four molecular subgroups that have distinctive genomic and epigenomic features. EECs predominate all molecular subgroups except the serous-like group, which contains 19% of G3 EECs and 5% of non-G3 EECs (Table 11.1). The genomic features of high-grade EECs generally more closely resemble EECs than SECs; for example, compared to SECs, G3 EECs exhibit higher rates of *ARID1A*, *PTEN*, *PIK3CA*, *CTCF*, *KRAS*, and *CTNNB1* mutations as well as *POLE* exonuclease domain mutations and MSI (Tables 11.3 and 11.4) [33,86,102]. Despite the overall similarity of most G3 EECs to EECs of other grades, a subset of high-grade EECs molecularly

Table 11.3 Molecular Aberrations Found in G3 EECs

Gene	Aberration	TCGA Aberration Frequency [86]	Range of Non-TCGA Reported Aberration Frequencies	Reference(s)
Mutations				
PTEN	Mutation	67% (26/39)	19–90%	[33] [a–d]
PIK3CA	Mutation	54% (21/39)	20–57%	[33] [a,c,d]
PIK3R1	Mutation	41% (16/39)	–	–
KRAS	Mutation	33% (11/39)	7–27%	[33] [a,c,d]
TP53	Mutation	31% (12/39)	3–43%	[33,102] [a,c]
ARID1A	Mutation	31% (12/39)	60%	[33]
RPL22	Mutation	28% (11/39)	–	–
CTCF	Mutation	20% (8/39)	–	–
CTNNB1	Mutation	18% (7/39)	14–20%	[33] [a,d]
FAM135B	Mutation	15% (6/39)	–	–
SIN3A	Mutation	15% (6/39)	–	–
RBMX	Mutation	13% (5/39)	–	–
FGFR2	Mutation	10% (4/39)	14%	[a]
PNN	Mutation	3% (1/39)	–	–
MSI Status				
MSI	MSI	64% (25/39)	36%	[102]

This table is filtered to significantly mutated (FDR CT < 0.02) genes identified by TCGA.
[a]*Cheung LW, et al., High frequency of PIK3R1 and PIK3R2 mutations in endometrial cancer elucidates a novel mechanism for regulation of PTEN protein stability. Cancer Discov 2011;1(2):170–85.*
[b]*Djordjevic B, et al., Clinical assessment of PTEN loss in endometrial carcinoma: immunohistochemistry outperforms gene sequencing. Mod Pathol 2012;25(5):699–708.*
[c]*Garcia-Dios DA, et al., High-throughput interrogation of PIK3CA, PTEN, KRAS, FBXW7 and TP53 mutations in primary endometrial carcinoma. Gynecol Oncol 2013;128(2): 327–34.*
[d]*McConechy MK, et al., Ovarian and endometrial endometrioid carcinomas have distinct CTNNB1 and PTEN mutation profiles. Mod Pathol 2014;27(1):128–34.*

resemble SECs, specifically in terms of high frequencies of copy number aberrations and frequent *TP53* mutations, and thus fall within the serous-like molecular subgroup (Table 11.1) [86], as discussed later in this chapter. In the remainder of this section, we describe subgroups comprised almost exclusively of EECs: copy number low (MSS), MSI (hypermutated), and *POLE* (ultramutated) tumors.

Table 11.4 Molecular Aberrations Found in SECs

Gene	Aberration	TCGA Aberration Frequency [86]	Range of Non-TCGA Reported Aberration Frequencies	Reference(s)
Mutations				
TP53	Mutation	90.7% (39/43)	60–100%	[23,33,88,89,91,102,124]
PIK3CA	Mutation	41.9% (18/43)	15–56%	[33,66,88–91] [a–c]
FBXW7	Mutation	30.2% (13/43)	17–29%	[88,89,91]
PPP2R1A	Mutation	27.9% (12/43)	15–43%	[23,33,88,89,91] [d–f]
CHD4	Mutation	16.3% (7/43)	10–19%	[88,89,91]
CSMD3	Mutation	11.6% (5/43)	8%	[89]
SLC9A11	Mutation	4.6% (2/43)	–	–
Copy Number Aberrations				
ERBB2	Amp	26% (14/53)	17–57%	[91] [g–m]
CCNE1	Amp	26% (14/53)	26–48%	[88,91] [n]
MYC	Amp	23% (12/53)	40%	[91]
SOX17	Amp	15% (8/53)	–	–
ZNF12	Amp	15% (8/53)	–	–
MCL1	Amp	13% (7/53)	–	–
PAX8	Amp	11% (6/53)	–	–
MECOM (TERT)	Amp	9% (5/53)	–	–
ERBB3	Amp	9% (5/53)	–	–
FGFR3	Amp	9% (5/53)	–	–
NEDD9	Amp	7% (4/53)	–	–
FGFR1	Amp	7% (4/53)	–	–
NF1	Loss	7% (4/53)	–	–
PARK2	Loss	2% (1/53)	–	–
RB1	Loss	2% (1/53)	–	–
MSI Status				
MSI	MSI	10% (4/41)	2%	[89]

(Continued)

Table 11.4 Molecular Aberrations Found in SECs (Continued)

Gene	Aberration	TCGA Aberration Frequency [86]	Range of Non-TCGA Reported Aberration Frequencies	Reference(s)

This table is filtered to significantly mutated (FDR CT < 0.02) and copy number aberrant (FDR CT < 0.15) genes identified by TCGA in non-ultramutated SECs and copy number cluster 4, respectively.
[a]Cheung LW, et al., High frequency of PIK3R1 and PIK3R2 mutations in endometrial cancer elucidates a novel mechanism for regulation of PTEN protein stability. Cancer Discov 2011;1(2):170–85.
[b]Hayes MP, Douglas W, Ellenson LH, Molecular alterations of EGFR and PIK3CA in uterine serous carcinoma. Gynecol Oncol 2009;113(3):370–3.
[c]Peterson LM, et al., Molecular characterization of endometrial cancer: a correlative study assessing microsatellite instability, MLH1 hypermethylation, DNA mismatch repair protein expression, and PTEN, PIK3CA, KRAS, and BRAF mutation analysis. Int J Gynecol Pathol 2012;31(3):195–205.
[d]McConechy MK, et al., Subtype-specific mutation of PPP2R1A in endometrial and ovarian carcinomas. J Pathol 2011;223(4): 567–73.
[e]Nagendra DC, Burke J. 3rd, Maxwell GL, Risinger JI, PPP2R1A mutations are common in the serous type of endometrial cancer. Mol Carcinogen 2012;51(10):826–31.
[f]Shih IM, et al., Somatic mutations of PPP2R1A in ovarian and uterine carcinomas. Am J Pathol 2011;178(4):1442–7.
[g]Konecny GE, et al., HER2 gene amplification and EGFR expression in a large cohort of surgically staged patients with nonendometrioid (type II) endometrial cancer. Brit J Cancer 2009;100(1):89–95.
[h]Morrison C, et al., HER-2 is an independent prognostic factor in endometrial cancer: Association with outcome in a large cohort of surgically staged patients. J Clin Oncol 2006;24(15):2376–85.
[i]Odicino FE, et al., HER-2/neu overexpression and amplification in uterine serous papillary carcinoma: comparative analysis of immunohistochemistry, real-time reverse transcription-polymerase chain reaction, and fluorescence in situ hybridization. Int J Gynecol Cancer 2008;18(1):14–21.
[j]Santin AD, et al., Amplification of c-erbB2 oncogene: a major prognostic indicator in uterine serous papillary carcinoma. Cancer 2005;104(7):1391–7.
[k]Fleming GF, et al., Phase II trial of trastuzumab in women with advanced or recurrent, HER2-positive endometrial carcinoma: a Gynecologic Oncology Group study. Gynecol Oncol 2010;116(1):15–20.
[l]Grushko TA, et al., An exploratory analysis of HER-2 amplification and overexpression in advanced endometrial carcinoma: a Gynecologic Oncology Group study. Gynecol Oncol 2008;108(1):3–9.
[m]Slomovitz BM, et al., Her-2/neu overexpression and amplification in uterine papillary serous carcinoma. J Clin Oncol 2004;22(15):3126–32.
[n]Kuhn E, Bahadirli-Talbott A, Shih IM, Frequent CCNE1 amplification in endometrial intraepithelial carcinoma and uterine serous carcinoma. Mod Pathol 2014;27(7):1014–19.

The Copy Number Low MSS Molecular Subgroup

The majority (98%) of tumors in the copy number low subgroup defined by TCGA were EECs (Table 11.1). In terms of tumor grade, this subgroup was largely comprised of lower grade (G1 and G2) EECs, which accounted for 91% of cases (Table 11.1) [86]. As the name implies, copy number low (MSS) ECs exhibit few copy number alterations and an MSS phenotype (Fig. 11.2); other characteristic features of this group include an almost complete lack of *TP53* mutations, and more frequent *CTNNB1* mutations than the MSI (hypermutated) subgroup (52% vs 20%, respectively) (Fig. 11.1) [86]. Copy number low (MSS) tumors also exhibit statistically significant rates of mutation (FDR CT < 0.02) in *PTEN* (77%), *PIK3CA* (53%), *CTNNB1* (52%), *ARID1A* (42%), *PIK3R1* (33%), *CTCF* (21%), *KRAS* (16%), *FGFR2* (13%), *CHD4* (12%),

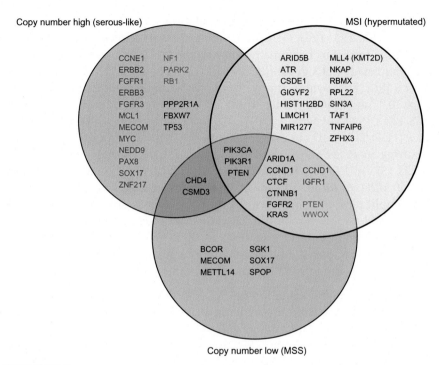

Copy number high (serous-like)

MSI (hypermutated)

CCNE1	NF1
ERBB2	PARK2
FGFR1	RB1
ERBB3	
FGFR3	PPP2R1A
MCL1	FBXW7
MECOM	TP53
MYC	
NEDD9	
PAX8	
SOX17	
ZNF217	

ARID5B	MLL4 (KMT2D)
ATR	NKAP
CSDE1	RBMX
GIGYF2	RPL22
HIST1H2BD	SIN3A
LIMCH1	TAF1
MIR1277	TNFAIP6
	ZFHX3

PIK3CA
PIK3R1
PTEN

CHD4
CSMD3

ARID1A	
CCND1	CCND1
CTCF	IGFR1
CTNNB1	
FGFR2	PTEN
KRAS	WWOX

BCOR	SGK1
MECOM	SOX17
METTL14	SPOP

Copy number low (MSS)

FIGURE 11.2

Significantly mutated (FDR CT < 0.02) (black), copy number amplified (Q < 0.15) (red), and copy number deleted (Q < 0.15) (blue) genes identified by TCGA in non-ultramutated molecular subgroups. Copy number amplified genes in the copy number high (serous-like) molecular subgroup are genes in TCGA's copy number cluster 4, and copy number amplified/deleted genes shared between the other two subgroups are genes in TCGA's copy number clusters 2 and 3 [86].

SPOP (10%), CSMD3 (10%), SOX17 (8%), and five other genes (mutated at frequencies < 8%) (Fig. 11.2) [86]. In addition to having a high frequency of PI3K pathway gene mutations, this subgroup exhibits mutual exclusivity of PIK3CA and PIK3R1 mutations but co-occurrence between mutations in these genes and PTEN mutations [86]. Finally, a strong mutual exclusivity between mutations in CTNNB1, KRAS, and SOX17 implicates KRAS and SOX17 aberrations in the deregulation of WNT signaling in copy number low (MSS) tumors [86].

The MSI (Hypermutated) Molecular Subgroup

The MSI (hypermutated) molecular subgroup was comprised exclusively of EECs, 37% of which were G3 EECs (Table 11.1). No difference in progression-free survival (PFS) was observed between patients in the MSI molecular subgroup and those in the copy number low (MSS) subgroup [86], which is in line with studies that noted a lack of prognostic significance for MSI status in EEC [105,106,125–127]. Consistent with early work showing that MSI in sporadic EC is caused by hypermethylation of the *MLH1* promoter [107,108], MSI (hypermutated) EECs exhibit heavy CpG methylation and *MLH1* hypermethylation [86]. EECs within this molecular subgroup exhibit a mutation rate of 18×10^{-6} mutations per Mb, which is 10-fold greater than that observed in MSS EECs. MSI (hypermutated) EECs have statistically significant mutation rates (FDR CT < 0.02) in 24 genes, with the highest frequencies of mutation occurring in *PTEN* (88%), *PIK3CA* (54%), *PIK3R1* (41%), *ARID1A* (37%), *RPL22* (37%), *KRAS* (35%), and *ZFHX3* (31%) [86]. The mutation frequencies of *ARID1A*, *KRAS*, and the PI3K pathway genes are consistent with the prior implication of these genes as drivers of EECs as discussed earlier in this section. MSI tumors exhibit mutual exclusivity of *PIK3CA* and *PIK3R1* mutations but co-occurrence of mutations in these genes with *PTEN* aberrations; the functional relevance of these aberrations is supported by high phospho-AKT expression and low PTEN expression in this subgroup [86].

A recurrent frameshift mutation (43delA) in the *RPL22* gene, which encodes a ribosomal protein subunit, occurs in ~52% of MSI (hypermutated) tumors but is not found in MSS tumors [86,104,128]. Similarly, MSI EECs exhibit more frequent mutations in *ZFHX3/ATBF1*, which encodes a transcription factor and putative tumor suppressor, than do MSS EECs [103]. Within one study patients (MSI and MSS combined) with *ZFHX3* mutations and/or allelic loss had significantly decreased recurrence-free survival compared to patients with wild-type *ZFHX3* [103]. The higher rates of mutation in *RPL22* and *ZFHX3/ATBF1* in MSI compared to MSS tumors implicate these genes as drivers that are targeted by MMR defects in EC. Other genes that may also be targets of MMR defects in EC include *c15orf40*, *TFAM*, and the *JAK1* kinase gene [128].

The *POLE* (Ultramutated) Molecular Subgroup

The *POLE* (ultramutated) molecular subgroup was comprised of EECs, of which 47% were G3 EECs (Table 11.1). *POLE* mutated EECs exhibit mutation rates among the highest reported (232×10^{-6} mutations per Mb) [129] and are therefore referred to as "ultramutated" [86]. Tumors in this molecular subgroup are typified by somatic mutations in the exonuclease domain of *POLE*, which encodes the catalytic subunit of polymerase epsilon, a replicative polymerase involved in nucleotide excision repair and base excision repair [130]; it is the acquisition of somatic mutations in *POLE* that are believed to underlie the ultramutated phenotype. Most EECs in this subgroup (76%; 13 out of 17) were mutated at *POLE* exonuclease domain hotspots (P286R and V411L), exhibited an increased C > A transversion frequency, and were MSS [86]. In addition to *POLE* mutations, ultramutated

EECs also exhibited statistically significant (FDR CT < 0.02) rates of mutation in 320 genes; the most frequently mutated of these genes were *PTEN* (94%), *FBXW7* (82%), *ARID1A* (76%), *PIK3CA* (71%), *PIK3R1* (65%), and *KRAS* (53%) [86].

THE GENOMICS OF SECs

Longstanding observations have shown that the mutational landscape of SECs is dominated by *TP53* mutations, which are found in 60–100% of cases (Table 11.4) and loss of p53 function is believed to play a crucial role in the development of the disease [102,124,131–134]. Stabilization of the p53 protein occurs at frequencies similar to *TP53* mutation [132,133,135] and is associated with loss of hormone receptor expression [132], a differentiating molecular feature of SECs compared to EECs [136,137]. *TP53*/p53 aberrations are early events in the development of SECs, as evidenced by the presence of *TP53* mutations and p53 expression in endometrial glandular dysplasia (EmGD) and endometrial intraepithelial carcinoma (EIC) [131–135], which are precursors to SECs.

Amplification and/or over-expression of the ERBB2/HER2 receptor tyrosine kinase occurs in 17–57% of SECs but is rare in other EC subtypes [138]. Candidate gene sequencing also implicated high rates of somatic mutations in the *PIK3CA*, *PIK3R1*, *PTEN*, and *PPP2R1A* genes in SEC (Table 11.4). Recent WES of SECs by our group and others led to the identification of other major candidate driver genes of SECs including *FBXW7*, *CHD4*, *SPOP*, and *TAF1* (Table 11.4) [86,88,89,91]. These observations have been confirmed in the comprehensive genomic analysis of SECs by TCGA and, as noted earlier, the integrated genomic analysis of SECs together with EECs showed that almost all SECs fall within a single molecular subgroup referred to as the copy number high (serous-like) subgroup (Table 11.1).

The Copy Number High (Serous-Like) Molecular Subgroup

Of the four molecular subgroups of EC described by TCGA, the copy number high (serous-like) subgroup had the worst outcome, as measured by PFS [86]. The majority (88%) of the serous-like tumors were SECs, G3 EECs, and mixed tumors (Table 11.1). Overarching genomic features that distinguish serous-like ECs from the other molecular subgroups are frequent *TP53* mutations and copy number alterations, a relatively low overall mutational load, low rates of MSI, and minimal DNA methylation changes [86]. In addition to *TP53*, seven other SMGs (FDR CT < 0.02) are found in serous-like tumors; these genes and their mutation frequencies include *PIK3CA* (47%), *FBXW7* (22%), *PPP2R1A* (22%), *PIK3R1* (13%), *CHD4* (13%), and *PTEN*

(10%) (Fig. 11.2) [86]. The PI3K pathway genes *PIK3CA, PIK3R1,* and *PTEN* are, collectively, mutated in 58% of serous-like ECs [86]. *PPP2R1A* (encoding a component of the PP2A phosphatase) and *FBXW7* (encoding the substrate recognition unit of the SCF^{FBXW7} ubiquitin ligase complex) are mutated at higher frequency in SEC and serous-like ECs than other non-ultramutated EC subtypes (Fig. 11.1). *CHD4* encodes a catalytic subunit of the NuRD chromatin remodeling complex and is mutated at comparable frequencies across non-ultramutated EC subgroups (Fig. 11.1).

Among SEC copy number alterations, amplification of the genomic region including *ERBB2 (HER2)* stands out as a frequent focal aberration (Table 11.4); statistically significant focal amplifications of *ERBB2 (HER2)* were found in 26% of serous-like ECs [86]. As discussed elsewhere [139], ERBB2 (HER2) is receiving considerable attention as a druggable target in SECs in both preclinical and clinical studies. Because activated ERBB2 (HER2) can signal through the PI3K pathway, it is noteworthy that *ERBB2 (HER2)* amplification and *PIK3CA* mutations often co-occur in serous-like tumors [112], raising the possibility that concurrent inhibition of ERBB2 (HER2) and PI3K might be clinically relevant in this genomic context.

Amplification of the genomic regions encompassing *CCNE1* and *MYC* were each reported in 23% of serous-like tumors [86]. *CCNE1* and *MYC* are oncoproteins that, respectively, regulate the cell cycle and transcription. Their turnover is mediated by FBXW7, a driver of SECs [86,88,89,91]. The mutual exclusivity of mutation/deletion of *FBXW7* with *CCNE1* amplification in SECs [88] suggests functional redundancy, although a tendency toward co-occurrence is observed in the serous-like molecular subgroup [112]. On the contrary, serous-like tumors tend toward mutual exclusivity for *MYC* amplification and *FBXW7* mutations, implying functional redundancy of these aberrations [112].

THE POTENTIAL CLINICAL UTILITY OF GENOMIC ABERRATIONS IN ECs

Importantly, the molecular classification of EECs and SECs may have prognostic significance. Among the molecular subtypes within the TCGA study, patients with *POLE* (ultramutated) EECs exhibit the longest PFS and those with serous-like tumors exhibit the shortest PFS [86]. It is unclear at this time whether the favorable prognosis of *POLE* (ultramutated) tumors is due to increased sensitivity of ultramutated tumors to chemotherapy or an alternate attribute of the tumors, such as high immune cell infiltration [140–142]. Although a favorable outcome in patients with *POLE* exonuclease domain mutated-EECs [23,97] and *POLE*-mutated SECs [143] has subsequently been confirmed by some, others have been unable to find a significant association between *POLE* mutation and progression-free and overall survival in EECs [144,145], suggesting that molecular profiling beyond *POLE* mutation status may be needed to explain the prognostic significance of the TCGA *POLE*-mutated subgroup. In this regard, a pragmatic molecular-based classification that can recapitulate the prognostic associations observed by TCGA, but without the need for WES, has recently been proposed [140]. In this approach, DNA MMR status is assessed first using immunohistochemistry (IHC), followed by an evaluation of *POLE* mutation status in MSS patients using targeted NGS, and finally an evaluation of p53 status by IHC in *POLE* wild-type patients [140]. Within this classification, patients with MSI, *POLE* mutations, or low p53 expression would represent a low-risk cohort that would receive minimal (vaginal brachytherapy or pelvic radiation) or no therapy; MSS, *POLE* wild-type, and p53 over-expressing patients would be high-risk and designated to receive chemotherapy and/or tumor-directed radiotherapy [140]. This approach reproduces the prognostic subgroups defined by TCGA and holds the potential to be used clinically for patient risk stratification [140]. This proposed model is in line with a study conducted by *Trans*PORTEC, an international consortium conducting translational research in high-risk ECs, which found that patients within an MSI subgroup or a *POLE*-mutant subgroup exhibit significantly fewer distant metastases and significantly increased five-year recurrence-free survival compared to patients within a *TP53* mutant subgroup or in a subgroup with no specific molecular profile [23].

THE GENOMICS OF UCSs/MMMTs

The genomic alterations that drive UCSs have been less well defined than those of EECs and SECs. Targeted sequencing of well-annotated cancer genes in UCSs revealed frequent mutations in *TP53* (23–88%), variable frequencies of mutation in *PIK3CA* (11–38%), *PTEN* (8–47%), *KRAS* (5–29%), *PIK3R1* (6–24%), and *PPP2R1A* (0–21%) as well as infrequent MSI [68] (Table 11.5). McConechy et al. noted that some UCSs resemble EECs, with mutations in *ARID1A*, *PTEN*, *PIK3CA*, and/or *KRAS*, whereas others resemble SECs, with *TP53* and *PPP2R1A* mutations [68]. Thus far, WES results have been reported for 17 UCSs [87], confirming frequent mutations in *TP53* (88%), *PTEN* (47%), *PIK3CA* (41%), *KRAS* (29%), and *PIK3R1* (18%) and also revealing frequent mutations in the chromatin remodeling genes *ARID1A* (23%), *ARID1B* (12%), *MLL3 (KMT2C)* (29%), *BAZ1A* (18%), and in the ubiquitin ligase complex genes *FBXW7* (23%) and *SPOP* (14%) [87]. Similar to SECs, *TP53* mutations are early events in the etiology of both the carcinoma

Table 11.5 Molecular Aberrations Found in UCSs

Gene	Aberration	Combined Aberration Frequency	Range of Reported Aberration Frequencies	Reference(s)
Mutations				
TP53	Mutation	53% (92/174)	23–88%	[33,61,64,67,68,87] [a]
MLL3 (KMT2C)	Mutation	29% (5/17)	–	[87]
PIK3CA	Mutation	27% (48/175)	11–38%	[33,66–68,87] [a,b]
PTEN	Mutation	26% (36/140)	8–47%	[33,68,87] [a,c,d]
FBXW7	Mutation	22% (10/46)	21–23%	[68,87]
CSMD3	Mutation	21% (6/29)	20.7%	[68]
ARID1A	Mutation	20% (18/88)	14–24%	[33,68,87]
ARID1B	Mutation	20% (12/59)	12–24%	[33,87]
BAZ1A	Mutation	18% (3/17)	–	[87]
MSH6	Mutation	18% (3/17)	–	[87]
PIK3R1	Mutation	17% (11/64)	6–24%	[68,87] [a]
KRAS	Mutation	15% (77/505)	5–29%	[33,64,67,68,87] [a,b]
BRCA2	Mutation	14% (3/22)	–	[87]
PPP2R1A	Mutation	11% (13/117)	0–21%	[33,68,87] [e]
ZFHX3	Mutation	10% (3/29)	–	[68]
FANCM	Mutation	9% (2/22)	–	[87]
SPOP	Mutation	8% (4/51)	3–14%	[68,87]
POLE*	Mutation	7% (2/29)	–	[68]
CHD4	Mutation	7% (2/29)	–	[68]
CTCF	Mutation	7% (2/29)	–	[68]
AKT3	Mutation	7% (2/29)	–	[68]
CTNNB1	Mutation	6% (5/87)	3–12%	[33,67,87]
MLH1	Mutation	6% (1/17)	–	[87]
ERBB3	Mutation	4% (1/22)	–	[87]
BRCA1	Mutation	4% (1/22)	–	[87]
MED12	Mutation	3% (1/29)	–	[68]
CCND1	Mutation	3% (1/29)	–	[68]
EP300	Mutation	3% (1/29)	–	[68]
PIK3R2	Mutation	3% (1/29)	–	[68]
GRLF1	Mutation	3% (1/29)	–	[68]
NRAS	Mutation	3% (1/31)	–	[67]
FGFR2	Mutation	3% (1/29)	–	[68]
BRAF	Mutation	2% (1/42)	–	[68]
Copy Number Aberrations				
URI1	Amp	40% (23/57)	–	f
ERBB2	Amp	18% (14/77)	14–20%	g,h
TP53	Loss	52% (21/40)	50–54%	[61,146]

(Continued)

Table 11.5 Molecular Aberrations Found in UCSs (Continued)

Gene	Aberration	Combined Aberration Frequency	Range of Reported Aberration Frequencies	Reference(s)
MSI Status				
MSI	MSI	17% (2/12)	–	[61]

Nonexonuclease POLE mutations.
[a]Cheung LW, et al., High frequency of PIK3R1 and PIK3R2 mutations in endometrial cancer elucidates a novel mechanism for regulation of PTEN protein stability. Cancer Discov 2011;1(2):170–85.
[b]Murray S, et al., Low frequency of somatic mutations in uterine sarcomas: a molecular analysis and review of the literature. Mutat Res 2010;686(1–2):68–73.
[c]Djordjevic B, et al., Clinical assessment of PTEN loss in endometrial carcinoma: immunohistochemistry outperforms gene sequencing. Mod Pathol 2012;25(5):699–708.
[d]Amant F, et al., PTEN mutations in uterine sarcomas. Gynecol Oncol 2002;85(1):165–9.
[e]Nagendra DC, Burke J. 3rd, Maxwell GL, Risinger JI, PPP2R1A mutations are common in the serous type of endometrial cancer. Mol Carcinogen 2012;51(10):826–31.
[f]Wang Y, Garabedian MJ, Logan SK, URI1 amplification in uterine carcinosarcoma associates with chemo-resistance and poor prognosis. Am J Cancer Res 2015;5(7):2320–9.
[g]Amant F, et al., ERBB-2 gene overexpression and amplification in uterine sarcomas. Gynecol Oncol 2004;95(3):583–7.
[h]Livasy CA, et al., EGFR expression and HER2/neu overexpression/amplification in endometrial carcinosarcoma. Gynecol Oncol 2006;100(1):101–6.

and sarcoma components of UCSs [62,146]. Currently, 57 cases of UCS are undergoing comprehensive genomic analysis by TCGA, which will provide the most complete view of their genomic landscapes and holds the potential to reveal additional aberrations that might explain their aggressive nature.

GENETIC PREDISPOSITION TO EC

Although the majority of ECs are sporadic and are driven by somatic genomic aberrations, a small fraction of cases (1.8–8.2%) are attributed to inherited genetic susceptibility [147–149]. Most hereditary ECs occur within Lynch syndrome families [150,151] and are linked to the inheritance of germline mutations in the MMR genes *MSH2* [150], *MLH1* [152,153], *MSH6* [154], and *PMS2* [150,155]. A small number of hereditary ECs are associated with Cowden syndrome [156–160] and Peutz–Jeghers syndrome [161–163], which, in most instances, are linked to germline mutations in *PTEN* [157,158] and *STK11* [161], respectively. A potential association between EC and hereditary breast-ovarian cancer [164] is attributed to tamoxifen use rather than genetic predisposition [165]. The implementation of NGS to search for genetic susceptibility to cancer has recently linked germline mutations in *POLD1* to an increased risk of colorectal and EC in families with so-called Polymerase Proofreading-Associated Polyposis [166].

CONCLUSION

The recent application of NGS to decode the mutational landscapes of EECs, SECs, and UCSs has revealed commonalities across conventional histopathological subtypes including aberrations in PI3K pathway genes (e.g., *PIK3CA*, *PIK3R1*, and *PTEN*), chromatin remodeling genes (e.g., *ARID1A* and *CHD4*), and ubiquitin ligase complex genes (e.g., *FBXW7* and *SPOP*), as well as distinguishing molecular features that can be used to categorize EECs and SECs into four discrete molecular subgroups. Collectively, the new insights into the genomics of EC that have come from systematic exome sequencing studies, as well as from TCGA's integrated genomic analysis, provide an unprecedented knowledge-base to generate hypotheses regarding the potential prognostic, predictive, and/or therapeutic significance of genomic alterations that are characteristic of ECs, particularly in regards to frequent genomic aberrations. However, it is important to note that there may also be low-frequency driver aberrations of ECs that have not yet been annotated as such, or have not yet been uncovered. This is particularly true for SECs and UCSs, for which relatively few tumors have been sequenced within any given study. Sequencing larger numbers of these rare tumors and indeed of each of the three molecular subgroups that are predominated by EECs could result in the identification of additional driver genes. In parallel with large-scale sequencing efforts, the use of so-called "mutation panels" to specifically screen for clinically actionable genomic aberrations in endometrial tumors provides a means to identify genomic targets that may be clinically meaningful for individual patients irrespective of the frequency at which they occur among ECs overall. In summary, since 2012, unprecedented strides have been made in our understanding of the genomic landscapes of primary ECs. The challenge now is to determine how this knowledge can be translated into the clinical setting, with the ultimate goal of improving outcome and quality of life for women with currently untreatable forms of EC.

GLOSSARY

amplification An increase in the number of copies of a genomic sequence
base excision repair A type of DNA repair that corrects small base (1–10 nucleotides) damage caused by oxidation, deamination, and alkylation
carcinoma A type of cancer caused by uncontrolled proliferation of epithelial cells
carcinosarcoma A biphasic malignant tumor comprised of carcinoma and sarcoma components
chemotherapy Chemicals used to kill replicating cells
chromatin remodeling Remodeling of DNA wrapped around core histones; typically used to control gene expression by either enabling or restricting polymerase interaction with DNA
clear cell endometrial cancer A rare histopathologic subtype of endometrial cancer that typically arises from atrophic (postmenopausal) endometrium with histopathology including large clear eosinophilic cells (basic cells that stain with the acidic dye eosin)

convolution test A method used to calculate summarized p-values

copy number aberration Any change in the number of segments of the genome

deletion A decrease in the number of copies of a genomic sequence

disease-free survival Length of time a patient lives, after diagnosis, without any indication or symptom of disease

driver gene A gene that is causally linked to cancer initiation and/or progression

endometrial cancer A type of cancer caused by uncontrolled proliferation of cells lining the uterus

endometrial glandular dysplasia A precursor lesion of serous endometrial cancer; characterized by dysplastic (abnormal) cells

endometrial intraepithelial carcinoma A noninvasive, cytologically malignant precursor lesion of serous endometrial cancer

endometrioid endometrial cancer The most common histopathologic subtype of endometrial cancer that typically arises from hyperplasia of the endometrium

false discovery rate The rate of predicted false positives

frameshift mutation A genomic deletion or insertion that removes or adds a number of nucleotides not divisible by three

genomics The study of the genome

hereditary endometrial cancer A disease caused by germline genomic abnormality and characterized by uncontrolled growth of uterine epithelial cells

heterologous Lacking correspondence; not homologous; not from a common origin

heterozygosity The presence of different alleles at one or more chromosomal loci

histopathology/histology A description of the microscopic structure of a tumor

homology Similarity resulting from a common origin

immunohistochemistry The use of antibodies to detect proteins in a section of tissue

methylation The addition of a methyl group (CH_3) to DNA

microsatellite instability A genetic abnormality in which the number of short, repeated sequences of DNA in a tumor tissue is altered because of a defect in DNA mismatch repair

mismatch repair A type of DNA repair that corrects base–base mismatches and insertion/deletion mispairs generated during DNA replication/recombination

mutation A nucleotide change in the genetic code

mutual exclusivity The state of not occurring simultaneously

next-generation sequencing Refers to high-throughput non-Sanger-based DNA sequencing technologies

nucleotide excision repair A type of DNA repair that corrects damage (bulky DNA adducts) caused by UV irradiation on a single strand of DNA

polymerase An enzyme that synthesizes polymers of nucleic acid

progression-free survival Length of time a patient lives after diagnosis without objective worsening of disease

radiotherapy Refers to the use of radiation as cancer treatment

recurrence/relapse Refers to the return of cancer following treatment or surgery

sarcoma A cancer arising from cells of mesenchymal origin; cancer arising in connective or nonepithelial tissue

serous endometrial cancer A rare histopathologic subtype of endometrial cancer that typically arises from atrophic (postmenopausal) endometrium, usually with well-formed papillae

somatic mutation A mutation that occurs in the nongermline (somatic) cells

The Cancer Genome Atlas (TCGA) An NIH-funded effort designed to systematically explore the entire spectrum of genomic changes in ~32 types of human cancer

tumor suppressor Refers to a gene that functions to prevent uncontrolled growth of cells

ubiquitin ligase complex Refers to a protein complex that contains an enzyme that adds ubiquitin to a substrate protein

whole exome sequencing An approach to decode the DNA sequence of all known protein-encoding genes

LIST OF ACRONYMS AND ABBREVIATIONS

ARID1A	AT rich interactive domain 1A (SWI-like)
BAZ1A	bromodomain adjacent to *zinc* finger domain, 1A
C15orf40	chromosome 15 open reading frame 40
CCEC	clear cell endometrial carcinoma
CCNE1	cyclin E1
CHD4	chromodomain helicase DNA binding protein 4
CSMD3	CUB and Sushi multiple domains 3
CT	convolution test
CTCF	CCCTC-binding factor (zinc finger protein)
CTNNB1	catenin (cadherin-associated protein), beta 1
EC	endometrial carcinoma
EEC	endometrioid endometrial carcinoma
EIC	endometrial intraepithelial carcinoma
EmGD	endometrial glandular dysplasia
ERBB2	erb-b2 receptor tyrosine kinase 2
FBXW7	F-box and WD repeat domain-containing 7, E3 ubiquitin protein ligase
FDR	false discovery rate
FGFR2	fibroblast growth factor receptor 2
G3	grade 3
IHC	immunohistochemistry
JAK1	Janus kinase 1
KRAS	Kirsten rat sarcoma viral oncogene homolog
MSH2	mutS homolog 2
MLH1	mutL homolog 1
MLL2 (KMT2B)	lysine (K)-specific methyltransferase 2B
MLL3 (KMT2C)	lysine (K)-specific methyltransferase 2C
MMMT	malignant mixed mesodermal (or Müllerian) tumor
MMR	DNA mismatch repair
MSH6	mutS homolog 6
MSI	microsatellite instable/instability
MSS	microsatellite stable/stability
MYC	v-myc avian myelocytomatosis viral oncogene homolog
NGS	next generation sequencing
PFS	progression-free survival
PIK3CA	phosphatidylinositol-4,5-bisphosphate 3-kinase, catalytic subunit alpha
PIK3R1	phosphoinositide-3-kinase, regulatory subunit 1 (alpha)
PMS2	PMS1 homolog 2, mismatch repair system component
POLD1	polymerase (DNA directed), delta 1, catalytic subunit
POLE	polymerase (DNA directed), epsilon, catalytic subunit
PPP2R1A	protein phosphatase 2, regulatory subunit A, alpha

PTEN	phosphatase and tensin homolog
RPL22	ribosomal protein L22
SEC	serous endometrial carcinoma
SMG	significantly mutated gene
SOX17	SRY (sex determining region Y)-box 17
SPOP	speckle-type POZ protein
STK11	serine/threonine kinase 11
TFAM	transcription factor A, mitochondrial
TP53	tumor protein p53
TCGA	The Cancer Genome Atlas
UCS	uterine carcinosarcoma
URI1	URI1, prefoldin-like chaperone
WES	whole exome sequencing
WNT	wingless-type MMTV integration site family
ZFHX3	zinc finger homeobox 3

References

[1] Ferlay J, et al. Cancer incidence and mortality worldwide: sources, methods and major patterns in GLOBOCAN 2012. Int J Cancer 2015;136(5):E359–86.

[2] Torre LA, et al. Global cancer statistics, 2012. CA Cancer J Clin 2015;65(2):87–108.

[3] Dedes KJ, et al. Emerging therapeutic targets in endometrial cancer. Nat Rev Clin Oncol 2011;8(5):261–71.

[4] Murali R, et al. Classification of endometrial carcinoma: more than two types. The Lancet Oncol 2014;15(7):e268–78.

[5] Gehrig PA, et al. Noninvasive papillary serous carcinoma of the endometrium. Obstet Gynecol 2001;97(1):153–7.

[6] Goff BA, et al. Uterine papillary serous carcinoma: patterns of metastatic spread. Gynecol Oncol 1994;54(3):264–8.

[7] Grice J, et al. Uterine papillary serous carcinoma: evaluation of long-term survival in surgically staged patients. Gynecol Oncol 1998;69(1):69–73.

[8] Hendrickson M, et al. Uterine papillary serous carcinoma: a highly malignant form of endometrial adenocarcinoma. Am J Surg Pathol 1982;6(2):93–108.

[9] Lee KR, et al. Recurrence in noninvasive endometrial carcinoma. Relationship to uterine papillary serous carcinoma. Am J Surg Pathol 1991;15(10):965–73.

[10] Sherman ME, et al. Uterine serous carcinoma. A morphologically diverse neoplasm with unifying clinicopathologic features. Am J Surg Pathol 1992;16(6):600–10.

[11] Williams KE, et al. Mixed serous-endometrioid carcinoma of the uterus: pathologic and cytopathologic analysis of a high-risk endometrial carcinoma. Int J Gynecol Cancer 1994;4(1):7–18.

[12] Gilks CB, et al. Poor interobserver reproducibility in the diagnosis of high-grade endometrial carcinoma. Am J Surg Pathol 2013;37(6):874–81.

[13] Han G, et al. Endometrial carcinomas with clear cells: a study of a heterogeneous group of tumors including interobserver variability, mutation analysis, and immunohistochemistry with HNF-1beta. Int J Gynecol Pathol 2015;34(4):323–33.

[14] Hoang LN, et al. Histotype-genotype correlation in 36 high-grade endometrial carcinomas. Am J Surg Pathol 2013;37(9):1421–32.

[15] Soslow RA. High-grade endometrial carcinomas - strategies for typing. Histopathology 2013;62(1):89–110.

[16] Cirisano FD, et al. Epidemiologic and surgicopathologic findings of papillary serous and clear cell endometrial cancers when compared to endometrioid carcinoma. Gynecol Oncol 1999;74(3):385–94.

[17] Chan JK, et al. Prognostic factors and risk of extrauterine metastases in 3867 women with grade 1 endometrioid corpus cancer. Am J Obstet Gynecol 2008;198(2) 216.e1–5.

[18] Farley JH, et al. Age-specific survival of women with endometrioid adenocarcinoma of the uterus. Gynecol Oncol 2000;79(1):86–9.

[19] Hirai M, et al. Prognostic factors relating to survival in uterine endometrioid carcinoma. Int J Gynaecol Obstet 1999;66(2):155–62.

[20] Rauh-Hain JA, et al. Mucinous adenocarcinoma of the endometrium compared with endometrioid endometrial cancer: a SEER analysis. Am J Clin Oncol 2014;39(1):43–8.

[21] Gottwald L, et al. Long-term survival of endometrioid endometrial cancer patients. Arch Med Sci 2010;6(6):937–44.

[22] Matthews RP, et al. Papillary serous and clear cell type lead to poor prognosis of endometrial carcinoma in black women. Gynecol Oncol 1997;65(2):206–12.

[23] Stelloo E, et al. Refining prognosis and identifying targetable pathways for high-risk endometrial cancer; a TransPORTEC initiative. Mod Pathol 2015;28(6):836–44.

[24] Matsuo K, et al. Time interval between endometrial biopsy and surgical staging for type I endometrial cancer: association between tumor characteristics and survival outcome. Obstet Gynecol 2015;125(2):424–33.

[25] Creasman WT, et al. Carcinoma of the corpus uteri. FIGO 26th Annual report on the results of treatment in gynecological cancer. Int J Gynaecol Obstet 2006;95(Suppl. 1):S105–43.

[26] Kuwabara Y, et al. Clinical characteristics of prognostic factors in poorly differentiated (G3) endometrioid adenocarcinoma in Japan. Jpn J Clin Oncol 2005;35(1):23–7.

[27] Hamilton CA, et al. Uterine papillary serous and clear cell carcinomas predict for poorer survival compared to grade 3 endometrioid corpus cancers. Br J Cancer 2006;94(5):642–6.

[28] Park JY, et al. Poor prognosis of uterine serous carcinoma compared with grade 3 endometrioid carcinoma in early stage patients. Virchows Arch 2013;462(3):289–96.

[29] Soslow RA, et al. Clinicopathologic analysis of 187 high-grade endometrial carcinomas of different histologic subtypes: similar outcomes belie distinctive biologic differences. Am J Surg Pathol 2007;31(7):979–87.

[30] Voss MA, et al. Should grade 3 endometrioid endometrial carcinoma be considered a type 2 cancer-A clinical and pathological evaluation. Gynecol Oncol 2012;124(1):15–20.

[31] Boruta II DM, et al. Uterine serous and grade 3 endometrioid carcinomas: is there a survival difference? Cancer 2004;101(10):2214–21.

[32] Alkushi A, et al. High-grade endometrial carcinoma: serous and grade 3 endometrioid carcinomas have different immunophenotypes and outcomes. Int J Gynecol Pathol 2010;29(4):343–50.

[33] McConechy MK, et al. Use of mutation profiles to refine the classification of endometrial carcinomas. J Pathol 2012;228(1):20–30.

[34] Benito V, et al. Pure papillary serous tumors of the endometrium: a clinicopathological analysis of 61 cases from a single institution. Int J Gynecol Cancer 2009;19(8):1364–9.

[35] Carcangiu ML, et al. Uterine papillary serous carcinoma: a study on 108 cases with emphasis on the prognostic significance of associated endometrioid carcinoma, absence of invasion, and concomitant ovarian carcinoma. Gynecol Oncol 1992;47(3):298–305.

[36] Idrees R, et al. Serous carcinoma arising in endometrial polyps: clinicopathologic study of 4 cases. Ann Diagn Pathol 2013;17(3):256–8.

[37] Kato DT, et al. Uterine papillary serous carcinoma (UPSC): a clinicopathologic study of 30 cases. Gynecol Oncol 1995;59(3):384–9.

[38] Lauchlan SC. Tubal (serous) carcinoma of the endometrium. Arch Pathol Lab Med 1981;105(11):615–18.

[39] McCluggage WG, et al. Uterine serous carcinoma and endometrial intraepithelial carcinoma arising in endometrial polyps: report of 5 cases, including 2 associated with tamoxifen therapy. Hum Pathol 2003;34(9):939–43.

[40] Yasuda M, et al. Endometrial intraepithelial carcinoma in association with polyp: review of eight cases. Diagn Pathol 2013;8:25.

[41] Bokhman JV. Two pathogenetic types of endometrial carcinoma. Gynecol Oncol 1983;15(1):10–17.

[42] Christopherson WM, et al. Carcinoma of the endometrium. II. Papillary adenocarcinoma: a clinical pathological study, 46 cases. Am J Clin Pathol 1982;77(5):534–40.

[43] Clement PB, et al. Non-endometrioid carcinomas of the uterine corpus: a review of their pathology with emphasis on recent advances and problematic aspects. Adv Anat Pathol 2004;11(3):117–42.

[44] Nordstrom B, et al. Endometrial carcinoma: the prognostic impact of papillary serous carcinoma (UPSC) in relation to nuclear grade, DNA ploidy and p53 expression. Anticancer Res 1996;16(2):899–904.

[45] Sutton GP, et al. Malignant papillary lesions of the endometrium. Gynecol Oncol 1987;27(3):294–304.

[46] Allen D, Bekkers R, Grant P, Hyde S. Adjuvant treatment, tumour recurrence and the survival rate of uterine serous carcinomas: a single-institution review of 62 women. S Afr J Gynaecol Oncol 2015;7(1):14–20.

[47] Scarfone G, et al. Clear cell and papillary serous endometrial carcinomas: survival in a series of 128 cases. Arch Gynecol Obstet 2013;287(2):351–6.

[48] Altman AD, et al. Canadian high risk endometrial cancer (CHREC) consortium: analyzing the clinical behaviour of high risk endometrial cancers. Gynecol Oncol 2015;139(2):269–74.

[49] Trope C, et al. Clear-cell and papillary serous cancer: treatment options. Best Pract Res Clin Obstet Gynaecol 2001;15(3):433–46.

[50] Christopherson WM, et al. Carcinoma of the endometrium: I. A clinicopathologic study of clear-cell carcinoma and secretory carcinoma. Cancer 1982;49(8):1511–23.

[51] Cirisano Jr. FD, et al. The outcome of stage I-II clinically and surgically staged papillary serous and clear cell endometrial cancers when compared with endometrioid carcinoma. Gynecol Oncol 2000;77(1):55–65.

[52] Giri PG, et al. Clear cell carcinoma of the endometrium: an uncommon entity with a favorable prognosis. Int J Radiat Oncol Biol Phys 1981;7(10):1383–7.

[53] Murphy KT, et al. Outcome and patterns of failure in pathologic stages I-IV clear-cell carcinoma of the endometrium: implications for adjuvant radiation therapy. Int J Radiat Oncol Biol Phys 2003;55(5):1272–6.

[54] Webb GA, et al. Clear cell carcinoma of the endometrium. Am J Obstet Gynecol 1987;156(6):1486–91.

[55] Abeler VM, et al. Clear cell carcinoma of the endometrium. Prognosis and metastatic pattern. Cancer 1996;78(8):1740–7.

[56] Arrastia CD, et al. Uterine carcinosarcomas: incidence and trends in management and survival. Gynecol Oncol 1997;65(1):158–63.

[57] Bartsich EG, et al. Carcinosarcoma of the uterus. A 50-year review of 32 cases (1917–1966). Obstet Gynecol 1967;30(4):518–23.

[58] Sartori E, et al. Carcinosarcoma of the uterus: a clinicopathological multicenter CTF study. Gynecol Oncol 1997;67(1):70–5.

[59] de Jong RA, et al. Molecular markers and clinical behavior of uterine carcinosarcomas: focus on the epithelial tumor component. Mod Pathol 2011;24(10):1368–79.

[60] Fujii H, et al. Frequent genetic heterogeneity in the clonal evolution of gynecological carcinosarcoma and its influence on phenotypic diversity. Cancer Res 2000;60(1):114–20.

[61] Jin Z, et al. Carcinosarcomas (malignant mullerian mixed tumors) of the uterus and ovary: a genetic study with special reference to histogenesis. Int J Gynecol Pathol 2003;22(4):368–73.

[62] Kounelis S, et al. Carcinosarcomas (malignant mixed mullerian tumors) of the female genital tract: comparative molecular analysis of epithelial and mesenchymal components. Human Pathol 1998;29(1):82–7.

[63] Schipf A, et al. Molecular genetic aberrations of ovarian and uterine carcinosarcomas – a CGH and FISH study. Virchows Archiv 2008;452(3):259–68.

[64] Wada H, et al. Molecular evidence that most but not all carcinosarcomas of the uterus are combination tumors. Cancer Res 1997;57(23):5379–85.

[65] Watanabe M, et al. Carcinosarcoma of the uterus: immunohistochemical and genetic analysis of clonality of one case. Gynecol Oncol 2001;82(3):563–7.

[66] Bashir S, et al. Molecular alterations of PIK3CA in uterine carcinosarcoma, clear cell, and serous tumors. Int J Gynecol Cancer 2014;24(7):1262–7.

[67] Growdon WB, et al. Tissue-specific signatures of activating PIK3CA and RAS mutations in carcinosarcomas of gynecologic origin. Gynecol Oncol 2011;121(1):212–17.

[68] McConechy MK, et al. In-depth molecular profiling of the biphasic components of uterine carcinosarcomas. J Pathol 2015;1(3):173–85.

[69] Silverberg SG, et al. Carcinosarcoma (malignant mixed mesodermal tumor) of the uterus. A Gynecologic Oncology Group pathologic study of 203 cases. I Int J Gynecol Pathol 1990;9(1):1–19.

[70] Macasaet MA, et al. Prognostic factors in malignant mesodermal (mullerian) mixed tumors of the uterus. Gynecol Oncol 1985;20(1):32–42.

[71] Barwick KW, et al. Malignant mixed mullerian tumors of the uterus. A clinicopathologic assessment of 34 cases. Am J Surg Pathol 1979;3(2):125–35.

[72] Ferguson SE, et al. Prognostic features of surgical stage I uterine carcinosarcoma. Am J Surg Pathol 2007;31(11):1653–61.

[73] Major FJ, et al. Prognostic factors in early-stage uterine sarcoma. A Gynecologic Oncology Group study. Cancer 1993;71(4 Suppl. 4):1702–9.

[74] Dinh TV, et al. Mixed mullerian tumors of the uterus: a clinicopathologic study. Obstet Gynecol 1989;74(3 Pt 1):388–92.

[75] Inthasorn P, et al. Analysis of clinicopathologic factors in malignant mixed Mullerian tumors of the uterine corpus. Int J Gynecol Cancer 2002;12(4):348–53.

[76] Spanos Jr. WJ, et al. Malignant mixed Mullerian tumors of the uterus. Cancer 1984;53(2):311–16.

[77] Bansal N, et al. Uterine carcinosarcomas and grade 3 endometrioid cancers: evidence for distinct tumor behavior. Obstet Gynecol 2008;112(1):64–70.

[78] Zhang C, et al. Uterine carcinosarcoma and high-risk endometrial carcinomas: a clinico-pathological comparison. Int J Gynecol Cancer 2015;25(4):629–36.

[79] Blom R, et al. Malignant mixed Mullerian tumors of the uterus: a clinicopathologic, DNA flow cytometric, p53, and mdm-2 analysis of 44 cases. Gynecol Oncol 1998;68(1):18–24.

[80] Doss LL, et al. Carcinosarcoma of the uterus: a 40-year experience from the state of Missouri. Gynecol Oncol 1984;18(1):43–53.

[81] Nakayama K, et al. Endometrial serous carcinoma: its molecular characteristics and histology-specific treatment strategies. Cancers (Basel) 2012;4(3):799–807.

[82] Callister M, et al. Malignant mixed Mullerian tumors of the uterus: analysis of patterns of failure, prognostic factors, and treatment outcome. Int J Radiat Oncol Biol Phys 2004;58(3):786–96.

[83] Spanos Jr. WJ, et al. Patterns of recurrence in malignant mixed mullerian tumor of the uterus. Cancer 1986;57(1):155–9.

[84] Yamada SD, et al. Pathologic variables and adjuvant therapy as predictors of recurrence and survival for patients with surgically evaluated carcinosarcoma of the uterus. Cancer 2000;88(12):2782–6.

[85] Koboldt DC, et al. The next-generation sequencing revolution and its impact on genomics. Cell 2013;155(1):27–38.

[86] The Cancer Genome Atlas Research Network Integrated genomic characterization of endometrial carcinoma. Nature 2013;497(7447):67–73.

[87] Jones S, et al. Genomic analyses of gynaecologic carcinosarcomas reveal frequent mutations in chromatin remodelling genes. Nat Commun 2014;5:5006.

[88] Kuhn E, et al. Identification of molecular pathway aberrations in uterine serous carcinoma by genome-wide analyses. J Natl Cancer Inst 2012;104(19):1503–13.

[89] Le Gallo M, et al. Exome sequencing of serous endometrial tumors identifies recurrent somatic mutations in chromatin-remodeling and ubiquitin ligase complex genes. Nat Genet 2012;44(12):1310–15.

[90] Liang H, et al. Whole-exome sequencing combined with functional genomics reveals novel candidate driver cancer genes in endometrial cancer. Genome Res 2012;22(11):2120–9.

[91] Zhao S, et al. Landscape of somatic single-nucleotide and copy-number mutations in uterine serous carcinoma. Proc Natl Acad Sci U S A 2013;110(8):2916–21.

[92] Bae HS, et al. Should endometrial clear cell carcinoma be classified as Type II endometrial carcinoma? Int J Gynecol Pathol 2015;34(1):74–84.

[93] DeLair D, et al. Molecular analysis of endometrial clear cell carcinoma by next generation sequencing: a study of 34 cases. Mod Pathol 2015;28:282a.

[94] Hoang LN, et al. Targeted mutation analysis of endometrial clear cell carcinoma. Histopathology 2015;66(5):664–74.

[95] Cohen Y, et al. AKT1 pleckstrin homology domain E17K activating mutation in endometrial carcinoma. Gynecol Oncol 2010;116(1):88–91.

[96] Hasegawa K, et al. Gynecologic Cancer InterGroup (GCIG) consensus review for clear cell carcinoma of the uterine corpus and cervix. Int J Gynecol Cancer 2014;24(9 Suppl. 3):S90–5.

[97] Meng B, et al. POLE exonuclease domain mutation predicts long progression-free survival in grade 3 endometrioid carcinoma of the endometrium. Gynecol Oncol 2014;134(1):15–19.

[98] Rudd ML, et al. Mutational analysis of the tyrosine kinome in serous and clear cell endometrial cancer uncovers rare somatic mutations in TNK2 and DDR1. BMC Cancer 2014;14:884.

[99] Rudd ML, et al. A unique spectrum of somatic PIK3CA (p110alpha) mutations within primary endometrial carcinomas. Clin Cancer Res 2011;17(6):1331–40.

[100] Urick ME, et al. PIK3R1 (p85alpha) is somatically mutated at high frequency in primary endometrial cancer. Cancer Res 2011;71(12):4061–7.

[101] Byron SA, et al. FGFR2 point mutations in 466 endometrioid endometrial tumors: relationship with MSI, KRAS, PIK3CA, CTNNB1 mutations and clinicopathological features. PLoS One 2012;7(2):e30801.

[102] Lax SF, et al. The frequency of p53, K-ras mutations, and microsatellite instability differs in uterine endometrioid and serous carcinoma: evidence of distinct molecular genetic pathways. Cancer 2000;88(4):814–24.

[103] Walker CJ, et al. Patterns of CTCF and ZFHX3 mutation and associated outcomes in endometrial cancer. J Natl Cancer Inst 2015;107(11):djv249.

[104] Novetsky AP, et al. Frequent mutations in the RPL22 gene and its clinical and functional implications. Gynecol Oncol 2013;128(3):470–4.

[105] Zighelboim I, et al. Microsatellite instability and epigenetic inactivation of MLH1 and outcome of patients with endometrial carcinomas of the endometrioid type. J Clin Oncol 2007;25(15):2042–8.

[106] MacDonald ND, et al. Frequency and prognostic impact of microsatellite instability in a large population-based study of endometrial carcinomas. Cancer Res 2000;60(6):1750–2.

[107] Esteller M, et al. MLH1 promoter hypermethylation is associated with the microsatellite instability phenotype in sporadic endometrial carcinomas. Oncogene 1998;17(18): 2413–17.

[108] Simpkins SB, et al. MLH1 promoter methylation and gene silencing is the primary cause of microsatellite instability in sporadic endometrial cancers. Hum Mol Genetics 1999;8(4):661–6.

[109] Esteller M, et al. hMLH1 promoter hypermethylation is an early event in human endometrial tumorigenesis. Am J Pathol 1999;155(5):1767–72.

[110] Diaz-Padilla I, et al. Mismatch repair status and clinical outcome in endometrial cancer: a systematic review and meta-analysis. Crit Rev Oncol Hematol 2013;88(1):154–67.

[111] Kinde I, et al. Evaluation of DNA from the Papanicolaou test to detect ovarian and endometrial cancers. Sci Transl Med 2013;5(167):167ra4.

[112] Cerami E, et al. The cBio Cancer Genomics Portal: an open platform for exploring multidimensional cancer genomics data. Cancer Discov 2012;2(5):401–4. http://www.cbioportal.org/index.do. [accessed September 2015].

[113] Guan B, et al. ARID1A, a factor that promotes formation of SWI/SNF-mediated chromatin remodeling, is a tumor suppressor in gynecologic cancers. Cancer Res 2011;71(21): 6718–27.

[114] Zhang ZM, et al. The clinicopathologic significance of the loss of BAF250a (ARID1A) expression in endometrial carcinoma. Int J Gynecol Cancer 2014;24(3):534–40.

[115] Wiegand KC, et al. Loss of BAF250a (ARID1A) is frequent in high-grade endometrial carcinomas. J Pathol 2011;224(3):328–33.

[116] Guan B, et al. Mutation and loss of expression of ARID1A in uterine low-grade endometrioid carcinoma. Am J Surg Pathol 2011;35(5):625–32.

[117] Werner HMJ, et al. ARID1A loss is prevalent in endometrial hyperplasia with atypia and low-grade endometrioid carcinomas. Mod Pathol 2013;26(3):428–34.

[118] Allo G, et al. ARID1A loss correlates with mismatch repair deficiency and intact p53 expression in high-grade endometrial carcinomas. Mod Pathol 2014;27(2):255–61.

[119] Bosse T, et al. Loss of ARID1A expression and its relationship with PI3K-Akt pathway alterations, TP53 and microsatellite instability in endometrial cancer. Mod Pathol 2013;26(11):1525–35.

[120] Huang HN, et al. Ovarian and endometrial endometrioid adenocarcinomas have distinct profiles of microsatellite instability, PTEN expression, and ARID1A expression. Histopathology 2015;66(4):517–28.

[121] Shen J, et al. ARID1A deficiency impairs the DNA damage checkpoint and sensitizes cells to PARP Inhibitors. Cancer Discov 2015;5(7):752–67.

[122] Liu Y, et al. Clinical significance of CTNNB1 mutation and Wnt pathway activation in endometrioid endometrial carcinoma. J Natl Cancer Inst 2014;106(9):dju245.

[123] Li JN, et al. Oncogenic K-ras stimulates wnt signaling in colon cancer through inhibition of GSK-3 beta. Gastroenterology 2005;128(7):1907–18.

[124] Tashiro H, et al. p53 gene mutations are common in uterine serous carcinoma and occur early in their pathogenesis. Am J Pathol 1997;150(1):177–85.

[125] Basil JB, et al. Clinical significance of microsatellite instability in endometrial carcinoma. Cancer 2000;89(8):1758–64.

[126] Cote ML, et al. A pilot study of microsatellite instability and endometrial cancer survival in white and African American women. Int J Gynecol Pathol 2012;31(1):66–72.

[127] Goodfellow PJ. MSI in endometrial cancer: prevalence and clinical applications. Int J Gynecol Cancer 2005;15(2):402–3.

[128] Kim TM, et al. The landscape of microsatellite instability in colorectal and endometrial cancer genomes. Cell 2013;155(4):858–68.

[129] Kandoth C, et al. Mutational landscape and significance across 12 major cancer types. Nature 2013;502(7471):333–9.

[130] Mjelle R, et al. Cell cycle regulation of human DNA repair and chromatin remodeling genes. DNA Repair (Amst) 2015;30:53–67.

[131] Jia L, et al. Endometrial glandular dysplasia with frequent p53 gene mutation: a genetic evidence supporting its precancer nature for endometrial serous carcinoma. Clin Cancer Res 2008;14(8):2263–9.

[132] Moll UM, et al. Uterine papillary serous carcinoma evolves via a p53-driven pathway. Hum Pathol 1996;27(12):1295–300.

[133] Sherman ME, et al. p53 in endometrial cancer and its putative precursors: evidence for diverse pathways of tumorigenesis. Hum Pathol 1995;26(11):1268–74.

[134] Zheng W, et al. A proposed model for endometrial serous carcinogenesis. Am J Surg Pathol 2011;35(1):e1–e14.

[135] Zheng W, et al. p53 overexpression and bcl-2 persistence in endometrial carcinoma: comparison of papillary serous and endometrioid subtypes. Gynecol Oncol 1996;61(2):167–74.

[136] Demopoulos RI, et al. Immunohistochemical comparison of uterine papillary serous and papillary endometrioid carcinoma: clues to pathogenesis. Int J Gynecol Pathol 1999;18(3):233–7.

[137] Sasano H, et al. Serous papillary adenocarcinoma of the endometrium. Analysis of proto-oncogene amplification, flow cytometry, estrogen and progesterone receptors, and immunohistochemistry. Cancer 1990;65(7):1545–51.

[138] Buza N, et al. HER2/neu in endometrial cancer: A promising therapeutic target with diagnostic challenges. Arch Pathol Lab Med 2014;138(3):343–50.

[139] El-Sahwi KS, et al. Development of targeted therapy in uterine serous carcinoma, a biologically aggressive variant of endometrial cancer. Exp Rev Anticancer Ther 2012;12(1):41–9.

[140] Talhouk A, et al. A clinically applicable molecular-based classification for endometrial cancers. Br J Cancer 2015;113(2):299–310.

[141] Hussein YR, et al. Clinicopathological analysis of endometrial carcinomas harboring somatic POLE exonuclease domain mutations. Mod Pathol 2015;28(4):505–14.

[142] van Gool IC, et al. POLE proofreading mutations elicit an antitumor immune response in endometrial cancer. Clin Cancer Res 2015;21(14):3347–55.

[143] Santin AD, et al. Improved survival of patients with hypermutation in uterine serous carcinoma. Gynecol Oncol Rep 2015;12:3–4.

[144] Billingsley CC, et al. Polymerase varepsilon (POLE) mutations in endometrial cancer: clinical outcomes and implications for Lynch syndrome testing. Cancer 2015;121(3):386–94.

[145] Church DN, et al. Prognostic significance of POLE proofreading mutations in endometrial cancer. J Natl Cancer Inst 2015;107(1):402.

[146] Taylor NP, et al. DNA mismatch repair and TP53 defects are early events in uterine carcinosarcoma tumorigenesis. Mod Pathol 2006;19(10):1333–8.

[147] Burleigh A, et al. Clinical and pathological characterization of endometrial cancer in young women: identification of a cohort without classical risk factors. Gynecol Oncol 2015;138(1):141–6.

[148] Hampel H, et al. Screening for Lynch syndrome (hereditary nonpolyposis colorectal cancer) among endometrial cancer patients. Cancer Res 2006;66(15):7810–17.

[149] Leenen CHM, et al. Prospective evaluation of molecular screening for Lynch syndrome in patients with endometrial cancer ≤ 70 years. Gynecol Oncol 2012;125(2):414–20.

[150] Dashti SG, et al. Female hormonal factors and the risk of endometrial cancer in Lynch syndrome. JAMA 2015;314(1):61–71.

[151] Wijnen J, et al. Familial endometrial cancer in female carriers of MSH6 germline mutations. Nat Genet 1999;23(2):142–4.

[152] Barrow E, et al. Cumulative lifetime incidence of extracolonic cancers in Lynch syndrome: a report of 121 families with proven mutations. Clin Genet 2009;75(2):141–9.

[153] Vasen HFA, et al. MSH2 mutation carriers are at higher risk of cancer than MLH1 mutation carriers: a study of hereditary nonpolyposis colorectal cancer families. J Clin Oncol 2001;19(20):4074–80.

[154] Hendriks YMC, et al. Cancer risk in hereditary nonpolyposis colorectal cancer due to MSH6 mutations: impact on counseling and surveillance. Gastroenterology 2004;127(1):17–25.

[155] ten Broeke SW, et al. Lynch syndrome caused by germline PMS2 mutations: delineating the cancer risk. J Clin Oncol 2015;33(4):319–25.

[156] Baker WD, et al. Endometrial cancer in a 14-year-old girl with Cowden syndrome: a case report. J Obstet Gynaecol Res 2013;39(4):876–8.

[157] Elnaggar AC, et al. Endometrial cancer in a 15-year-old girl: a complication of Cowden syndrome. Gynecol Oncol Case Rep 2012;3:18–19.

[158] Schmeler KM, et al. Endometrial cancer in an adolescent: a possible manifestation of Cowden syndrome. Obstet Gynecol 2009;114(2 Pt 2):477–9.

[159] Starink TM, et al. The Cowden syndrome – a clinical and genetic study in 21 patients. Clin Genetics 1986;29(3):222–33.

[160] Tan MH, et al. A clinical scoring system for selection of patients for PTEN mutation testing is proposed on the basis of a prospective study of 3042 probands. Am J Hum Genetics 2011;88(1):42–56.

[161] Banno K, et al. Hereditary gynecological tumors associated with Peutz-Jeghers syndrome (Review). Oncol Lett 2013;6(5):1184–8.

[162] Kondi-Pafiti A, et al. Endometrial carcinoma and ovarian sex cord tumor with annular tubules in a patient with history of Peutz-Jeghers syndrome and multiple malignancies. Eur J Gynaecol Oncol 2011;32(4):452–4.

[163] Noriega-Iriondo MF, et al. High-grade endometrial stromal sarcoma as the initial presentation of an adult patient with Peutz-Jeghers syndrome: a case report. Hered Cancer Clin Pract 2015:13.

[164] Pennington KP, et al. BRCA1, TP53, and CHEK2 germline mutations in uterine serous carcinoma. Cancer 2013;119(2):332–8.

[165] Segev Y, et al. The incidence of endometrial cancer in women with BRCA1 and BRCA2 mutations: an international prospective cohort study. Gynecol Oncol 2013;130(1):127–31.

[166] Palles C, et al. Germline mutations affecting the proofreading domains of POLE and POLD1 predispose to colorectal adenomas and carcinomas. Nat Genet 2013;45(2):136–44.

CHAPTER 12

Sentinel Lymph Node Mapping Procedures in Endometrial Cancer

D.M. Boruta II[1,2]

[1]Harvard Medical School, Boston, MA, United States [2]Massachusetts General Hospital, Boston, MA, United States

CONTENTS

INTRODUCTION

Endometrial cancer is the most common gynecologic malignancy in the United States with nearly 55,000 estimated cases in 2015 [1]. Most women (>90%) present with early-stage, uterine-confined disease and have an excellent prognosis for cure with standard surgical management including hysterectomy and bilateral salpingo-oophorectomy. When present, metastatic disease is often lymphatic and best detected by comprehensive surgical staging, including bilateral pelvic and paraaortic lymphadenectomy [2].

Despite the failure of two randomized controlled trials to demonstrate a survival benefit associated with staging lymphadenectomy, the role of comprehensive surgical staging remains controversial [3,4]. Current recommendations concerning lymphatic staging vary from consideration of systematic comprehensive dissection for any woman diagnosed to selective dissection or even sampling depending upon the presence or absence of various risk factors [5]. Although the presence of metastatic disease in lymph nodes is a highly significant prognostic factor and frequently used in tailoring optimal adjuvant therapy, dissection entails risks of injury to adjacent nerves and blood vessels, as well as for morbidity including lymphedema and lymphocyst formation [6]. Given the rate of lymphatic metastasis is relatively low but when present, critically important, the surgeon is confronted with whether to perform staging and potentially over-treat with surgery, or risk under-treatment in not identifying the presence of potentially deadly, but curable, metastatic disease.

Translational Advances in Gynecologic Cancers. DOI: http://dx.doi.org/10.1016/B978-0-12-803741-6.00012-4

SENTINEL LYMPHATIC MAPPING

Sentinel lymph node mapping and biopsy has replaced systematic, regional lymphatic dissections in the management of breast cancer, melanoma, and more recently vulvar cancer [7]. The technique relies upon the concept of sequential drainage from a tumor to lymph nodes in an orderly pattern. A marker, usually either a dye or radiolabeled colloid, most commonly technetium-99 (Tc-99), is injected in the tumor and identified in the first ("sentinel") lymph node to which it drains. In theory, this lymph node is the first at risk for harboring metastatic disease and likewise, the absence of metastatic disease in a sentinel lymph node should predict absence in the remainder of the regional lymph nodes. Limitation of resection of lymphatic material to a sentinel node or nodes, as compared to systematic clearance of a regional lymphatic bed, should limit the morbidity of the procedure. Identification of one or a small number of sentinel lymph nodes also allows for thorough inspection focused on this tissue by pathology, potentially facilitating detection of low-volume metastasis. In the setting of aberrant lymphatic pathways, sentinel lymph node mapping may identify an involved node in a region outside of the boundaries usually sampled. Thus, sentinel lymphatic mapping has been proposed as a potential "win-win" scenario as part of endometrial cancer management [8].

TECHNIQUE IN ENDOMETRIAL CANCER

Since its description by Burke in 1996, with blue dye injection at the cervix, ongoing investigation and development have led to increased use of sentinel lymph node mapping in endometrial cancer, as well as modification and study of the technique [9].

Sentinel lymph node mapping and biopsy in breast cancer or melanoma is facilitated by relative ease of access directly to the tumor for injection and lymphatic drainage that is most often unilateral. In the case of endometrial cancer, the tumor location within the uterine cavity makes access for direct injection cumbersome. A discrete tumor may not be visualized or may blanket a wide portion of the endometrium. Furthermore, lymphatic drainage of such a lesion from within the midline located uterus with lymphatic drainage pathways to both the pelvis and paraaortic regions must be anticipated to be potentially bilateral and complex. This has prompted calls for caution in the adoption of sentinel lymphatic mapping in endometrial cancer management [10].

The cervix has been the most often used site for injection of a tracer material (Fig. 12.1). Injection at the cervix has the advantage of being simple and

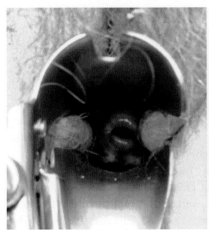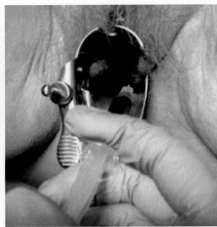

FIGURE 12.1
Injection of the uterine cervix with tracer material for sentinel lymphatic mapping.

consistent from patient to patient. Potential drainage of fundal endometrial tumors via lymphatic pathways along the gonadal vasculature, as opposed to the uterine vasculature more closely associated with the cervix and lower uterus, however, understandably prompts concern regarding use of the cervix as a proxy injection site for a tumor located within the uterine cavity. Alternative to the cervix, hysteroscopic injection of the endometrium or directly into visible tumor, and injection of the subserosa of the uterus have both been explored.

Injection of Tc-99 into the endometrium during hysteroscopy was compared to cervical injection in a study of 100 consecutive patients with endometrial cancer [11]. Sentinel lymph node detection rate was 96% by cervical injection compared to 78% by hysteroscopic injection. Cervical injection, however, did not result in detection of any sentinel paraaortic lymph nodes as compared to detection in 56% of cases with hysteroscopic injection. Paraaortic lymphadenectomy was performed in 99 of 100 women and metastases found in eight. Five of these women had metastatic disease within a detected sentinel lymph node, one had a negative pelvic sentinel lymph node, and sentinel lymph nodes were not identified in the remaining two. This finding is reflective of the relatively low rate of isolated paraaortic lymphatic spread noted in the absence of pelvic lymphatic disease as has previously been published [12,13].

The diagnostic performance of the sentinel lymph node mapping for the assessment of nodal status in patients with endometrial cancer was examined in a meta-analysis of 26 studies including 1101 mapping procedures [14]. While the overall rate of sentinel lymph node detection and pooled sensitivity of the procedure was 78% and 93%, respectively, the location of tracer injection was a significant variable. Cervical injection was correlated with an increase in detection rate ($p = 0.031$). Alternatively, hysteroscopic injection was associated with a decrease in detection rate ($p = 0.045$) and subserosal injection with a decrease in the sensitivity of sentinel lymph node biopsy ($p = 0.049$), if they were not combined with cervical injection.

In addition to studies exploring feasibility of different injection sites, alternatives to radioactive isotopes as tracer materials have been studied. Sentinel lymph node detection rate was compared in a study of 188 women with endometrial cancer where either Tc-99 or blue dye was injected into the cervix and blue dye into the fundal subserosa [15]. Although the rate of detection of at least one sentinel lymph node as well as the rate of bilateral detection did not differ between those groups (95.1% vs 87.7% and 79.5% vs 66.6%, respectively; all $p > 0.05$), the rate of detection from fundal subserosal injection of blue dye alone compared to cervical Tc-99 injection was significantly lower (74.4% vs 91.5%; $p < 0.05$). In another study including 42 women with grade 1 endometrial cancer, sentinel lymph node detection rates did not appear to improve with the addition of fundal injections of blue dye compared to cervical injection of radioisotope and blue dye [16].

More recently, indocyanine green dye (ICG) has been described as a tracer material for injection in sentinel lymph node mapping (Fig. 12.2). Although not approved by the FDA for subdermal injection, numerous studies have established its feasibility for use in mapping sentinel lymph nodes in women with endometrial cancer [15,17–24]. ICG has numerous advantages

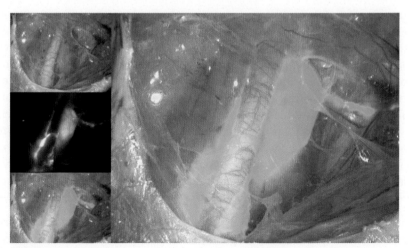

FIGURE 12.2
Indocyanine green and near-infrared imaging to visualize the sentinel lymph node.

compared to traditional tracer materials. It fluoresces when exposed to near-infrared light and is more easily visualized than blue dyes, especially when surrounding tissues are blood stained. Compared to radiolabeled colloid, ICG is simpler to inject and detect, not requiring assistance from nuclear medicine specialists or the use of radioactivity detectors. Initial reports of visualization of ICG for sentinel lymphatic mapping entailed use of a robotic-assisted laparoscopic platform, but its use with a newly available nonrobotic laparoscope has recently been reported [17,23].

A comparison of sentinel lymph mapping using Tc-99, ICG, or blue dye was prospectively completed in 100 women with endometrial cancer [18]. The tracer material was injected into the cervix both submucosally and deep. The detection rate overall for at least one side sentinel lymph node was 92% compared to bilateral detection in 76% of women. The detection rate with ICG as a tracer material was higher than with blue dye for both overall (87% vs 71%, respectively; $p = 0.005$) and bilateral (65% vs 43%, respectively; $p = 0.005$). The detection rate for ICG did not differ significantly from those with Tc-99 (87% vs 88%, respectively for overall; $p = 0.83$; and 65% vs 71%, respectively for bilateral; $p = 0.36$). No complications or allergic reactions occurred. The authors concluded that given the poor performance of blue dye, it could be omitted when ICG and Tc-99 are used in combination.

Additional studies corroborate the finding that ICG is associated with higher sentinel lymph node detection rates in comparison to use of blue dye [21,24]. An early publication describing ICG lymphatic mapping in 35 women undergoing robotic-assisted laparoscopic lymphadenectomy for treatment of endometrial cancer demonstrated bilateral detection of pelvic or paraaortic lymph nodes in 77% and 97% of women using blue dye versus ICG, respectively ($p = 0.03$) [24]. Another study described treatment of 71 women that underwent robotic-assisted laparoscopic hysterectomy and sentinel lymph node mapping for endometrial cancer ($n = 64$) or complex atypical hyperplasia ($n = 7$) [21]. Bilateral detection of sentinel lymph nodes was noted in 78.9% of women injected with ICG compared to 42.4% of those injected with blue dye ($p = 0.02$). Of additional interest in this study was the finding that on multivariate analysis body mass index was negatively correlated with success of bilateral mapping, but when stratified for tracer material, the association was only significant for blue dye.

Finally, detection of ICG for sentinel lymph node mapping in endometrial cancer has been reported using a near-infrared fluorescence imaging system allowing for a traditional (nonrobotic-assisted) laparoscopic approach for detection of ICG for sentinel lymph node mapping [17]. Overall and bilateral detection rate was 96% and 88% in this series of 50 women (42 endometrial and 8 cervical cancers).

PATHOLOGIC ULTRASTAGING

In addition to identification of the lymph node or nodes expected to be at the greatest risk of harboring metastatic disease, sentinel lymph node mapping facilitates pathologic ultrastaging of these nodes, further assisting in accurate detection of metastatic disease even when present in low volumes. Traditional pelvic and/or paraaortic lymphadenectomy specimens consist of a mixture of fatty and lymphatic tissue with a variable number of lymph nodes. These represent a challenge for pathology evaluation and likely contribute to false-negative interpretation. Providing pathology with a more discrete specimen allows for more detailed examination.

The impact of pathologic ultrastaging was described in a study of 643 women with endometrial cancer, wherein at least one sentinel lymph node was detected in 508 women after cervical injection of blue dye [25]. The sentinel nodes were examined by routine hematoxylin and eosin (H&E), and if negative, by additional sections and immunohistochemistry (IHC). Specifically, two adjacent 5 μm sections were cut from each paraffin block at each of two levels 50 μm apart. At each level, one side was stained with H&E and the other with IHC using an anticytokeratin. Findings were reported as positive for the presence of isolated tumor cells (ITCs, defined as single cells or small clusters of cells ≤0.2 mm in largest dimension), micrometastasis (defined as tumor deposits measuring larger than 0.2–2 mm or less in largest dimension), or macrometastases (tumor cell deposits larger than 2.0 mm). Cells staining positive for cytokeratin alone were considered negative for metastatic disease. Positive sentinel nodes were noted in 64 women (12.6%). Routine H&E detected 35 women (6.9%), while ultrastaging detected an additional 23 women (4.5%) with either micrometastases (4 women) or ITC (19 women). Of additional interest, 6 women (1.2%) had metastatic disease in a nonsentinel lymph node [25].

In another study including 304 women with presumed low- or intermediate-risk endometrial cancer, the impact of sentinel lymph node mapping and ultrastaging on therapeutic management was evaluated [26]. In this retrospective study, 156 women underwent sentinel lymph node assessment after cervical injection of both Tc-99 and blue dye followed by complete pelvic lymphadenectomy. Sentinel lymph nodes were detected in 136 women and these underwent pathologic ultrastaging. Another 95 women underwent pelvic lymphadenectomy without a sentinel mapping procedure, and no lymph node assessment was performed in 53 women. Lymph nodes were considered positive if either a tumor cell cluster was present as a macrometastasis (more than 2 mm) or micrometastasis (between 0.2 and 2 mm). No cases demonstrated ITCs alone. Sentinel lymph node biopsies resulted in diagnosis of metastases in 22 women (7 macrometastases (31.8%) and 15 micrometastases (68.2%)). Of the 15 sentinel lymph nodes with micrometastases, 11 (73.3%) were detected by pathologic ultrastaging. The false-negative rate of sentinel lymph node biopsy was 0%. Metastatic lymphatic disease was detected in 16.2% (22/136) of women that underwent a successful sentinel lymph node mapping procedure compared to 6.1% (7/115) of women that underwent complete pelvic lymphadenectomy either alone ($n = 95$) or following an unsuccessful sentinel lymph node mapping procedure ($n = 20$)($p = 0.03$).

Data concerning the potential clinical significance of findings from lymph node ultrastaging, including the presence of low-volume metastases (micrometastases or ITC) or cytokeratin expression, are limited. The potential significance of cytokeratin expression was assessed in a study of 304 pelvic lymph nodes from 46 women with endometrial cancer including 36 with stage I disease and 10 with stage IIIc disease [27]. Tissue sections were stained with H&E and underwent IHC using antibodies against cytokeratin, CA125, and macrophage-related antigen. Cytokeratin expression was present in all 13 lymph nodes with known metastatic disease from the 10 stage IIIc patients as well as in 20 of 66 additional nodes interpreted as negative for metastases. Similar expression was noted in 37 of 225 lymph nodes (16.4%) obtained from 14 of the 36 women with stage I disease. Recurrent pelvic disease was diagnosed in 5 of the 14 women (35.7%) with cytokeratin expression, whereas no recurrences were noted in the 22 women whose lymph nodes did not demonstrate cytokeratin staining.

The rate of micrometastases in lymph nodes of 47 women that underwent surgical staging for endometrial cancer and whose pathology were previously reported negative was examined in another retrospective study [28]. Using IHC with cytokeratin staining, 7 women (14.9%) were found to have micrometastases. With a mean follow-up time of 55.5 ± 13.3 months, recurrence occurred in two women found to have micrometastases compared to none in the group without micrometastases (recurrence-free survival of 71% vs 100%, respectively; $p = 0.0004$).

PROSPECTIVE EVALUATION

Although the feasibility and potential value of sentinel lymph node mapping procedures in endometrial cancer have been established, integration into universal management is being approached with caution. Prospective study and careful consideration of the procedure's potential limitations are necessary. Appropriately, sentinel lymph node mapping in women with endometrial cancer is currently being examined in multiple prospective trials [29].

At present, only one prospective trial (SENTI-ENDO) assessing the detection rate and diagnostic accuracy of sentinel lymph node mapping in women with endometrial cancer has been published [30]. In SENTI-ENDO, 133 women with endometrial cancer at nine centers underwent cervical dual-injection with Tc-99 and blue dye prior to sentinel lymph node biopsy. Of note, no complications related to these injections or the sentinel lymph node biopsy occurred. This was followed by complete pelvic lymphadenectomy, either by laparoscopy or laparotomy, allowing the ability of sentinel lymph nodes to predict pathologic pelvic node status to be assessed. Importantly, the negative predictive value (NPV) of pelvic sentinel nodes was determined both regarding each hemipelvis and regarding the patient as a whole (the study's primary and secondary objectives, respectively). The sentinel lymph node biopsy was considered a true positive if at least one sentinel lymph node had metastatic disease. When a nonsentinel lymph node contained metastases, but the sentinel lymph node did not, the biopsy was considered to be a false negative.

A sentinel lymph node was identified in 111 of 125 evaluable women, with a detection rate of 77% and 76% in the left and right pelvis, respectively, and a per patient rate of 89% [30]. A paraaortic sentinel lymph node was detected in 5% of women. Among the women with successful pelvic sentinel lymph node identification, detection in bilateral hemipelvises occurred in 69% and in only one hemipelvis in 31% of cases. Although NPV was 100% when considering the right and left hemipelvis as single units, three false-negative cases were encountered when considering the patient as the unit of analysis for an NPV of 97% (95% CI 91–99). These occurred in cases where a sentinel lymph node was detected in only one hemipelvis and was negative, but a nonsentinel lymph node from the contralateral pelvis ($n = 2$) or paraaortic region ($n = 1$) was involved with metastasis. Of additional interest, of 19 women (17.1%) found to have metastatic disease in a sentinel lymph node, 9 cases (47%) were identified as a result of pathologic ultrastaging with IHC and serial sectioning. As such, 10% of women with "low-risk" and 15% of women with "intermediate-risk" endometrial cancer were upstaged based upon the sentinel lymph node biopsy result.

Long-term results of the SENTI-ENDO study reported recurrence-free survival and the impact of sentinel lymph node biopsy on adjuvant therapy [31]. With a median follow-up of 50 months (range: 3–77), recurrence-free survival was 84.7% and no difference was noted between women with or without a positive sentinel lymph node ($p = 0.05$). Women with a positive sentinel lymph node were more often treated with external beam radiotherapy compared to women with a negative biopsy (78.6% vs 30.3%, respectively; $p = 0.0001$). Chemotherapy was similarly administered at a greater frequency to women with a positive sentinel lymph node biopsy (50% vs 12.5%, respectively; $p = 0.009$).

INCORPORATION INTO AN ALGORITHM

The collective results of SENTI-ENDO demonstrate both potential pitfalls of a sentinel lymph node mapping procedure as well as the powerful influence it can have on management decisions in women with endometrial cancer. Investigators from Memorial Sloan-Kettering Cancer Center (MSKCC) have argued for the importance of incorporating sentinel lymph node mapping into a broader surgical algorithm [32]. In a study of 498 women that underwent sentinel lymph node mapping with a cervical injection of blue dye, the false-negative rate of the surgical algorithm in detecting metastatic endometrial cancer was examined. The surgical algorithm included peritoneal and serosal evaluation and washings; retroperitoneal evaluation including excision of all mapped sentinel lymph nodes and suspicious nodes regardless of mapping; and if there was no mapping on a hemipelvis, performance of a side-specific pelvic, common iliac, and interiliac lymph node dissection. Paraaortic lymph node dissection was performed at the discretion of the surgeon.

At least one sentinel lymph node was successfully identified in 81% (401/498) of women [32]. The sentinel lymph node correctly diagnosed 40 of 47 women found to have nodal metastatic disease who had at least one sentinel lymph node mapped, resulting in an observed 15% false-negative rate. However, when the surgical algorithm was retrospectively applied, the observed false-negative rate for detection of metastatic nodal disease was reduced to 2%. With use of the algorithm, only one patient, with an isolated positive right paraaortic lymph node and negative ipsilateral sentinel and pelvic lymph node dissection, would have been misidentified as not having metastatic disease. The authors concluded that the surgical algorithm provided a "potential middle ground in the controversy of endometrial cancer surgical staging by providing a reasonably low false-negative rate, while sparing complete bilateral LND in the majority of cases." Further, they acknowledged "isolated paraaortic lymphatic spread remains a known but small risk in the majority of patients."

An additional study by investigators from MSKCC sought to determine whether use of the surgical algorithm incorporating sentinel lymph node mapping in treatment of women with endometrial cancer would affect the frequency of cases diagnosed with metastatic lymphatic disease [33]. The study included 507 women undergoing a minimally invasive staging procedure not requiring conversion to laparotomy from January 1, 2008, to December 31, 2010. Over this period of time, the use of the sentinel lymph node mapping surgical algorithm in endometrial cancer management increased significantly from 23% in 2008, to 52% in 2009, to 71% in 2010 ($p < 0.001$). As a result, the performance of comprehensive pelvic and paraaortic lymph node dissection significantly decreased from 65% in 2008, to 35% in 2009 and to 23% in 2010 ($p < 0.001$). All the while, the number of women diagnosed with stage IIIc disease remained similar: 7% in 2008, 7.9% in 2009, and 7.5% in 2010 ($p = 1.0$). Of further interest was that median operative times and the number of lymph nodes retrieved both decreased over the time periods (218 min and 20 in 2008, 198 min and 10 in 2009, and 176.5 min and 7 in 2010; both $p < 0.001$). The authors concluded that "incorporation of a modified staging approach utilizing the SLN mapping algorithm reduces the need for standard lymphadenectomy and does not appear to adversely affect the rate of stage IIIC detection."

CONCLUSION

Sentinel lymph node mapping appears to be an attractive compromise in the approach to detection of lymphatic metastases in women with endometrial cancer. The most recent NCCN (National Comprehensive Cancer Network) Clinical Practice Guidelines in Oncology for Uterine Neoplasms lists sentinel lymphatic mapping as a procedure that "may be considered (category 3) in selected patients" as part of surgical staging [34]. While universal performance of comprehensive lymph node dissection maximizes detection of metastases at the cost of potentially significant morbidity, abandonment of or even selective performance of lymph node dissection may result in missed opportunities for cure as effective adjuvant therapies are identified. The availability of accurate information regarding lymph node status in women with endometrial cancer is critical to planning appropriate postoperative management. Sentinel lymph node mapping appears to offer an opportunity to capture this information while minimizing surgical morbidity. Additional study concerning the potential clinical significance of low-volume metastatic disease detected on pathologic ultrastaging of sentinel lymph nodes is critically necessary in order to optimize management of women with endometrial cancer.

References

[1] Siegel RL, Miller KD, Jemal A. Cancer statistics, 2015. CA Cancer J Clin 2015;65(1):5–29.

[2] Pecorelli S. Revised FIGO staging for carcinoma of the vulva, cervix, and endometrium. Int J Gynaecol Obstet 2009;105(2):103–4.

[3] ASTEC Study Group Kitchener H, Swart AM, Qian Q, Amos C, Parmar MK. Efficacy of systematic pelvic lymphadenectomy in endometrial cancer (MRC ASTEC trial): a randomised study. Lancet 2009;373(9658):125–36.

[4] Benedetti Panici P, Basile S, Maneschi F, Alberto Lissoni A, Signorelli M, Scambia G, et al. Systematic pelvic lymphadenectomy vs. no lymphadenectomy in early-stage endometrial carcinoma: randomized clinical trial. J Natl Cancer Inst 2008;100(23):1707–16.

[5] SGOCPECW Group Burke WM, Orr J, Leitao M, Salom E, Gehrig P, et al. Endometrial cancer: a review and current management strategies: part I. Gynecol Oncol 2014;134(2):385–92.

[6] Abu-Rustum NR, Alektiar K, Iasonos A, Lev G, Sonoda Y, Aghajanian C, et al. The incidence of symptomatic lower-extremity lymphedema following treatment of uterine corpus malignancies: a 12-year experience at Memorial Sloan-Kettering Cancer Center. Gynecol Oncol 2006;103(2):714–18.

[7] Slomovitz BM, Coleman RL, Oonk MH, van der Zee A, Levenback C. Update on sentinel lymph node biopsy for early-stage vulvar cancer. Gynecol Oncol 2015;138(2):472–7.

[8] Kitchener HC. Sentinel-node biopsy in endometrial cancer: a win-win scenario? Lancet Oncol 2011;12(5):413–14.

[9] Burke TW, Levenback C, Tornos C, Morris M, Wharton JT, Gershenson DM. Intraabdominal lymphatic mapping to direct selective pelvic and paraaortic lymphadenectomy in women with high-risk endometrial cancer: results of a pilot study. Gynecol Oncol 1996;62(2):169–73.

[10] Frumovitz M, Coleman RC, Soliman PT, Ramirez PT, Levenback CF. A case for caution in the pursuit of the sentinel node in women with endometrial carcinoma. Gynecol Oncol 2014;132(2):275–9.

[11] Niikura H, Kaiho-Sakuma M, Tokunaga H, Toyoshima M, Utsunomiya H, Nagase S, et al. Tracer injection sites and combinations for sentinel lymph node detection in patients with endometrial cancer. Gynecol Oncol 2013;131(2):299–303.

[12] Chiang AJ, Yu KJ, Chao KC, Teng NN. The incidence of isolated para-aortic nodal metastasis in completely staged endometrial cancer patients. Gynecol Oncol 2011;121(1):122–5.

[13] Abu-Rustum NR, Chi DS, Leitao M, Oke EA, Hensley ML, Alektiar KM, et al. What is the incidence of isolated paraaortic nodal recurrence in grade 1 endometrial carcinoma? Gynecol Oncol 2008;111(1):46–8.

[14] Kang S, Yoo HJ, Hwang JH, Lim MC, Seo SS, Park SY. Sentinel lymph node biopsy in endometrial cancer: meta-analysis of 26 studies. Gynecol Oncol 2011;123(3):522–7.

[15] Sawicki S, Lass P, Wydra D. Sentinel lymph node biopsy in endometrial cancer—comparison of 2 detection methods. Int J Gynecol Cancer 2015;25(6):1044–50.

[16] Abu-Rustum NR, Khoury-Collado F, Pandit-Taskar N, Soslow RA, Dao F, Sonoda Y, et al. Sentinel lymph node mapping for grade 1 endometrial cancer: is it the answer to the surgical staging dilemma? Gynecol Oncol 2009;113(2):163–9.

[17] Plante M, Touhami O, Trinh XB, Renaud MC, Sebastianelli A, Grondin K, et al. Sentinel node mapping with indocyanine green and endoscopic near-infrared fluorescence imaging in endometrial cancer. A pilot study and review of the literature. Gynecol Oncol 2015;137(3):443–7.

[18] How J, Gotlieb WH, Press JZ, Abitbol J, Pelmus M, Ferenczy A, et al. Comparing indocyanine green, technetium, and blue dye for sentinel lymph node mapping in endometrial cancer. Gynecol Oncol 2015;137(3):436–42.

[19] Favero G, Pfiffer T, Ribeiro A, Carvalho JP, Baracat EC, Mechsner S, et al. Laparoscopic sentinel lymph node detection after hysteroscopic injection of technetium-99 in patients with endometrial cancer. Int J Gynecol Cancer 2015;25(3):423–30.

[20] Buda A, Bussi B, Di Martino G, Di Lorenzo P, Palazzi S, Grassi T, et al. Sentinel lymph node mapping with near-infrared fluorescent imaging using indocyanine green: a new tool for laparoscopic platform in patients with endometrial and cervical cancer. J Minim Invasive Gynecol 2015;23(2):265–9.

[21] Sinno AK, Fader AN, Roche KL, Giuntoli II RL, Tanner EJ. A comparison of colorimetric versus fluorometric sentinel lymph node mapping during robotic surgery for endometrial cancer. Gynecol Oncol 2014;134(2):281–6.

[22] Jewell EL, Huang JJ, Abu-Rustum NR, Gardner GJ, Brown CL, Sonoda Y, et al. Detection of sentinel lymph nodes in minimally invasive surgery using indocyanine green and near-infrared fluorescence imaging for uterine and cervical malignancies. Gynecol Oncol 2014;133(2):274–7.

[23] Rossi EC, Ivanova A, Boggess JF. Robotically assisted fluorescence-guided lymph node mapping with ICG for gynecologic malignancies: a feasibility study. Gynecol Oncol 2012;124(1):78–82.

[24] Holloway RW, Bravo RA, Rakowski JA, James JA, Jeppson CN, Ingersoll SB, et al. Detection of sentinel lymph nodes in patients with endometrial cancer undergoing robotic-assisted staging: a comparison of colorimetric and fluorescence imaging. Gynecol Oncol 2012;126(1):25–9.

[25] Kim CH, Soslow RA, Park KJ, Barber EL, Khoury-Collado F, Barlin JN, et al. Pathologic ultrastaging improves micrometastasis detection in sentinel lymph nodes during endometrial cancer staging. Int J Gynecol Cancer 2013;23(5):964–70.

[26] Raimond E, Ballester M, Hudry D, Bendifallah S, Darai E, Graesslin O, et al. Impact of sentinel lymph node biopsy on the therapeutic management of early-stage endometrial cancer: Results of a retrospective multicenter study. Gynecol Oncol 2014;133(3):506–11.

[27] Yabushita H, Shimazu M, Yamada H, Sawaguchi K, Noguchi M, Nakanishi M, et al. Occult lymph node metastases detected by cytokeratin immunohistochemistry predict recurrence in node-negative endometrial cancer. Gynecol Oncol 2001;80(2):139–44.

[28] Erkanli S, Bolat F, Seydaoglu G. Detection and importance of micrometastases in histologically negative lymph nodes in endometrial carcinoma. Eur J Gynaecol Oncol 2011;32(6):619–25.

[29] Darin MC, Gómez-Hidalgo NR, Westin SN, Soliman PT, Escobar PF, Frumovitz M, et al. Role of indocyanine green (ICG) in sentinel node mapping in gynecologic cancer: Is fluorescence imaging the new standard? J Minim Invasive Gynecol 2015;23(2):186–93. http://dx.doi.org/10.1016/j.jmig.2015.10.011.

[30] Ballester M, Dubernard G, Lecuru F, Heitz D, Mathevet P, Marret H, et al. Detection rate and diagnostic accuracy of sentinel-node biopsy in early stage endometrial cancer: a prospective multicentre study (SENTI-ENDO). Lancet Oncol 2011;12(5):469–76.

[31] Darai E, Dubernard G, Bats AS, Heitz D, Mathevet P, Marret H, et al. Sentinel node biopsy for the management of early stage endometrial cancer: long-term results of the SENTI-ENDO study. Gynecol Oncol 2015;136(1):54–9.

[32] Barlin JN, Khoury-Collado F, Kim CH, Leitao Jr. MM, Chi DS, Sonoda Y, et al. The importance of applying a sentinel lymph node mapping algorithm in endometrial cancer staging: beyond removal of blue nodes. Gynecol Oncol 2012;125(3):531–5.

[33] Leitao Jr. MM, Khoury-Collado F, Gardner G, Sonoda Y, Brown CL, Alektiar KM, et al. Impact of incorporating an algorithm that utilizes sentinel lymph node mapping during minimally invasive procedures on the detection of stage IIIC endometrial cancer. Gynecol Oncol 2013;129(1):38–41.

[34] National Comprehensive Cancer Network. Uterine neoplasms (Version 2.2016), <http://www.nccn.org/professionals/physician_gls/pdf/uterine.pdf> [accessed 29.11.15].

CHAPTER 13

The Role of Radiation in Uterine Cancer

A.L. Russo[1,2]

[1]Harvard Medical School, Boston, MA, United States [2]Massachusetts General Hospital, Boston, MA, United States

CONTENTS

INTRODUCTION

Radiation therapy (RT) is a necessary adjuvant, and occasionally, definitive treatment for uterine cancer. Advances in radiation technology as well as in the understanding of radiobiology have allowed us to deliver more precise, effective, and less toxic treatment to patients. RT exists in a multitude of forms including external beam radiation utilizing either photons, electrons or protons, as well as radioactive isotopes used in brachytherapy. In essence, radiation is carefully designed, highly targeted delivery of high energy X-rays intended to kill tumor cells. This chapter will describe essential radiation biology necessary to understand how radiation works and will review the specific role of radiation in both early and locally advanced staged uterine cancer. It will also discuss the specific forms of radiation in use and cover expected radiation-induced toxicities. It should serve as a basic framework from which translational researchers can understand the issues pertaining to radiation treatment of this disease in the hopes of improving patient outcomes and minimizing toxicity.

RADIOBIOLOGY AND RADIOSENSITIZERS

Radiation kills cells by inducing single- and double-stranded DNA breaks and ultimately leading to cell death through various mechanisms including mitotic catastrophe, defined as when a cell dies attempting the next mitosis, and apoptosis. The four "R's" are biologic factors that influence a tumor's response to fractionated radiation. These are described in Table 13.1. In considering drugs to combine with radiation for enhanced radiosensitivity, the

Translational Advances in Gynecologic Cancers. DOI: http://dx.doi.org/10.1016/B978-0-12-803741-6.00013-6

Table 13.1 The Four R's of Radiobiology

Four R's	Time to Occurrence	
1. Repair	Hours	Fractionating RT allows normal cells time to repair damage while tumor cells often cannot due to damaged repair pathways
2. Reassortment	Hours	After RT, living cells reassort into different phases of the cell cycle, ideally into late G2 and M (radiosensitive phases)
3. Reoxygenation	Hours–days	Hypoxic cells become reoxygenated ~24 h after RT
4. Repopulation	Weeks	After RT, surviving clonogens proliferate at an accelerated rate

four R's should also be considered. Cells are most sensitive to radiation in the mitosis (M) and G2 phases of the cell cycle and most radioresistant in late S phase [1]. The presence of oxygen acts to enhance the effects of radiation and the oxygen enhancement ratio (OER) also varies with phases of the cell cycle. The OER is the ratio of a radiation dose under hypoxic conditions to aerated conditions to produce the same biologic effect [2]. The presence of molecular oxygen fixes free radicals which ultimately enhances DNA damage. Hence, hypoxic tumors respond poorly to RT, and means to enhance oxygen delivery to tumors has been attempted clinically, for example, with hypoxic-cell radiosensitizers such as misonidazole [3,4]. Forcing cells into the radio-sensitive phase of the cell cycle or increasing the presence of oxygen prior to irradiating can enhance cell kill. Radiation induces single-strand DNA breaks and double-strand breaks (dsDNA breaks) which are repaired by nonhomologous recombination and homologous recombination. Cisplatin is commonly employed as a radiosensitizing agent as it is synergistic with radiation by producing additional intra- and interstrand DNA crosslinks leading to DNA strand breakage during replication. In uterine cancer, the combination of cisplatin and radiation is currently being studied in two randomized trials for patients with locally advanced disease (PORTEC-3 and GOG 258).

Ataxia telangiectasia mutated (ATM) and ATM- and Rad3-related (ATR) are primary signal transducers in mediating the DNA damage response that occurs after radiation. Inhibition of ATM and ATR are thought to enhance the response to radiation by preventing repair of the damage induced by radiation. Preclinical data combining ATR and ATM inhibitors with radiation in the treatment of endometrial cells demonstrate an enhanced response to radiation with either inhibitor present when assessed by clonogenic survival [5]. There is significant potential to identify further radiosensitizing agents, in particular ones that are specific for tumor cells.

RADIATION IN EARLY STAGE UTERINE CANCER

Early stage uterine cancer is treated primarily with surgery followed by adjuvant radiation to eradicate micrometastatic residual disease. There are two methods of employing radiation in this setting. For very early stage disease, where there is little concern for lymphatic spread, vaginal brachytherapy (VB) is performed, or occasionally in combination with chemotherapy for high-risk histologies (grade 3 endometrioid adenocarcinoma, papillary serous and clear cell). If there is a higher risk of lymph node involvement, external pelvic radiation is often recommended. Four large randomized trials have demonstrated the need for adjuvant radiation to lower the recurrence risk in certain high-intermediate risk patients. GOG 99, which compared pelvic radiation to observation in early stage patient undergoing total abdominal hysterectomy, bilateral salpingo-oophorectomy, and pelvic lymphadenectomy, identified advanced age, deep 1/3 myometrial invasion, and lymphovascular invasion as risk factors for recurrence [6]. The high-intermediate risk group from this study had a recurrence rate of 26% in the observation group compared to 6% in patients receiving pelvic radiation. A similar study, PORTEC-1, compared pelvic RT to observation in women who had not undergone pelvic lymph node dissection and noted similar high risk features and similar rates of recurrence in the observation arm [7]. The two other randomized trials, MRC ASTEC and the Norwegian trial, found similar results (Table 13.2) [8,9].

Once pelvic radiation was established as a standard of care for high-intermediate risk early stage uterine cancer, investigators attempted to de-escalate

Table 13.2 Randomized Controlled Trials of Adjuvant Pelvic Radiation for Early Stage Uterine Cancer

First Author/ Study Name	Risk Factors Identified	Locoregional Recurrence Rate	
		Pelvic RT (%)	Observation (%)
GOG 99	All patients	3	12
	HIR: grade 2 or 3, LVI, outer 1/3 MMI, age >70 with any RF, age 50–70 with 2 RF, age <50 with 3 RF	6	26
PORTEC-1	>50% MMI grade 1, grade 2 any MMI, grade 3 <50% MMI	4	14
Aalders/ Norwegian[a]	Grade 3, >50% MMI, LVI	2	7
	Grade 3 and >50% MMI	5	20
MRC ASTEC	IA/IB grade 3, IC all grades, papillary serous, clear cell	6.4 (Crude)	2.9 (Crude)

[a]On the observation arm of the Norwegian trial, all patients received VB.

treatment to reduce acute and late radiation-induced toxicities. Additionally, it was noted that 70% of locoregional recurrences were located in the vaginal vault or vagina [10]. This prompted the PORTEC-2 trial which compared pelvic radiation to VB alone in patients over the age of 60 with stage IC grade 1 or 2 disease or stage IB grade 3 disease or stage 2A, any age (FIGO 1988 staging) [11]. The results of this study justified the use of VB alone for these patients. The ongoing PORTEC-4 trial is intended to attempt to further de-escalate treatment and compares VB versus observation in high-intermediate risk patients.

For patients with early stage, high grade disease, the combination of VB and chemotherapy is often recommended to target both the systemic risk of disseminated disease and the high vaginal recurrence risk. Preliminary data from GOG 249 comparing pelvic radiation to VB and three cycles of carboplatin/paclitaxel is encouraging, however final results are forthcoming.

Early Stage Uterine Papillary Serous Carcinoma

Early stage uterine papillary serous carcinoma (UPSC) can also effectively be managed with a combination of systemic chemotherapy and intravaginal brachytherapy. In a series of 77 women with stage I–II UPSC treated at Memorial Sloan Kettering Cancer Center, 14% of patients relapsed with a 5-year disease-free survival of 88% and overall survival (OS) of 91% [12]. Given the aggressive nature of UPSC, a combination of pelvic radiation and chemotherapy is often employed. A prospective phase II study of combination chemoradiation followed by adjuvant chemotherapy for patients with stage I–IIIA UPSC showed this aggressive treatment strategy to be effective with 5-year local control of 87% and OS of 85% and tolerable with 20% grade 3/4 toxicities [13].

LOCALLY ADVANCED UTERINE CANCER

The role of radiation in locally advanced uterine cancer is also well established although controversies exist regarding sequencing with chemotherapy and utility for some patients. Unfortunately, GOG 122, which compared chemotherapy to whole abdominal radiation, invoked now outdated techniques, but established systemic chemotherapy as the standard of care for locally advanced disease because of the significant improvement in progression-free survival (PFS) and OS with chemotherapy [14]. However, in patients treated on the chemotherapy arm of GOG 122, 18% had pelvic recurrences as the first site of recurrence, suggesting the need to combine chemotherapy with local radiation. Furthermore, GOG 122 included patients with stage IV disease, those with gross residual disease up to 2 cm, and high risk, non-endometrioid, histologies, further making results difficult to interpret. An Italian randomized trial of patients with high risk stage I, II, and III patients with endometrioid, adenocanthoma or adenosquamous histologies compared adjuvant chemotherapy to adjuvant pelvic RT and found no difference in PFS or OS [15]. A combined analysis of two randomized trials (NSGO-9501/EORTC-55991 and MaNGO ILIADE-III trials) comparing adjuvant radiation to adjuvant sequential chemotherapy and radiation showed a significant reduction in relapse or death, including cancer-specific survival in patients receiving combination chemotherapy and radiation [16]. In addition to randomized data, there is convincing phase II data to support combination radiation and chemotherapy for locally advanced disease. RTOG 9708 was a phase II trial of concurrent chemoradiation followed by adjuvant chemotherapy for patients with high risk stage I, stage II, and pelvic-confined extrauterine disease [17]. Four-year overall and disease-free survival was 85% and 81%, respectively, and 77% and 72% for patients with stage III disease. This compares favorably to the 55% 5-year OS rate from GOG 122, while recognizing the limitations of comparison among studies.

Several retrospective studies also support combination chemotherapy and radiation for locally advanced uterine cancer and demonstrate efficacy of various sequencing approaches. A multicenter retrospective analysis of 265 patients with stage IIIC endometrial cancer compared outcomes of patients treated with chemotherapy alone, radiation alone, or combination therapy and found a significantly higher risk of recurrence and death in patients treated with chemotherapy alone compared to those treated with chemotherapy and radiation [18]. This study also gave credence to the "sandwich" approach of chemotherapy followed by radiation and then followed by further chemotherapy, however only 38% of patients in the analysis were treated with the "sandwich approach." A retrospective series of 71 patients with stage IIIC endometrial adenocarcinoma treated with radiation ± chemotherapy or chemotherapy alone showed significant improvement in pelvic relapse free survival, disease-specific survival (DSS), and OS within the patients receiving radiation [19]. In this study a mix of sequencing approaches was utilized including concurrent chemoradiation as well as sequential radiation followed by chemotherapy. A retrospective study from Memorial Sloan Kettering showed feasibility and efficacy using an approach similar to RTOG 9708 of concurrent chemoradiation followed by adjuvant chemotherapy with excellent 5-year freedom from relapse of 79% and OS of 85% in patients with stage III disease [20].

In summary, both prospective and retrospective data suggest RT plays an important role in local control and possibly survival for patients with locally advanced uterine cancer. There are two large ongoing trials further investigating the utility of radiation in addition to chemotherapy which will help to elucidate the most effective adjuvant approach. PORTEC-3 randomizes patients with high risk early stage disease and stage IIIA/C disease with no gross residual to pelvic radiation alone versus concurrent chemoradiation with two cycles of cisplatin followed by four cycles of adjuvant carboplatin and paclitaxel. GOG 258, alternatively, is comparing chemotherapy alone versus concurrent chemoradiation followed by adjuvant chemotherapy of carboplatin/paclitaxel in patients with stage III or IVA endometrial cancer or stage I/II serous or clear cell histology.

RADIATION DELIVERY METHODS

External Beam Radiation Therapy

Improved radiation delivery methods have enhanced tumor and regional targeting and have contributed to the reduction of significant radiation-induced toxicities. Most patients receiving external beam radiation therapy (EBRT) undergo either three-dimensional conformal radiation therapy (3D-CRT) or intensity modulated radiation therapy (IMRT). Radiation planning begins with a CT simulation where specially trained radiation therapists place the patient in the treatment position and a high quality CT-scan is obtained typically with 2.5 mm slice thickness and often intravenous, oral, and vaginal contrast to assist in target delineation. The radiation oncologist then determines the isocenter, or center of intended radiation, and the radiation therapists ensure a reproducible patient set up for daily treatments by marking x, y, and z coordinates on the patient. The radiation oncologist contours three important volumes on the CT scan as well as all nearby organs at risk. A gross target volume (GTV) encompasses any gross disease visible either on CT scan or by palpation; the clinical target volume (CTV) is defined as any region of microscopic tumor spread and for gynecologic cancers often includes regional lymph node basins at risk; the planning target volume (PTV) is a margin added to the CTV to account for daily intrapatient movement or set-up variation from day to day.

Pelvic external beam radiation can be delivered with 3D-CRT or IMRT. The 3D-CRT approach typically involves utilizing a 4-field box approach, meaning that radiation beams are entering anterior-to-posterior, posterior-to-anterior, right lateral, and left lateral (Fig. 13.1). Traditional borders using a 4-field box approach including treating superiorly to the level of the L4/L5 interspace, inferiorly to the bottom of the obturator foramen, and laterally to 1–2 cm outside of the pelvic brim on an anterior/posterior field. On a lateral field this includes covering the sacrum to S3 and anteriorly to the anterior aspect of the pubic symphysis. Further conformality can be achieved with the use of IMRT which allows for improved dose shaping by modulating the

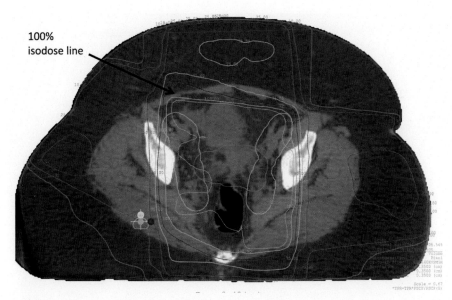

100%
isodose line

FIGURE 13.1

Axial CT scan showing a four-field pelvis 3D conformal radiation treatment plan in a patient receiving adjuvant pelvic radiation. Colored lines are isodose lines representing the dose received with the green line (black arrow) representing the region receiving full dose to 45 Gy.

radiation dose in small volumes but also increases the total low dose volume receiving radiation (Fig. 13.2).

The use of EBRT must be carefully considered for each individual patient, as there can be significant acute and late toxicity associated with EBRT. Acute risks of EBRT can include loose stools or diarrhea, skin irritation, dysuria, urinary frequency, vaginal irritation, and decreased blood counts. The acute side effects typically improve within 2 weeks of completion of EBRT and most are managed symptomatically. It is rare and not desirable to interrupt a course of radiation due to side effects, as treatment breaks correspond to decreased efficacy of treatment. Late toxicity can include significant damage to bowel or bladder requiring surgery (5%), pelvic insufficiency fractures (5–10%), lymphedema (10–20%), radiation induced malignancy (<1%), and vaginal narrowing or shortening (high risk without the use of a vaginal dilator) [21–24]. Late side effects can develop months to years after completion of therapy and can be permanent.

There are several techniques that can be utilized to decrease the risk of radiation-induced toxicity. While prospective comparative studies are awaiting final analysis, retrospective studies comparing 3D-CRT to IMRT have repeatedly demonstrated a significant reduction in late gastrointestinal and

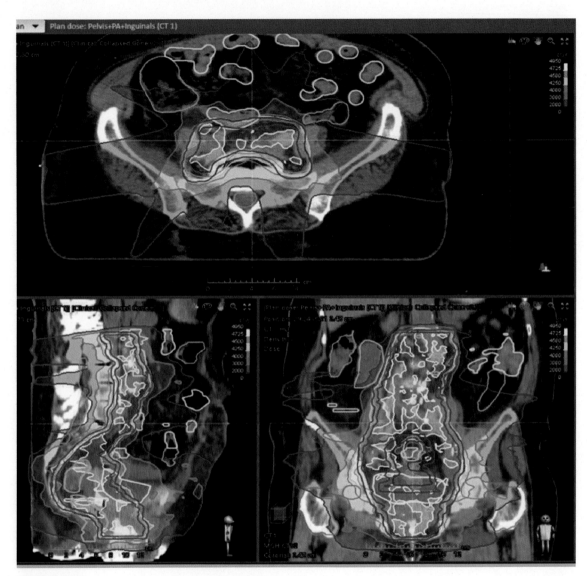

FIGURE 13.2

Dose distribution of an intensity modulated radiation (IMRT) treatment plan to the pelvis and para-aortics prescribed to 45 Gy. The pink line represents the CTV. Light blue area represents the 30 Gy region. Darker blue area represents the 20 Gy region. Note the "low dose bath" of IMRT.

genitourinary toxicity with the use of IMRT [25]. Additionally two prospective phase II trials investigating toxicity after IMRT in the adjuvant setting have shown exceedingly low rates of gastrointestinal toxicity and have helped to establish parameters for limiting bone marrow dose and toxicity [26,27]. IMRT is also particularly useful when carrying the treatment to very high

Small bowel

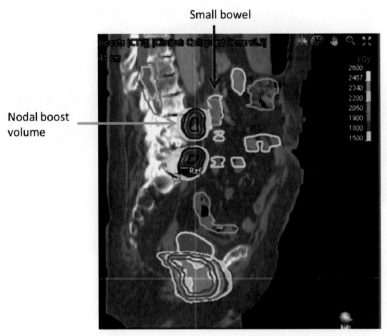

Nodal boost
volume

FIGURE 13.3
IMRT nodal and vaginal boost. Note the rapid dose fall off from the nodal boost volume (*blue arrow*) and proximity of the adjacent small bowel (*red arrow*).

doses to eradicate gross disease. This technology allows sculpting of dose and rapid dose fall off around critical organs, for example, as seen in Fig. 13.3 with dose fall off anteriorly to spare the immediately adjacent small bowel. Another technique to minimize toxicity is to treat patients in a prone, rather than supine, position. This allows the bowel of the abdominal cavity to fall away from the targeted area of interest (Fig. 13.4). Treatment in the prone position significantly lowers the volume of small bowel irradiated and may reduce gastrointestinal toxicity [28,29].

Vaginal Brachytherapy

For the majority of patients with stage I uterine cancer, VB is the only radiation treatment indicated. VB differs significantly from pelvic external beam radiation as rather than photons from an external source, a radioisotope such as iridium-192 is used which undergoes nuclear decay to emit radiation. The target volume for VB is limited to the vagina and vaginal cuff, where 70% of uterine cancers recur. The method of delivery is through a vaginal cylinder (Fig. 13.5). Cylinders come in various sizes ranging from 2 to 4 cm, with most patients having a vaginal diameter of 3 cm. Prior to the first treatment

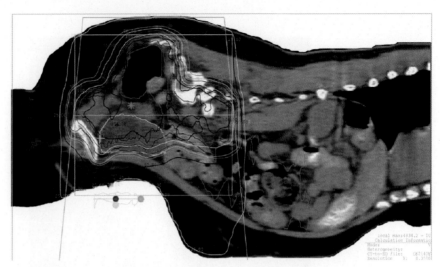

FIGURE 13.4
Prone treatment setup in a patient receiving 3D conformal pelvic radiation. Prone positioning allows bowel contents to fall away from radiation treatment field.

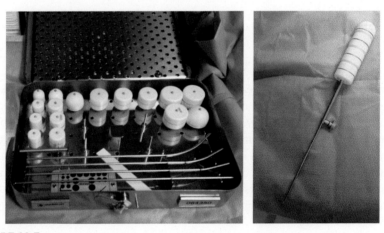

FIGURE 13.5
Left. HDR cylinder kit for VB. Cylinders come in different diameters from 2 to 3.5 cm and in adjustable lengths. Right: Vaginal cylinder. End of metal rod is hollow for radioisotope to pass.

a cylinder fitting is performed in which the radiation oncologist determines the most appropriate diameter and the length of the vagina. The largest cylinder that comfortably fits is the size chosen to minimize dose to the vaginal mucosa and maximize dose prescribed to 5 mm depth. There is a range of accepted doses and fractionation schedules for treatment with VB alone

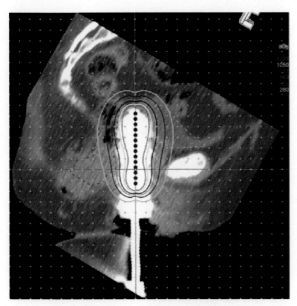

FIGURE 13.6

Isodose distribution of a patient receiving vaginal cylinder brachytherapy. Red line represents the 100% isodose line which is 700 cGy for this patient.

in the adjuvant setting; however, the most commonly used regimen is treating to 21 Gy in three fractions of 7 Gy per fraction to 5 mm depth. The upper half of the vagina is treated for most patients, however many physicians will treat the full length of the vagina in patients with high risk features including extensive lymphovascular invasion or high grade disease due to the risk of retrograde lymphatic spread and distal vaginal recurrences [30]. Fig. 13.6 shows a typical isodose distribution for a patient receiving VB. A catheter tube is attached to the hollow end of the cylindrical tube and connects to the high dose rate (HDR) remote after-loader, which eliminates radiation exposure to staff (Fig. 13.7).

VB is a well-tolerated treatment with few side effects. During the course of treatment patients may experience vaginal irritation, spotting, or burning with urination. The most important side effect is the risk of vaginal narrowing for which the use of a vaginal dilator is recommended. Patients are typically instructed to use a vaginal dilator daily for 6 weeks after completion of radiation for 10 s/day and subsequently twice weekly for life. Use of a dilator helps to maintain vaginal patency needed for follow up pelvic exams as well as for patients who wish to have vaginal intercourse. A randomized trial on VB fractionation regimens showed decreased vaginal shortening as well as decreased mucosal atrophy and bleeding in patients treated with a lower

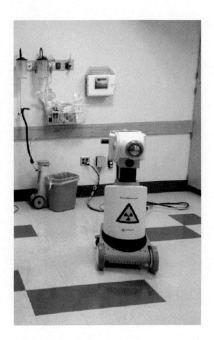

FIGURE 13.7
HDR remote after-loader. Iridium-192 is stored in this machine and travels through a catheter into the vaginal cylinder for treatment. The after-loader is computer-controlled from outside of the room and eliminates risk of exposure to staff.

dose per fraction and lower total dose [31]. A retrospective study found more vaginal stenosis when a higher total dose was used and a large proportion of vagina was treated [32]. Use of a vaginal dilator was thought to be protective. A Cochrane Database analysis showed no evidence of decreased stenosis or improved quality of life with dilator use during the course of radiation but did show lower rates of self-reported stenosis with frequent dilator use after radiation has been completed [33].

Whole Abdominal Radiation Therapy

While whole abdominal radiation therapy (WART) has largely been abandoned by many radiation oncologists, there remains utility in a select group of patients with uterine cancer. WART may be considered in patients with resected extrauterine abdomen-confined disease with no gross residual or patients with recurrent abdomen-confined disease involving the peritoneum. A recent study of patients treated with WART at Massachusetts General Hospital included patients with locally advanced endometrial cancer with one or more sites of abdomen-confined extrauterine disease who underwent primary surgery followed by chemotherapy [34]. Patients were excluded if they had residual macroscopic disease, distant metastases, or positive cytology only. In this series, 20 patients received WART from 2000 to 2011 using 3D-CRT radiation to doses of 20–30 Gy with kidney and liver shielding. Treatment was well tolerated with no disease-free patients developing toxicity greater than grade 2. The 3-year relapse-free survival and OS was 57% and 62%, respectively. The majority of recurrences were in patients with papillary serous histology and stage IVB abdomen-confined disease, suggesting the ideal cohort for treatment with WART is patients with endometrioid histology and stage III–IVA disease. The peritoneal cavity was the most common site of failure and this is thought possibly to be from the kidney and liver shielding needed to protect these organs.

IMRT for the treatment of WART may be a treatment modality that allows for improved target coverage while still respecting liver and kidney tolerances. Preliminary data on the use of IMRT for WART demonstrates feasibility and safety with initial phase I studies showing excellent planning target volume coverage and no grade 4 or 5 toxicity [35]. Further studies are warranted to investigate the improved target coverage and decreased dose to kidney/liver with the use of IMRT for WART in select patients with locally advanced uterine cancer. Fig. 13.8 shows the dose distribution using IMRT WART.

Proton Therapy

Proton therapy is a unique form of external radiation that creates rapid dose fall off after radiation hits its target and therefore allows for very precise dose delivery and sparing of adjacent critical structures. The physical characteristics of protons that allow for dose fall off are characterized by the Bragg peak; a plot of energy loss of ionizing radiation as it travels through matter. The peak of the proton occurs immediately before the particle comes to a stop and deposits its energy.

Proton therapy is widely used for tumors of the central nervous system and head and neck, as well as in pediatric patients where the target volume is in close proximity to critical structures, such as the optic nerve or brain stem. The use in gynecologic malignancies is primarily limited to patients enrolled on clinical trials. The advantage for patients with uterine cancer requiring adjuvant pelvic and para-aortic radiation is the anterior sparing of bowel which theoretically will reduce acute and late bowel toxicities. There are few proton centers available worldwide. In the United States, there are currently 14 centers; therefore access to protons is quite limited. More recently pencil beam scanning proton therapy has become more widely available which allows for more conformal treatments by using many small pencil beams to cover the target volume. In contrast to photons, protons deposit dose over a very narrow range with minimal radiation transmitted beyond

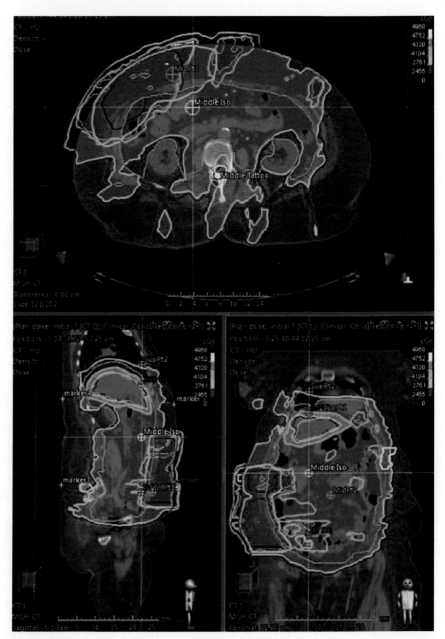

FIGURE 13.8

Isodose distribution for a patient undergoing whole abdominal radiation using IMRT. Note the large volume treated and sparing of kidneys.

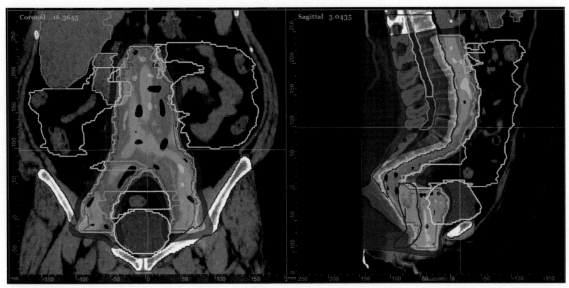

FIGURE 13.9
Coronal and sagittal images of pelvic and para-aortic radiation using pencil beam scanning proton beam. Orange is the target volume receiving 45 Gy. Blue region represents the entry dose of 25 Gy. There is minimal to no dose anterior or lateral to the target volume.

that point. Pencil beam scanning allows coverage of a desired target volume using one single posterior entry point and thereby limits dose to adjacent normal organs anterior and lateral to the volume of interest (Fig. 13.9). At Massachusetts General Hospital there is currently an open protocol utilizing pencil beam scanning proton therapy to treat patients with resected node positive uterine cancer requiring adjuvant radiation of the pelvic and para-aortic lymph nodes.

MEDICALLY INOPERABLE UTERINE CANCER

There is a subset of patients with early stage medically inoperable uterine cancer, for whom RT alone is an effective definitive treatment. Several retrospective series have shown very good local control and cancer-specific survival after treatment with radiation alone [36–39]. In a retrospective series of 74 patients, of those alive after 3 years, only 14% recurred [38]. The majority of patients were treated with 45 Gy to the pelvis followed by three brachytherapy uterine insertions to a total of 20.5 Gy. A total of 17% of patients were treated with intracavitary brachytherapy alone to a dose of 35 Gy divided in three implants. In another series of women treated with radiation alone, the DSS at 3 years was 73% [39]. In a similar fashion, the majority of patients received a combination of external beam radiation combined with brachytherapy. Investigators have

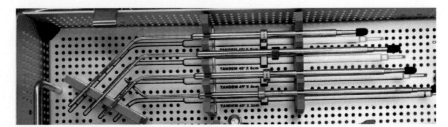

FIGURE 13.10
An intrauterine tandem, which comes in different lengths, can be placed in the uterus for patients with medically inoperable uterine cancer. The end of the tandem tube is hollow allowing passage of the radioisotope.

also demonstrated preliminary promising results using stereotactic body radiotherapy as an alternative to intracavitary brachytherapy [37].

The approach at our institution for women with uterine-confined disease has been treatment with high-dose rate iridium-192 intracavitary uterine tandem (Fig. 13.10). Tandem alone treatments are a relatively easy and effective way to treat these medically complex patients. This is an outpatient one-hour procedure not requiring anesthesia and is often repeated for six fractions. For patients with concern for extrauterine disease or high grade histology, we recommend combination treatment with external beam pelvic radiation followed by intrauterine tandem placement.

CONCLUSION

In conclusion, there are many roles that radiation plays in treating uterine cancer and a multitude of radiation techniques in use to treat all stages of disease. There are many exciting areas of collaboration between radiation oncology and basic and translational research that utilize the principles of radiobiology to improve outcomes and minimize treatment related-toxicities. There remains a great need to combine targeted therapies and radiosensitizers both in preclinical and clinical studies to enhance care for patients with uterine cancer.

LIST OF ACRONYMS AND ABBREVIATIONS

3D-CRT three-dimensional conformal radiation therapy
CTV   clinical target volume
DSS   disease-specific survival
EBRT   external beam radiation therapy
GTV   gross target volume
GOG   Gynecologic Oncology Group
IMRT   intensity modulated radiation therapy

OER	oxygen enhancement ratio
OS	overall survival
PTV	planning target volume
PFS	progression-free survival
RTOG	Radiation Therapy Oncology Group
UPSC	uterine papillary serous carcinoma
VB	vaginal brachytherapy

References

[1] Hall EJ, Brown JM, Cavanagh J. Radiosensitivity and the oxygen effect measured at different phases of the mitotic cycle using synchronously dividing cells of the root meristem of Vicia faba. Radiat Res 1968;35(3):622–34. PubMed PMID: 5675172.

[2] Palcic B, Skarsgard LD. Reduced oxygen enhancement ratio at low doses of ionizing radiation. Radiat Res 1984;100(2):328–39. PubMed PMID: 6494444.

[3] Ash DV, Peckham MJ, Steel GG. The quantitative response of human tumours to radiation and misonidazole. Br J Cancer 1979;40(6):883–9. PubMed PMID: 526430. Pubmed Central PMCID: 2010139.

[4] Hockel M, Schlenger K, Aral B, Mitze M, Schaffer U, Vaupel P. Association between tumor hypoxia and malignant progression in advanced cancer of the uterine cervix. Cancer Res 1996;56(19):4509–15. PubMed PMID: 8813149.

[5] Teng PN, Bateman NW, Darcy KM, Hamilton CA, Maxwell GL, Bakkenist CJ, et al. Pharmacologic inhibition of ATR and ATM offers clinically important distinctions to enhancing platinum or radiation response in ovarian, endometrial, and cervical cancer cells. Gynecol Oncol 2015;136(3):554–61. PubMed PMID: 25560806. Pubmed Central PMCID: 4382918.

[6] Keys HM, Roberts JA, Brunetto VL, Zaino RJ, Spirtos NM, Bloss JD, et al. A phase III trial of surgery with or without adjunctive external pelvic radiation therapy in intermediate risk endometrial adenocarcinoma: a Gynecologic Oncology Group study. Gynecol Oncol 2004;92(3):744–51. PubMed PMID: 14984936.

[7] Creutzberg CL, van Putten WL, Koper PC, Lybeert ML, Jobsen JJ, Warlam-Rodenhuis CC, et al. Surgery and postoperative radiotherapy versus surgery alone for patients with stage-1 endometrial carcinoma: multicentre randomised trial. PORTEC Study Group. Post Operative Radiation Therapy in Endometrial Carcinoma. Lancet 2000;355(9213):1404–11. PubMed PMID: 10791524.

[8] Group AES, Blake P, Swart AM, Orton J, Kitchener H, Whelan T, et al. Adjuvant external beam radiotherapy in the treatment of endometrial cancer (MRC ASTEC and NCIC CTG EN.5 randomised trials): pooled trial results, systematic review, and meta-analysis. Lancet 2009;373(9658):137–46. PubMed PMID: 19070891. Pubmed Central PMCID: 2646125.

[9] Aalders J, Abeler V, Kolstad P, Onsrud M. Postoperative external irradiation and prognostic parameters in stage I endometrial carcinoma: clinical and histopathologic study of 540 patients. Obstet Gynecol 1980;56(4):419–27. PubMed PMID: 6999399.

[10] Creutzberg CL, van Putten WL, Koper PC, Lybeert ML, Jobsen JJ, Warlam-Rodenhuis CC, et al. Survival after relapse in patients with endometrial cancer: results from a randomized trial. Gynecol. Oncol 2003;89(2):201–9. PubMed PMID: 12713981.

[11] Nout RA, Smit VT, Putter H, Jurgenliemk-Schulz IM, Jobsen JJ, Lutgens LC, et al. Vaginal brachytherapy versus pelvic external beam radiotherapy for patients with endometrial cancer of high-intermediate risk (PORTEC-2): an open-label, non-inferiority, randomised trial. Lancet 2010;375(9717):816–23. PubMed PMID: 20206777.

[12] Desai NB, Kiess AP, Kollmeier MA, Abu-Rustum NR, Makker V, Barakat RR, et al. Patterns of relapse in stage I-II uterine papillary serous carcinoma treated with adjuvant intravaginal radiation (IVRT) with or without chemotherapy. Gynecol Oncol 2013;131(3):604–8. PubMed PMID: 24055615.

[13] Jhingran A, Ramondetta LM, Bodurka DC, Slomovitz BM, Brown J, Levy LB, et al. A prospective phase II study of chemoradiation followed by adjuvant chemotherapy for FIGO stage I-IIIA (1988) uterine papillary serous carcinoma of the endometrium. Gynecol Oncol 2013;129(2):304–9. PubMed PMID: 23385150.

[14] Randall ME, Filiaci VL, Muss H, Spirtos NM, Mannel RS, Fowler J, et al. Randomized phase III trial of whole-abdominal irradiation versus doxorubicin and cisplatin chemotherapy in advanced endometrial carcinoma: a Gynecologic Oncology Group Study. J Clin Oncol 2006;24(1):36–44. PubMed PMID: 16330675.

[15] Maggi R, Lissoni A, Spina F, Melpignano M, Zola P, Favalli G, et al. Adjuvant chemotherapy vs radiotherapy in high-risk endometrial carcinoma: results of a randomised trial. Br J Cancer 2006;95(3):266–71. PubMed PMID: 16868539. Pubmed Central PMCID: 2360651.

[16] Hogberg T, Signorelli M, de Oliveira CF, Fossati R, Lissoni AA, Sorbe B, et al. Sequential adjuvant chemotherapy and radiotherapy in endometrial cancer--results from two randomised studies. Eur J Cancer 2010;46(13):2422–31. PubMed PMID: 20619634. Pubmed Central PMCID: 3552301.

[17] Greven K, Winter K, Underhill K, Fontenesci J, Cooper J, Burke T. Final analysis of RTOG 9708: adjuvant postoperative irradiation combined with cisplatin/paclitaxel chemotherapy following surgery for patients with high-risk endometrial cancer. Gynecol Oncol 2006;103(1):155–9. PubMed PMID: 16545437.

[18] Secord AA, Geller MA, Broadwater G, Holloway R, Shuler K, Dao NY, et al. A multicenter evaluation of adjuvant therapy in women with optimally resected stage IIIC endometrial cancer. Gynecol Oncol 2013;128(1):65–70. PubMed PMID: 23085460.

[19] Klopp AH, Jhingran A, Ramondetta L, Lu K, Gershenson DM, Eifel PJ. Node-positive adenocarcinoma of the endometrium: outcome and patterns of recurrence with and without external beam irradiation. Gynecol Oncol 2009;115(1):6–11. PubMed PMID: 19632709.

[20] Milgrom SA, Kollmeier MA, Abu-Rustum NR, Tew WP, Sonoda Y, Barakat RR, et al. Postoperative external beam radiation therapy and concurrent cisplatin followed by carboplatin/paclitaxel for stage III (FIGO 2009) endometrial cancer. Gynecol Oncol 2013;130(3):436–40. PubMed PMID: 23800696.

[21] Kuku S, Fragkos C, McCormack M, Forbes A. Radiation-induced bowel injury: the impact of radiotherapy on survivorship after treatment for gynaecological cancers. Br J Cancer 2013;109(6):1504–12. PubMed PMID: 24002603. Pubmed Central PMCID: 3777000.

[22] Tokumaru S, Toita T, Oguchi M, Ohno T, Kato S, Niibe Y, et al. Insufficiency fractures after pelvic radiation therapy for uterine cervical cancer: an analysis of subjects in a prospective multi-institutional trial, and cooperative study of the Japan Radiation Oncology Group (JAROG) and Japanese Radiation Oncology Study Group (JROSG). Int J Radiat Oncol, Biol, Phys 2012;84(2):e195–200. PubMed PMID: 22583605.

[23] Shih KK, Folkert MR, Kollmeier MA, Abu-Rustum NR, Sonoda Y, Leitao Jr. MM, et al. Pelvic insufficiency fractures in patients with cervical and endometrial cancer treated with postoperative pelvic radiation. Gynecol Oncol 2013;128(3):540–3. PubMed PMID: 23262211.

[24] Todo Y, Yamamoto R, Minobe S, Suzuki Y, Takeshi U, Nakatani M, et al. Risk factors for postoperative lower-extremity lymphedema in endometrial cancer survivors who had treatment including lymphadenectomy. Gynecol Oncol 2010;119(1):60–4. PubMed PMID: 20638109.

[25] Chen LA, Kim J, Boucher K, Terakedis B, Williams B, Nickman NA, et al. Toxicity and cost-effectiveness analysis of intensity modulated radiation therapy versus 3-dimensional

conformal radiation therapy for postoperative treatment of gynecologic cancers. Gynecol Oncol 2015;136(3):521–8. PubMed PMID: 25562668.

[26] Barillot I, Tavernier E, Peignaux K, Williaume D, Nickers P, Leblanc-Onfroy M, et al. Impact of post operative intensity modulated radiotherapy on acute gastro-intestinal toxicity for patients with endometrial cancer: results of the phase II RTCMIENDOMETRE French multi-centre trial. Radiother Oncol 2014;111(1):138–43. PubMed PMID: 24630537.

[27] Klopp AH, Moughan J, Portelance L, Miller BE, Salehpour MR, Hildebrandt E, et al. Hematologic toxicity in RTOG 0418: a phase 2 study of postoperative IMRT for gynecologic cancer. Int J Radiat Oncol, Biol, Phys 2013;86(1):83–90. PubMed PMID: 23582248.

[28] Adli M, Mayr NA, Kaiser HS, Skwarchuk MW, Meeks SL, Mardirossian G, et al. Does prone positioning reduce small bowel dose in pelvic radiation with intensity-modulated radio-therapy for gynecologic cancer? Int J Radiat Oncol, Biol, Phys 2003;57(1):230–8. PubMed PMID: 12909238.

[29] Ghosh K, Padilla LA, Murray KP, Downs LS, Carson LF, Dusenbery KE. Using a belly board device to reduce the small bowel volume within pelvic radiation fields in women with post-operatively treated cervical carcinoma. Gynecol Oncol 2001;83(2):271–5. PubMed PMID: 11606083.

[30] Ng TY, Perrin LC, Nicklin JL, Cheuk R, Crandon AJ. Local recurrence in high-risk node-negative stage I endometrial carcinoma treated with postoperative vaginal vault brachytherapy. Gynecol Oncol 2000;79(3):490–4. PubMed PMID: 11104626.

[31] Sorbe B, Straumits A, Karlsson L. Intravaginal high-dose-rate brachytherapy for stage I endometrial cancer: a randomized study of two dose-per-fraction levels. Int J Radiat Oncol, Biol, Phys 2005;62(5):1385–9. PubMed PMID: 16029797.

[32] Park HS, Ratner ES, Lucarelli L, Polizzi S, Higgins SA, Damast S. Predictors of vaginal stenosis after intravaginal high-dose-rate brachytherapy for endometrial carcinoma. Brachytherapy 2015;14(4):464–70. PubMed PMID: 25887343.

[33] Miles T, Johnson N. Vaginal dilator therapy for women receiving pelvic radiotherapy. Cochrane Database Syst Rev 2014;9:CD007291. PubMed PMID: 25198150.

[34] Rochet N, Kahn RS, Niemierko A, Delaney TF, Russell AH. Consolidation whole abdomen irradiation following adjuvant carboplatin-paclitaxel based chemotherapy for advanced uterine epithelial cancer: feasibility, toxicity and outcomes. Radiat Oncol 2013;8:236. PubMed PMID: 24125168. Pubmed Central PMCID: 3842773.

[35] Rochet N, Sterzing F, Jensen AD, Dinkel J, Herfarth KK, Schubert K, et al. Intensity-modulated whole abdominal radiotherapy after surgery and carboplatin/taxane chemotherapy for advanced ovarian cancer: phase I study. Int J Radiat Oncol, Biol, Phys 2010;76(5):1382–9. PubMed PMID: 19628341.

[36] Gerszten K, Faul C, Kelley J, Selvaraj R, King GC, Mogus R, et al. Twice-daily high-dose-rate brachytherapy for medically inoperable uterine cancer. Brachytherapy 2006;5(2):118–21. PubMed PMID: 16644466.

[37] Jones R, Chen Q, Best R, Libby B, Crandley EF, Showalter TN. Dosimetric feasibility of stereotactic body radiation therapy as an alternative to brachytherapy for definitive treatment of medically inoperable early stage endometrial cancer. Radiat Oncol 2014;9:164. PubMed PMID: 25059785. Pubmed Central PMCID: 4118162.

[38] Podzielinski I, Randall ME, Breheny PJ, Escobar PF, Cohn DE, Quick AM, et al. Primary radiation therapy for medically inoperable patients with clinical stage I and II endometrial carcinoma. Gynecol Oncol 2012;124(1):36–41. PubMed PMID: 22015042.

[39] Wegner RE, Beriwal S, Heron DE, Richard SD, Kelly JL, Edwards RP, et al. Definitive radiation therapy for endometrial cancer in medically inoperable elderly patients. Brachytherapy 2010;9(3):260–5. PubMed PMID: 20122872.

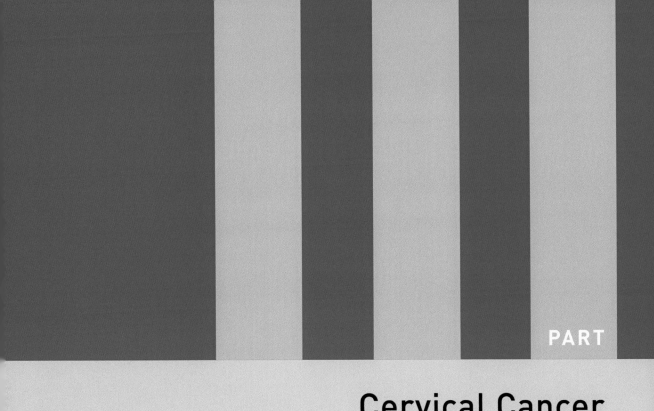

PART

Cervical Cancer

CHAPTER 14

Novel Therapies for Advanced Cervical Cancer

K.S. Tewari

University of California, Irvine, CA, United States

CONTENTS

INTRODUCTION TO CERVICAL CANCER

Cervical cancer is second only to breast and lung cancers as a cause of cancer-related mortality with approximately 529,800 new cases diagnosed annually worldwide and 275,100 deaths [1]. During 2016, the American Cancer Society estimates that there will be 12,990 new cases and 4120 deaths in the United States [2]. The median age at diagnosis is 47 years. The causative agent responsible for invasive cervical cancer and its precursor, cervical intraepithelial neoplasia (CIN), is oncogenic subtypes of the human papillomavirus (HPV), with HPV 16 and HPV 18 being most common [3,4]. HPV infection is a global epidemic and causes not only cervical cancer, but also carcinoma of the vulva, vagina, anus, penis, and oropharynx, as well as anogenital warts. In developed countries such as the United States, screening programs using cervical cytology with and without high-risk HPV DNA testing have resulted in a dramatic decrease in the incidence and mortality due to cervical cancer. Ultimately, widespread adoption of prophylactic HPV vaccination is expected to reduce incidence and mortality rates further.

Typically, microinvasive disease manifests approximately 12–15 years following initial HPV infection. During this period, the precursor, CIN1–3, is detectable and can be eradicated via the immune system among women under age 30 years, or through local therapy including ablative (e.g., CO_2 laser vaporization and cryotherapy) and excisional biopsy (e.g., large loop excision of the transformation zone and cold knife conization) [3]. Women with early invasive disease may be cured via extrafascial hysterectomy (International Federation of Gynecology and Obstetrics (FIGO) stage IA_1) or radical hysterectomy plus bilateral pelvic lymphadenectomy with or without adjuvant chemoradiation (FIGO stages IA_2–IB_1). Women of child-bearing age with early-stage, small lesions (i.e., 2 cm maximal diameter) strongly desirous

Translational Advances in Gynecologic Cancers. DOI: http://dx.doi.org/10.1016/B978-0-12-803741-6.00014-8

of future fertility may be candidates for radical trachelectomy with lymphadenectomy. Each of the surgical procedures used for early-stage cervical cancer can be performed via laparotomy, laparoscopy, or robotic-assisted laparoscopy.

The vast majority of unscreened patients present with locally advanced disease (FIGO IB$_2$–IVA). In the past, these tumors were treated with pelvic radiotherapy (50.4 Gy) plus intracavitary low-dose-rate cesium-137 brachytherapy (30–35 Gy) using an intrauterine tandem with intravaginal ovoids [3]. Following the publication in 1999–2000 of five pivotal randomized phase III trials which documented significantly improved survival outcomes with the addition of chemotherapy to the radiotherapy protocols [5–9], the National Cancer Institute (NCI) issued a rare Clinical Announcement strongly recommending the adoption of chemoradiation protocols for locally advanced cervical carcinoma [10]. Chemotherapy, typically weekly cisplatin 40 mg/m^2, acts as a radiosensitizer and also may eradicate occult metastatic tumor foci [11].

Despite the proven efficacy of chemoradiation for locally advanced disease, recurrence remains problematic, with relapse rates reported ranging from 20% to over 70%, particularly with FIGO IIIB–IVA disease. Patients manifesting with local central recurrences and a negative metastatic workup through physical examination, PET/CT imaging, and intraoperative exploration inclusive of aorto-caval lymphadenectomy submitted for intraoperative rapid mandatory frozen section analysis may be salvaged by total pelvic exenteration [3].

Unfortunately, in most cases, local failure is accompanied by distant relapse, and for years cisplatin-based palliative systemic chemotherapy served as the mainstay of treatment. For patients with recurrent disease who are not candidates for pelvic exenteration as well as for those who present with metastatic cervical cancer (i.e., FIGO stage IVB), response rates (RRs) up to 36% using the cisplatin–paclitaxel doublet were achieved; however, most were short-lived, with rapid deterioration of quality of life (QoL) and death with 7–12 months being the rule [3]. Importantly, this poor prognostic population is comprised predominantly of preirradiated patients with diminished bone marrow reserve, many of whom have impaired renal function due to pelvic sidewall extension of recurrent tumor and/or radiation fibrosis, and are suffering from malnutrition, cancer-related and/or neuropathic pain, and multiple medical comorbidities.

Furthermore, the widespread implementation of cisplatin-based chemoradiation protocols for the treatment of locally advanced disease during the first decade of the 21st century had rendered platinum-based chemotherapy doublets less effective in treating recurrence due to acquired drug resistance [3]. Therefore, a high unmet need for novel therapies exists within this population of patients (Table 14.1).

Table 14.1 Novel Therapies for Recurrent/Persistent and Metastatic Cervical Carcinoma

Anti-Angiogenesis	Immunotherapy	Signal Transduction	Other Molecules of Interest
Bevacizumab	Adoptive T-cell therapy	PI3K/AKT/mTOR	Anti-EGF
Cediranib	ADXS-HPV vaccine	PARP inhibitor	WEE1 checkpoint blockade
Pazopanib	Anti-PD-1/PD-L1	Notch signaling	Cyclooxygenase-2 inhibitors
TNP-470	Dendritic cell vaccine		Proteasome inhibitors
Angiopoietin axis blockade	Chimeric T-cell antigen		Demethylating, histone deacetylase inhibitors
Tumor-VDAs	Bispecific T-cell engager		Micro RNAs (miRNA) and RNA interfering (siRNA)
			Stem cell targeting
			Gene therapy

VDA, vascular disrupting agent; EGF, epidermal growth factor; PD-1, programmed death 1; PD-L1, programmed death ligand 1; PI3K/Akt/mTOR, phosphatidylinositol-3-kinase-(protein kinase B)-mammalian target of rapamycin.

THE ANGIOGENESIS STORY

Rationale to Target Tumor Angiogenesis in Recurrent/Persistent and Metastatic Cervical Cancer

Prior to the NCI's mandate in 2013 to merge the Gynecologic Oncology Group (GOG) with the National Surgical Adjuvant Breast and Bowel Project and Radiation Therapy Oncology Group cooperative groups into NRG Oncology, GOG protocol 240 had been designed to address the clinical problem described above for women with incurable recurrent/persistent and metastatic cervical cancer. Previously, the GOG had performed eight randomized phase III trials in this population [12–19] and although platinum-based chemotherapy doublets had emerged as the standard of care, the results in terms of duration of response, survival, and QoL (when measured) were very poor [2]. When the eighth trial (i.e., GOG protocol 204) was closed for futility, the pool from which platinum-based chemotherapy doublets had been obtained had dried up [20–23]. In addition to studying nonplatinum chemotherapy doublets (specifically, topotecan ($0.75\,\text{mg/m}^2$ days 1–3) plus paclitaxel ($175\,\text{mg/m}^2$ day 1) as a strategy to circumvent acquired drug resistance to platinum-based salvage regimens, GOG 240 also investigated the efficacy and tolerability of adding the fully humanized antiangiogenesis monoclonal antibody, bevacizumab (15 mg/kg), to both the topotecan-paclitaxel doublet and the cisplatin (50 mg/m^2) plus paclitaxel (135 or 175 mg/m^2) control doublet [24–29].

To generalize its mechanism of action, bevacizumab sequesters the vascular endothelial growth factor (VEGF) ligand and thereby prevents it from binding to and activating the transmembrane VEGF receptor (VEGFR), which in turn initiates the intracellular signal transduction cascade which leads to increased tumor angiogenesis [3] (Fig. 14.1). As such, VEGF is a highly specific mitogen for vascular endothelial cells and through alternative splicing from a single VEGF gene, five VEGF isoforms are generated, including the prototype, VEGFA (also denoted as VDGFA165), as well as VEGFB, VEGFC, VEGFD, and placenta growth factor (PlGF) [30]. Although primarily found as homodimeric polypeptides, naturally occurring heterodimers of VEGDFA and PlGF have reported. VEGF expression is potentiated in response to hypoxia, activated oncogenes (e.g., HPV E6 and E7), and several cytokines. VEGFs act through three structurally related VEGFR tyrosine kinases, denoted as VEGFR1 (Flt1, expressed in monocytes and macrophages), VEGFR2 (Flk1, vascular endothelial cells), and VEGFR3 (Flt4, lymphatic endothelial cells) [30]. VEGFR1 signaling appears elusive, and although it is widely expressed, its kinase activity is not required for endothelial cell function. Rather it appears that VEGFR1 is in negative regulation of VEGFR2 biology, and through binding of VEGF, it regulates monocyte migration during inflammation. Interestingly, VEGFR1 ligands, VEGFB, and PlGF have distinct functions, including fatty acid transport and regulation of pathologic angiogenesis. With VEGFR2 being the main VEGFR on endothelial cells, its activation through VEGFA is essential for endothelial cell development, physiology, and pathology [31]. Receptor binding to VEGFR2 induces endothelial cell proliferation, promotes cell migration, and inhibits apoptosis, with the end results being angiogenesis induction, permeabilization of blood vessels, and regulation of vasculogenesis. Finally VEGFC/VEGFR3 are critical regulators of lymphendothelial function [31]. VEGFR3 also binds VEGFD.

Importantly, deregulated VEGF expression underlies the carcinogenesis process of solid tumors by promoting tumor angiogenesis. Indeed, VEGF has emerged as an important therapeutic target in multiple solid tumors, including colorectal carcinoma, lung cancer, glioblastoma, renal cell carcinoma, breast cancer, and ovarian cancer. The hypothesis that antiangiogenesis strategies could lead to a survival advantage among women with advanced cervical cancer (defined as recurrent/persistent disease following local therapy not amenable to exenteration and metastatic disease on presentation) is based on clinical, pathologic, molecular, and therapeutic rationales.

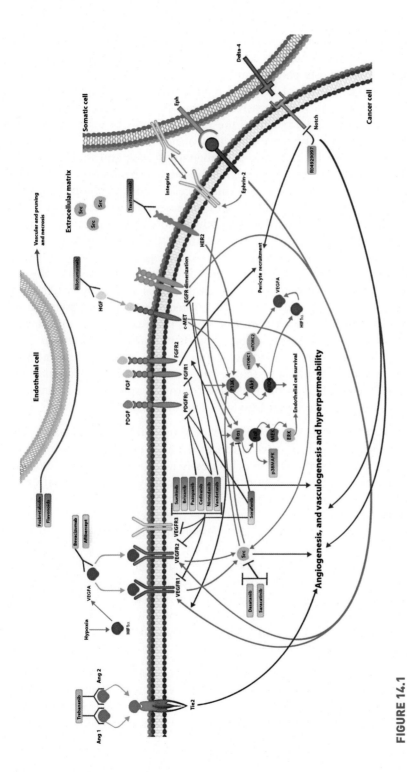

FIGURE 14.1

The angiogenesis map. *Used with permission from: Liu FW, Cripe J, Tewari KS. Oncology 2015;29:350–60. The angiogenesis map created by Liu FW from a concept by Tewari KS.*

The *clinical rationale* to target angiogenesis in cervical cancer is based upon investigations into abnormal screening results. Specifically, the vascular markings observed on colposcopy among women with abnormal cervical cytology (i.e., Papanicolaou tests) represent harbors of angiogenesis that often are indicative of microinvasive disease. Over 100 years ago, Goldman, in 1911 [32], published a paper on the vascular supply of malignant tumors and found a distinct difference from that of normal tissues. In 1927, Lewis [33] suggested that the diagnosis of malignant disease and the type of cancer could be readily made from studying the vascular pattern. When considering the cervix, in 1956, Koller [34] introduced a refined colpophotographic method to obtain a precise evaluation of the vascular patterns in normal and pathologic conditions. Using Koller's method, Kolstad [35] published a series of images with detailed descriptions of many of the vascular hallmarks (e.g., punctuation, mosaicism, and atypical vessels) recognized today as being consistent with oncogenic HPV infection and development of CIN and invasive carcinoma. In a study concerning atypical vessels and neovascularization in cervical neoplasia, Sillman et al. [36] noted that atypical vessels are not present with CIN3 and rarely with carcinoma in situ. Although they may be associated with microinvasion, atypical vessels were required in cases of frank invasion. Indeed, when a microinvasive carcinoma develops, the morphologic features of the terminal vessels become more complex and are markedly twisted with irregular branching, as newly formed capillaries run parallel to the surface and may proliferate into large mosaic fields [36].

The *pathologic rationale* to support the development of an antiangiogenesis therapeutic platform in cervical cancer is based on the expression of the endothelial cell antigen CD31 which lines newly formed blood vessels and can be detected by immunohistochemistry in frankly invasive cervical cancer (Fig. 14.2). The microvessel count within the primary tumor (i.e., microvessel density (MVD)) was first reported to be negatively associated with prognosis by Wiggins et al. in 1995 [37]. Other investigations confirmed these early observations [38–40], and in a study of intratumoral MVD of 166 cases of FIGO stage IB cervical cancer, Obermair et al. [41] reported that the 5-year survival rate was 89.7% for the 102 patients with an MVD <20/field, and 63.0% for the 64 patients with an MVD >20/field (log rank $p < 0.0001$) (Fig. 14.3). In a multivariate Cox model, MVD, lymph node metastases, tumor size, and application of radiation therapy were found to be independent prognostic factors for survival [41].

The *molecular rationale* in support of angiogenesis blockade in cervical cancer is related directly to viral properties. While HPV is unable to effect neoplastic transformation of infected cells when it remains in its double-stranded circular native episomal form, viral integration (necessary for the

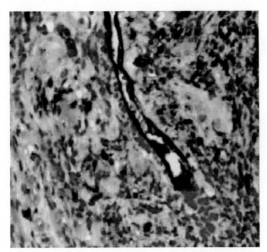

FIGURE 14.2
Immunohistochemistry for the endothelial cell antigen, CD31, in a woman with early-stage invasive squamous cell carcinoma.

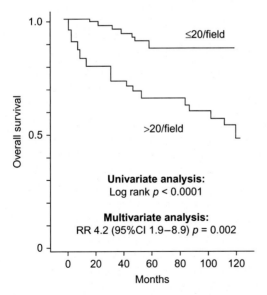

FIGURE 14.3
Intratumoral MVD in 166 patients with stage IB cervical cancer demonstrating significantly worse prognosis among those with >20/field. *From Obermair A, Wanner C, Bigi S, et al. Tumor angiogenesis in stage IB cervical cancer: correlation of microvessel density with survival. Am J Obstet Gynecol 1998;178:314–19.*

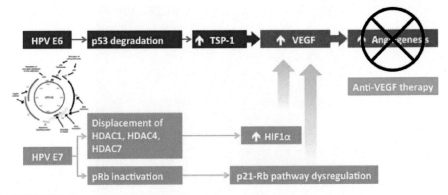

FIGURE 14.4

A molecular rationale to target angiogenesis in cervical cancer. Following oncogenic HPV integration into host DNA from native episomal double-strand DNA form, repression of transcription of viral oncogenes, E6 and E7, is relieved due to the integration event occurring in the reading frame of negative regulator, E2. The resulting oncoproteins, E6 and E7, proceed to degrade and/or inactive cellular tumor gene products, p53 and pRb, respectively. Ultimately, the molecular cascade results in increased expression of HIF1-α and production of VEGF. Binding of the VEGF ligand to the VEGFR leads to tumor angiogenesis.

development of invasive carcinoma) occurs in the viral *E2* reading frame. With lack of E2 expression, the viral oncogenes *E6* and *E7* are transcribed and these proteins degrade and/or inactivate the cellular tumor suppressor gene products, p53 and retinoblastoma protein (pRb), respectively [3,4]. Ultimately this molecular cascade leads to increased hypoxia-inducible factor-1 (HIF1-α) and thrombospondin-1 expression which in turn results in increased VEGF expression and tumor angiogenesis (Fig. 14.4). Using the human cervical cancer cell lines C-33A and HeLa, Tang et al. have shown that HPV 16 E6- and E7-transfected cervical cancer cells express increased HIF1-α and VEGF expression. Transfected cells stimulated in vitro capillary and tubule formation and these angiogenic effects were abolished through cotransfection with HIF1-α siRNA or treatment with reservatol [42]. Because several oncogenes including mutant *ras*, EGF receptor, ErbB2/Her2 c-*myc*, and v-*src* upregulate VEGF expression, Lopez-Ocejo et al. [43] studied the ability of HPV 16 E6 to directly stimulate VEGF expression and reported that the oncoprotein upregulates VEGF promoter-Luc (luciferase) in a p53-independent manner. Similarly, Walker et al. [44] have also reported a mechanism of HPV 16 E7-mediated increased levels of VEGF expression that is pRb-independent. Thus the molecular cascade that is initiated following viral integration leads to increased VEGF expression and ultimately tumor angiogenesis through p53- and pRb-dependent and -independent pathways.

Finally, the *therapeutic rationale* that led to the proof of concept of an antiangiogenesis approach to cervical cancer is found in a series of three independent clinical trials and a small case series that suggested clinical efficacy in this disease. Fumagillin, an antibiotic secreted by the fungus *Aspergillus fumigatus fresenius*, was shown to inhibit endothelial cell proliferation and tumor-induced neovascularization [45]. Because fumagillin was associated with an intolerable toxicity profile, synthetic analogues were developed, with TNP-470 being the most potent angiogenesis inhibitor with the least side effects. TNP-470 was shown to inhibit endothelial cell migration, proliferation, and capillary-like tube formation in in vitro studies [46]. The angio-inhibitory properties were also demonstrated in vivo through the induction of avascular zones in chorioallantoic membrane assays and by the suppression of the number and length of new blood vessels in rat cornea in the presence of growth factors [47–49]. In a phase I study of TNP-470 in 18 evaluable women (median age 48 years) with inoperable, recurrent, or metastatic squamous cell carcinoma of the cervix, the drug was administered at a starting dose of $9.3\,mg/m^2$ over 60 minutes every other day for 28 days followed by a 14-day rest period [50]. The intermediate dose level of $60\,mg/m^2$ was identified for further study, with reversible neurotoxicity being dose limiting [50]. Among three patients with initially progressive disease, treatment with TNP-470 resulted in stable disease for 5, 7.7, and over 19 months [50].

Importantly, one patient with bilateral pulmonary metastases experienced complete resolution of metastatic disease [50]. This patient received TNP-470 for 22 months and at the time when her case was highlighted in a *Letter* to the *New England Journal of Medicine* in 1998 [51] there had been no evidence of recurrent disease 8 months after TNP-470 had been discontinued.

Antiangiogenesis therapy in cervical cancer targeting the VEGF axis using bevacizumab was first reported by Wright et al. [52] in a small case series of six patients in 2006. In 2009 and 2010, Monk et al. reported separately on two important phase II trials targeting VEGF-dependent angiogenesis in recurrent cervical cancer. GOG protocol 227C studied single-agent bevacizumab at $15\,mg/kg$ in 46 women with recurrent or persistent squamous cell carcinoma of the cervix [53]. Thirty-eight patients (82.6%) had received prior pelvic radiotherapy as well as either one ($n = 34$, 73.9%) or two ($n = 12$, 26.1%) prior cytotoxic regimens for recurrent disease [53]. Grade 3 or 4 adverse events included hypertension ($n = 7$), thromboembolism ($n = 5$), gastrointestinal toxicity ($n = 4$), anemia and cardiovascular ($n = 2$ each), and vaginal bleeding, neutropenia, and fistula ($n = 1$ each) [53]. One grade 5 infection was observed. Eleven patients (23.9%) survived progression-free for at least 6 months, and five patients (10.9%) experienced partial responses [53]. The median duration of response was 6.21 moths (range, 2.83–8.28 months), and the median progression-free survival (PFS) and overall survival (OS) were 3.4

months (95% CI, 2.53–4.58 months) and 7.29 months (95% CI, 6.11–10.41 months), respectively [53]. When compared with the GOG's cervical cancer database of cytotoxic single-agent compounds studied in the phase II setting of treatment failure with one prior regimen, bevacizumab compared favorably in GOG 227C, justifying its advancement into the phase III arena.

The randomized phase II open-label study reported by Monk et al. of the oral tyrosine kinase inhibitor (TKI) of VEGFR pazopanib (800 mg daily), vs the oral dual anti-EGFR and anti-HER2/neu TKI lapatinib (1500 mg daily) in patients with recurrent/persistent cervical cancer who had been treated with at least one prior regimen for recurrent/metastatic disease was actually initiated as a three-arm 1:1:1 randomized study but the third arm of combined therapy (pazopanib 400 or 800 mg daily plus lapatinib 1000 or 1500 mg daily) was dropped when the futility boundary was breached [54]. Of the 230 patients enrolled, 152 were randomly assigned to the monotherapy arms. Anti-VEGF therapy using pazopanib improved the primary endpoint PFS (HR = 0.66; 90% CI, 0.48–0.91, $p = 0.013$) and also OS (HR = 0.67; 90% CI 0.46–0.99; $p = 0.045$) over anti-EGF therapy using lapatinib [54]. The median OS was 50.7 and 39.1 weeks, and the objective RRs were 9% and 5%, respectively [54]. Notable toxicities attributed to pazopanib included grade 3 diarrhea (11%), with grade 4 toxicities occurring in 12% [54]. An updated report demonstrated no OS benefit with prolonged follow-up [55].

Clinical Practice Ramifications of GOG Protocol 240

As described earlier, GOG 240 was originally conceived as a phase III trial to compare the efficacy and tolerability of a nonplatinum chemotherapy doublet to the platinum–paclitaxel control which had emerged from the GOG's eighth phase III randomized trial in the recurrent/persistent and metastatic cervical cancer setting, GOG protocol 204 [19]. Although GOG 204 to this day remains the phase III experience to have recruited the largest number of patients (>500), it was closed by the NCI's data safety monitoring board (DSMB) for futility when it was determined at interim analysis that none of the investigation arms (cisplatin–topotecan, cisplatin–gemcitabine, or cisplatin–vinorelbine) would outperform the reference arm of cisplatin–paclitaxel [19]. Therefore, with the exception of the GOG's phase II trial of cisplatin-pemetrexed (ongoing during 2009) [56–58], when GOG 204 was reported, the platinum-based chemotherapy doublets had been exhausted. Moreover, with the aforementioned widespread adoption of cisplatin-based chemoradiation protocols for locally advanced disease, identification of a nonplatinum doublet for the treatment of recurrent and metastatic disease seemed attractive. At that time the combination of gemcitabine plus docetaxel was being studied in the SCOTCERV phase II study and data would not be forthcoming for several years [59]. Therefore, based on laboratory data by Bahadori et al.

[60] which suggested synergy between topotecan and microtubule-interfering agents such as paclitaxel and the phase II experience by Tiersten et al. [61] in which topotecan plus paclitaxel appeared to be active and tolerable in a preirradiated recurrent cervix population, the combination of topotecan plus paclitaxel emerged as a frontrunner to be studied in GOG 240.

Ultimately, the opportunity to *also* study anti-VEGF therapy using bevacizumab in GOG 240 was approved by both the GOG's Cervix and Protocol Development Committees, the NCIs Cervical Cancer Task Force, Gynecologic Cancer Steering Committee, Central Institutional Review Board, and Cancer Evaluation and Therapy Program, and Roche/Genentech which would supply bevacizumab. Assuming an absence of interaction between the factors being investigated (nonplatinum chemotherapy doublet and antiangiogenesis therapy), GOG 240 randomized using a 2 × 2 factorial design to determine whether one or both factors could reduce the hazard of death by 30% [62]. In this prospective, international, phase III randomized, four-arm clinical trial, 452 patients were stratified according to performance status, prior to cisplatin with radiotherapy, and recurrent/persistent vs metastatic disease prior to randomization and were treated every 21 days until voluntary patient withdrawal, disease progression, unacceptable toxicity, or complete response [62] (Fig. 14.5). An important consideration concerning the eligibility criteria in

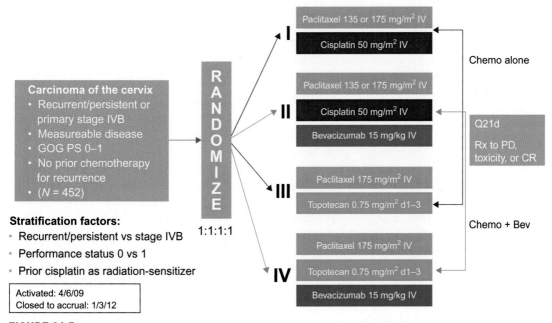

FIGURE 14.5

Schema for GOG protocol 240, the phase III randomized trial studying antiangiogenesis therapy and a nonplatinum chemotherapy doublet in recurrent/persistent and metastatic cervical cancer.

GOG 240 was that the patient population was required to undergo optimization of medical comorbidities, including correction of any existing impaired renal function, malnutrition, neuropathic or tumor-related pain, and poor performance status.

The primary endpoints of GOG 240 included OS and tolerability. Secondary endpoints were PFS, RR, and health-related QoL as determined by patient-reported outcomes. The major exploratory objectives were the prospective evaluation of previously identified pooled prognostic factors known as the Moore criteria [63], and the prevalence and impact on survival of current tobacco use. The translational endpoint centered on the potential identification of circulating tumor cells (CTCs) in patient serum and their impact (if found) on survival endpoints. The trial was activated on April 9, 2009, with a single interim analysis planned at 173 events, and 346 deaths required at final analysis to inform a study designed to improve median OS in this population from 12 months (based on the prior GOG 204 study) to at least 16 months.

At the preplanned interim analysis in February 2012, the topotecan–paclitaxel doublet (with or without bevacizumab) was found to be not superior to the cisplatin–paclitaxel doublet (with or without bevacizumab). The database was frozen again on December 12, 2012, at 271 deaths (the time determined by the NCI's DSMB at first interim). During the second analysis it was determined that the trial had met its primary endpoint and the NCI issued a Press Release in March 2013 indicating that the addition of bevacizumab to chemotherapy was associated with significantly improved OS (17.0 months vs 13.3 months; hazard of death 0.71; 98% CI, 0.54–0.95; $p = 0.004$) [64] (Fig. 14.6).

A significant improvement in PFS was also observed (8.2 months vs 5.9 months; hazard of progression 0.67; 95% CI, 0.54–0.82; $p = 0.002$) [62] (Fig. 14.7). The RR was also significantly higher among patients who received chemotherapy plus bevacizumab (48% vs 36%, relative probability of response 1.35; 95% CI, 1.08–1.68; $p = 0.008$) [62]. In an analysis of prognostic factors, it was shown that the survival benefit conferred by bevacizumab was sustained even when disease was present in the previously irradiated pelvis, a region previously held to be a "sanctuary site" for recurrent disease following definitive chemoradiation. Interestingly, tumors with glandular histology (e.g., adenocarcinoma and adenosquamous histology) did not appear to respond as favorably to antiangiogenesis therapy as squamous cell carcinoma, but this was likely due to an underpowering of this histologic subtype as a consequence of its relative lower prevalence than squamous cell carcinoma lesions.

Although the arms administering bevacizumab were associated with higher rates of grade 2+ hypertension, this was easily managed and no patients were

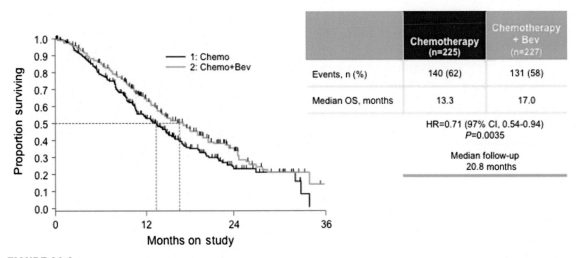

FIGURE 14.6

Kaplan–Meier survival curves of OS in GOG protocol 240 demonstrating the survival advantage conferred by the arms administering the antiangiogenesis drug, bevacizumab. *From Tewari KS, Sill MW, Long HJ III, et al. Improved survival with bevacizumab in advanced cervical cancer. N Engl J Med 2014;370:734–43.*

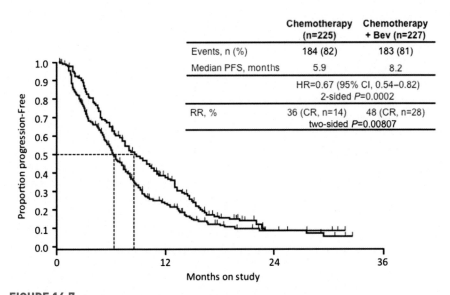

FIGURE 14.7

Kaplan–Meier survival curves of PFS in Gynecologic Oncology. Group protocol 240 demonstrating superiority of chemotherapy plus bevacizumab vs chemotherapy alone. Antiangiogenesis therapy was also associated with significantly improved RRs by RECIST criteria. *From Tewari KS, Sill MW, Long HJ III, et al. Improved survival with bevacizumab in advanced cervical cancer. N Engl J Med 2014;370:734–43.*

taken off investigational therapy due to this adverse event [62]. The fistula rate was 8.6% among patients receiving bevacizumab, all of whom had been preirradiated, but unlike the gastrointestinal perforations associated with bevacizumab in patients with colorectal and ovarian carcinoma, fistula development did not constitute a surgical emergency, and none led to sepsis or death [62]. Finally, venous thromboembolism occurred in 8% of patients receiving chemotherapy plus bevacizumab as compared to 1% of patients treated with chemotherapy alone [62]. Immediately following the Press Release and submission of Dear Investigator and Dear Patient Letters, the NCI and Genentech made arrangements for provision of bevacizumab to patients who were on the chemotherapy alone arms.

GOG 240 realized a high previously unmet clinical need for women with recurrent/persistent and metastatic cervical cancer. In March 2013, the American Society of Clinical Oncology (ASCO) made an exception to their embargo policy and released the GOG 240 abstract concerning the findings of bevacizumab into the public domain, 3 months ahead of the meeting. ASCO still accepted the abstract in the General Plenary (abstract #3) and it was featured in the ASCO Press Briefing on June 2, 2013.

In the months following presentation of these data, Penson et al. reported that patients treated on chemotherapy plus bevacizumab did not experience a significant deterioration in health-related QoL as measured by the Functional Assessment of Cancer Therapy for Cervical Cancer Trial Outcome Index (FACT-Cx TOI), the GOG Neurotoxicity subscale and the Brief Pain Inventory [65]. Using these three previously validated instruments, patient-reported outcomes were assessed prior to treatment cycles 1, 2, and 5 and 6 and 9 months following cycle 1. Patients who completed baseline QoL assessment and at least one further assessment were evaluable ($n = 390$), and compared to patients treated with chemotherapy alone, those who received chemotherapy plus bevacizumab-reported FACT-Cx TOI scores that were an average of 1.2 points lower (98.75% CI, −4.1 to 1.7; $p = 0.30$) [65]. Similarly there is no significant difference in neurotoxicity or brief pain inventory scores between the two groups, although there was noted a nonsignificant trend among patients treated with bevacizumab to report less neurotoxicity and be less likely to report pain, the latter possibly attributed to enhanced tumor shrinkage rates demonstrated with antiangiogenesis therapy resulting in patients feeling better [65].

The primary manuscript was published in the February 20, 2014, issue of the *New England Journal of Medicine* [62], and 1 month following publication, the United Kingdom's Cancer Drug Fund approved bevacizumab for women in England diagnosed with advanced cervical cancer [66,67]. The manufacturer, Genentech, submitted an application to the United States Food and Drug Administration (US FDA) to expand the label of bevacizumab to include cervical cancer, noting an even higher OS improvement of 3.9 months (as compared to 3.7 months in the primary manuscript). The (although favorable) discrepancy between the OS improvement reported by Genentech to the US FDA and what was reported by the GOG at ASCO was due to the collection of additional data from reporting institutions that had been delinquent at the time of the NCI DSMB's second analysis. The application was selected by the US FDA for Priority Review [68] and on August 14, 2014, bevacizumab received regulatory approval for use with cisplatin–paclitaxel or with topotecan–paclitaxel for the treatment of recurrent/persistent and/or metastatic cervical carcinoma [69]. The National Comprehensive Cancer Network (NCCN) Cervical Cancer Treatment Guidelines were updated in 2014 to include both bevacizumab-containing triplet regimens studied in GOG 240 and listed both as Category 1 which indicates that based on overwhelming evidence and unanimous NCCN consensus that the intervention is appropriate [70].

Following the protocol-specified 348 deaths, the final OS analysis by intent-to-treat demonstrated continued separation of the curves beyond a maximum follow-up of 50 months (16.8 months vs 13.3 months; HR = 0.765; 95% CI, 0.62–0.95; p = 0.0068) [71]. The survival benefit attributed to antiangiogenesis therapy in the experimental arms was not attenuated by having bevacizumab made available to patients assigned to chemotherapy alone following the NCI Press Release issued upon completion of the second analysis. PFS was also updated at this time (8.2 months vs 6.0 months; hazard of progression 0.684; 95% CI, 0.56–0.84; p = 0032), as was RR (49% vs 36%, p = 0.0032) [71]. Bevacizumab was approved in December 2014 by Swissmedic for women with cervical cancer in Switzerland [72] and following a positive opinion issued in March 2015 by the European Union's Committee for Medicinal Products for Human Use (CHMP) [73,74], the drug was approved for cervical cancer by the European Medicines Agency on April 8, 2015 [75]. During 2015, many other countries, including Canada, Mexico, Costa Rica, Panama and all of South America, including Brasil, as well as Australia, Turkey, Lebanon, Israel, and five other Middle Eastern Countries, Morocco, South Africa, Hong Kong, Korea, Japan, Malaysia, Singapore, and India, also approved bevacizumab for the recurrent and metastatic cervical cancer population.

While HPV infection and dysplasia remain ubiquitous, the burden of invasive cervical cancer is felt principally in the developing world. Although it is not disputed that resources in the developing world should be concentrated on cervical cancer prophylactic vaccination and screening programs, the positive clinical and societal impact of such programs will not be enjoyed for at least one generation after implementation. For this reason, there remains an obligation to help those patients today struggling with recurrent and metastatic disease. A Markov decision tree based on GOG 240 trial data was created by Minion et al. using the 2013 MediCare Services Drug Payment Table and Physician Fee Schedule [76]. Noting that the median number of cycles in the experimental arm(s) of GOG 240 was seven and the serious adverse event rate peaked at 8.6% (i.e., fistula), the estimated cost of therapy with bevacizumab was noted to be approximately 13.2 times that for chemotherapy alone, adding $73,791 per 3.5 months of life gained [76]. This resulted in an incremental cost-effectiveness ratio (ICER) of $24,597/quality adjusted life months (QALmonth) [76]. With a 75% reduction in the cost of bevacizumab, the ICER reduces to $6737/QALmonth [76]. The Markov chain revealed that increased costs are primarily related to the cost of drug and not the management of bevacizumab-induced adverse events. Through the future availability of biosimilars, dramatic declines in the ICER can be achieved making anti-VEGF therapy cost-effective in both the developing and industrialized world.

The previously described Moore criteria [63] were prospectively evaluated and ultimately validated during the conduct of GOG 240 as one of two major (i.e., tertiary) endpoints. Originally identified by pooling 20 clinical factors extracted from three prior phase III randomized trials of the GOG [15–17], Moore et al. [63] developed a scoring system comprised of the five strongest factors: performance status >0, time to recurrence <12 months, prior platinum exposure, pelvic disease, and African-American ethnicity. When weighted equally, successive significant deteriorations in prognosis were reported in the original paper when moving from low-risk (0 or 1 factor) to mid-risk (2 or 3 factors) to high-risk (4 or 5 factors) populations [63]. The same effect was observed in GOG 240 among the entire study population [77]. Although the validated Moore criteria were not able to guide chemotherapy backbone selection, the placement of GOG 240 study population into low-risk, mid-risk, and high-risk groups based on treatment with and without bevacizumab provided interesting observations. Specifically, the hazard ratios of death for treating with bevacizumab in low-risk, mid-risk, and high-risk subsets were 0.96 (95% CI, 0.51–1.83; $p = 0.9087$), 0.673 (95% CI, 0.51–0.91; $p = 0.0094$), and 0.536 (95% CI, 0.32–0.905; $p = 0.0196$), respectively [77]. These data indicate that low-risk patients by the Moore scoring system do not derive a survival benefit from treatment with bevacizumab. Thus, for low-risk patients at significant risk for adverse events (e.g., fistula formation among a previously irradiated patient), the Moore criteria can be used to justify omitting bevacizumab given no survival gains. The case is very different for mid-risk and high-risk patients, with both groups benefiting from bevacizumab, not only in terms of OS but also with significant gains in PFS and higher RRs. In fact, the high-risk group derives the greatest benefit from antiangiogenesis therapy [77], and it is interesting to note that during the development of GOG 240, it had been suggested by some to exclude the high-risk patients from enrollment and shunt them to best supportive care. Fortunately, the need to first prospectively validate the Moore criteria was recognized before high-risk patients were declared ineligible. The Moore criteria constitutes the first prospectively validated scoring system in cervical cancer and can be used as a clinical instrument to counsel patients and families regarding their likelihood of response and months of survival gain (if any) when considering the addition of bevacizumab to chemotherapy. It should be noted that African-American ethnicity may represent a surrogate for lack of access to health care since several studies have documented equivalent outcomes to Caucasians when the playing field is level (i.e., universal health care coverage within the Kaiser Permanente health care system, military health care coverage, etc.). As such, the Moore criteria may retain their prognostic capabilities even in populations where African-Americans are not present in appreciable numbers.

The second tertiary endpoint of GOG 240 involves the prevalence and impact on survival of concurrent tobacco use. Previously, Waggoner et al. showed that among patients treated for locally advanced cervical cancer, compared with nonsmokers, median survival was 15 months shorter for reported smokers and 20 months shorter for cotinine-derived smokers ($p < 0.01$) [78]. In that original study, a significant increase in the risk of death was observed for reported smokers (HR = 1.51; 95% CI, 1.01–2.27; $p = 0.04$) and cotinine-derived smokers (HR = 1.57; 95% CI, 1.03–2.38; $p = 0.04$) [78]. Currently, data are being extracted and analyzed from GOG 240 smoking questionnaires [79].

The principal translational endpoint of GOG 240 was to determine whether CTCs were present in the study population and detectable as minimally invasive "liquid biopsies." According to theory, highly angiogenic tumors have a leaky vasculature, through which cancer cells could be shed and shunted. It followed then that the presence of CTCs may identify subsets of tumors vulnerable to antiangiogenesis therapies. Immunomagnetic separation using a microfluidic-based CTC chip containing 78,000 EpCAM-coated microposts was performed on 8.5 mL whole blood specimens drawn precycle 1 and 36 days postcycle 1 [80]. Captured CTCs were stained to identify DNA content, epithelial cells, and nonspecifically bound leukocytes. The CTC chip employed was a second-generation Herringbone Chip developed at Massachusetts General Hospital and so named because the herringbone pattern provides increased surface area for CTCs to bind. Not only were CTCs identified in GOG 240 patients (median precycle 1 CTC count was 7 CTCs/8.5 mL whole blood), but CTC counts significantly decreased during treatment, with those patients experiencing the greater declines in CTCs having a lower risk of dying (HR = 0.87; 95% CI, 0.79–0.95) [80]. An exploratory analysis suggests a potential role of CTCs as a *predictive biomarker* in this disease as patients with high levels of CTCs treated with bevacizumab had a lower risk of progression (HR = 0.59; 95% CI, 0.36–0.96) compared to those with high CTC levels who did not receive bevacizumab [80]. In other words, the survival curve described by high CTC counts shifts to the right with intervening anti-VEGF therapy.

The GOG 240 database continues to be mined for clinically useful information. Approved ongoing ancillary data studies that have been presented in part at scientific meetings include identification of risk factors for the development of fistula among patients treated with bevacizumab (e.g., prior pelvic irradiation, persistent disease in the pelvis, current tobacco use, preexisting hypertension) [81], the pooling of glandular histology from GOG 240 and prior phase III studies to compare survival following systemic therapy with squamous cell carcinoma [82], and clinical factors associated with complete responders and long-term survivors from GOG 240 [83], along with postprogression survival following treatment with bevacizumab.

Antiangiogenesis Therapy Beyond GOG 240

While a 3.7-month (or FDA-reported 3.9-month) improvement in OS is not very long, it still represents a proverbial "foot in the door," opening up a potential therapeutic window of opportunity through which patients deriving benefit from antiangiogenesis therapy can be treated with other novel agents prior to progression. Molecules of interest include other classes of antiangiogenesis/antivascular agents, immunotherapies, as well as other small molecule inhibitors of critical signal transduction pathways involved in cervical cancer tumorigenesis [84–98]. Recently, Symonds et al. [99] reported on CIRCCa, the randomized, phase II, double-blind, placebo-controlled trial involving 69 women with recurrent cervical cancer. The investigators demonstrated significantly improved PFS using daily administration of cediranib (20 mg), a VEGFR1–3 oral TKI, plus six cycles of carboplatin (AUC [area under the curve] 5) plus paclitaxel (175 mg/m^2) (HR = 0.58; 80% CI 0.4–0.85; $p = 0.032$) [99] (Fig. 14.8). Although CIRCCa was not powered for survival, the RR of 64% is the highest reported for any regimen in this disease [99,100]. Unlike GOG 240, febrile neutropenia was a significant problem in CIRCCa, as was cediranib-induced diarrhea that significantly impacted one QoL measure [99,100]. Fistula, however, was not reported.

The use of carboplatin in CIRCCa deserves comment [101–105]. Although the Japanese Clinical Oncology Group (JCOG) have demonstrated significantly noninferiority in JCOG 0505 (a phase III randomized trial comparing cisplatin–paclitaxel to carboplatin–paclitaxel), they also reported that among platinum-naïve patients, the cisplatin–paclitaxel arm was superior [106]. While the carboplatin–paclitaxel doublet is certainly easier to administer, caution should be exercised in women who have not seen platinum previously (e.g., those treated with radiotherapy alone for locally advanced disease and those who present with FIGO IVB tumors), as well as the elderly and those

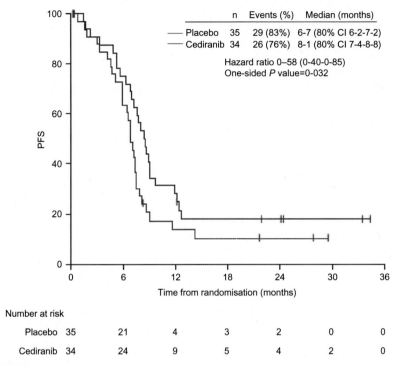

	n	Events (%)	Median (months)
Placebo	35	29 (83%)	6-7 (80% CI 6-2-7-2)
Cediranib	34	26 (76%)	8-1 (80% CI 7-4-8-8)

Hazard ratio 0–58 (0-40-0-85)
One-sided *P* value=0-032

Number at risk

Placebo	35	21	4	3	2	0	0
Cediranib	34	24	9	5	4	2	0

FIGURE 14.8

Improved PFS with the incorporation of cediranib to chemotherapy for recurrent or metastatic cervical cancer. *From Symonds RP, Gourley C, Davidson S, et al. Cediranib combined with carboplatin and paclitaxel in patients with metastatic or recurrent cervix cancer (CIRCCa): a randomised, double blind, placebo-controlled phase 2 trial. Lancet Oncol 2015;16:1515–24.*

previously treated with extended field radiation for whom carboplatin may prove to be intolerable due to vastly diminished bone marrow reserves [100].

In addition to studying different agents that target the VEGF-dependent angiogenesis axis, future trials may also choose to incorporate molecules that target the angiopoietin/Tie2 angiogenesis pathway. This signaling pathway is comprised of two tyrosine kinase receptors (Tie1 and Tie2) which are preferentially expressed on vascular endothelium, and three known Tie2 ligands (Ang1, Ang2, and Ang4) [107]. Ang1 binding to Tie2 activates downstream signaling pathways which promote blood vessel stability through enhancing interactions between perivascular cells and endothelium and increased endothelial cell survival, both of which lead to a more stable vasculature with decreased permeability [107]. Importantly, Ang2 expression is tightly regulated and is primarily synthesized and secreted by endothelial cells at sites of vascular remodeling in response to proinflammatory stimuli, other proangiogenic cytokines (e.g., insulin-like growth factor 1, platelet-derived growth factor B), or hypoxia [107]. The peptibody, trebananib, is an active inhibitor of the angiopoietin pathway [107].

Finally, an antivascular strategy designed to disrupt established tumor blood vessels involves the use of tumor-vascular disrupting agents (VDAs). These agents include flavonoid compounds such as ASA404 and tubulin-binding agents including combretastatin A-4 phosphate and AVE8062 [108]. Unlike antiangiogenesis agents which predominantly inhibit neovascularization with greatest activity at the tumor periphery and in small lesions, tumor-VDAs prune existing vasculature and inhibit tumor blood flow, resulting in extensive necrosis in the tumor core [108].

IMMUNOTHERAPEUTIC APPROACHES

HPV-Targeted Tumor-Infiltrating T-Cells (Adoptive T-Cell Therapy)

Cervical cancer has been strongly associated with failure to mount a strong HPV-specific type 1 T-helper and cytotoxic T-lymphocyte response. In addition, the lack of CD8+ T-cells migrating into hypoxic tumor core and induction of HPV subtype-specific regulatory T-cells have also been implicated in the pathogenesis [88,93]. It has been shown that the ratio between tumor-infiltrating CD8+ T-cells and co-infiltrating CD4+Foxp3+regulatory T-cells is an independent prognostic factor for OS in cervical cancer [88,93].

Adoptive T-cell therapy involves the isolation and ex vivo expansion of tumor-specific T-cells. In patients with B-cell malignancies and malignant melanoma, complete clinical responses have been achieved following infusion of autologous tumor-reactive T-cells. Stevanovic et al. developed a method for generating T-cell cultures from HPV-positive cancers and recently reported on nine patients with metastatic cervical cancer treated with a single infusion of tumor-infiltrating lymphocytes selected for HPV E6 and E7 reactivity (HPV-TIL) (median 81×10^9 T-cells, range $33-159 \times 10^9$) [109]. Infusion was preceded by nonmyeloablative conditioning with lymphocyte-depleting chemotherapy and followed by high-dose bolus aldesleukin [109]. Two patients with no HPV reactivity did not respond to treatment and three of six patients with HPV reactivity demonstrated objective tumor responses. One patient had a 39% partial response and two patients with widespread metastases (one with chemotherapy-refractory HPV 16+ squamous cell carcinoma and one with chemoradiation-refractory HPV 18+ adenocarcinoma) experienced complete tumor responses that were ongoing at the time of publication, 22 and 15 months after treatment [109]. This study represented a proof of concept that cellular therapy can mediate complete, durable regression of an epithelial malignancy (Fig. 14.9).

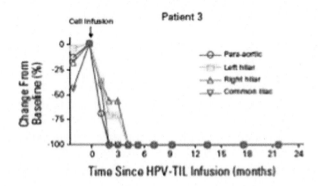

FIGURE 14.9
Complete response in a patient with metastatic cervical cancer treated with tumor-infiltrating T-cells selected for HPV E6 and E7 reactivity (HPV-TILs). *From Stevanovic S, Draper LM, Langhan MM, et al. Complete regression of metastatic cervical cancer after treatment with human papillomavirus-targeted tumor-infiltrating T-cells. J Clin Oncol 2015;33:1543–50.*

Therapeutic Vaccination

While successful prophylactic cervical cancer vaccines have been derived from cDNA of the L1 virus capsid, because appreciable levels of L1 and/or L2 capsid antigens are not expressed by infected basal epithelial cells and cervical cancer cells, therapeutic HPV vaccines targeting antigens other than L1 and L2 are needed. Both HPV oncoproteins, E6 and E7, represent ideal targets for therapeutic vaccine development. Technologies to create therapeutic HPV vaccines include live vector-based vaccines, peptide/protein-based vaccines, nucleic acid-based vaccines, and whole cell vaccines. Live vectors can be bacterial (e.g., *Listeria monocytogenes*) or viral (e.g., Newcastle disease virus) [88,93]. Nucleic acid vaccines may be DNA-based or created from naked RNA replicons, while whole cell vaccines may be derived from dendritic cells or even tumor cells [88,93].

In their construction of an attenuated *L. monocytogenes strain* for use in cancer immunotherapy, Wallecha et al. [110] acknowledge that this gram-positive bacterium serves as an ideal vaccine vector by virtue of its ability to trigger potent cellular immune responses in an infected host due to its ability to survive in both phagocytic and cytosolic compartments. However, the use of a live attenuated bacterium represents a very complex method of immunotherapy as over millennia *L. monocytogenes* has evolved to infect humans and during this same period humans have evolved to prevent and reject this infection.

Although insertion of a heterologous gene in the bacterial chromosome either by homologous recombination or by phage-specific insertion represents a strategy through which *L. monocytogenes* can be genetically modified to express heterologous antigens in vivo, Wallecha's team elected to transform the bacterium with a plasmid carrying a foreign antigen stating that the plasmid-based strategy has the advantage of multicopy expression but relies on complementation for the maintenance of the plasmid in vivo [110]. The result was an irreversibly attenuated and highly immunogenic *L. monocytogenes* platform which was then used to create a live attenuated *L. monocytogenes-* (*Lm-*) based immunotherapy (AXAL, previously called ADXS-HPV or ADXS11-001) that secretes an antigen–adjuvant fusion (Lm-LLO) protein consisting of a truncated fragment of the Lm protein listeriolysin O (LLO) fused to HPV 16 E7 oncoprotein [111].

At the 2014 Annual Meeting of the ASCO, Basu et al. [112] presented final results of a randomized phase II trial involving 110 women from India with recurrent cervical cancer who had progressed following treatment with chemotherapy and/or radiation therapy. Patients were randomized to receive ADXS-HPV (1×10^9 cfu $\times$ 3 or 4 doses) with and without cisplatin ($40\,mg/m^2$) [112]. The final 12-month OS was 36% and 18-month OS was 28% [112]. The overall RR was 11% and included six complete responders and six partial responders [112]. The average duration of response in both treatment groups was 10.5 months. The addition of cisplatin did not significantly improve OS or RRs. Grade 3 adverse events related to study drug occurred in 2% [112].

This was followed by a second phase II trial (GOG protocol 0265, NCT01266460) in which the vaccine was studied in women with squamous and nonsquamous persistent/recurrent or metastatic cervical cancer who had previously received one or more lines of therapy for recurrent/metastatic disease. In stage 1, 26 patients were enrolled of whom 31% had received prior bevacizumab, and in stage 2, 83% of the 24 women enrolled had received prior anti-VEGF therapy with bevacizumab. Importantly, in stage 1, the 12-month OS was 38.5% ($n = 10$), indicating that the vaccine could move forward into the second stage. Of particular interest was one patient previously treated with chemotherapy plus bevacizumab who achieved a complete response that has lasted 11 months following second line treatment with three doses of axalimogene filolisbac on GOG-265. This patient's postbevacizumab recurrence had manifested along the periaortic nodal chain and had resolved on PET/CT within 6 months of vaccine administration. These data were reported by Huh et al. at the 2016 ASCO [113].

Currently in development is AIM2CERV (NCT02853604), a phase 3, placebo-controlled randomized trial of adjuvant axalimogene filolisbac for high-risk locally advanced squamous cell carcinoma, adenocarcinoma, and adenosquamous carcinoma of the cervix following definitive chemoradiotherapy plus brachytherapy. Randomization will be 2:1 with induction therapy consisting of 1×10^9 cfu or placebo q3wks $\times$ 3 doses, followed by maintenance 1×10^9 cfu or placebo q8wks x 5 doses or until recurrence. High-risk patients will be those with one or more of the following: (1) FIGO stage IB2, IIA2, IIB with positive pelvic nodes (biopsy or MRI/CT/PET); (2) FIGO IIA, IIIB, IVA; and (3) any stage with positive para-aortic nodes (biopsy or MRI/CT/PET).

Breaking Immune Tolerance Through PD-1/PD-L1 Blockade

The emerging paradigm in the new landscape of immunotherapy involves reawakening silenced immune responses by *inhibiting the inhibitors* that are responsible for paralyzing T-cells and creating a state of immune tolerance. In point of fact, breaking of immune tolerance may represent a powerful strategy to overcome immune suppression and result in a more robust intervention than short-lived immune-activating approaches. Cytotoxic T-lymphocyte antigen 4 (CTLA-4) attenuates the early activation of naïve and memory T-cells. Ipilimumab is a checkpoint blocking antibody directed against CTLA-4 and has received regulatory approval in advanced melanoma [114]. Programmed cell death protein 1 (PD-1) is primarily involved in modulating T-cell activity in peripheral tissues via interaction with its ligands, PD-L1 and PD- L2, and can be inhibited by checkpoint blocking antibodies such as nivolumab and pembrolizumab, both of which have also received US FDA approval in advanced melanoma [115–121]. Although both CTLA-4 and PD-1 function as negative regulators, their roles in immune modulation are nonredundant and do not overlap.

T-cell activation may result in expression of the PD-1 receptor. The programmed death-ligand 1 (PD-L1) can be induced by the entire spectrum of immune cells and nonhematopoietic cell types, while the PD-L2 ligand is found on activated macrophages and dendritic cells exclusively [122]. Simultaneous engagement of both ligands by PD-1 results in transmittal of negative immunoregulatory signals to T-cells, resulting in decreased effector response and T-cell tolerance [122]. Importantly, the PD-1/PD-L1 interactions inhibit a wide range of immune responses directed against tumors, pathogens, and even, self-antigens. PD-L1 is expressed on the cell surface in most human cancers and correlates with poor prognosis in many solid tumors, most likely by providing an immune escape mechanism through downregulation of tumor-specific T-cell responses [122]. This phenomenon has been previously shown to be indicative of chronic antigen stimulation and T-cell exhaustion. Some investigators have speculated that PD-1 expression is even more pronounced in advanced stage cervical cancer and that blockade of PD-1 could have therapeutic potential in this patient population [123].

The PD-1/PD-L1 pathway plays a critical role in attenuating T-cell responses and promoting T-cell tolerance during chronic viral infections. In a study by Yang et al. [124], PD-1 and PD-L1 expression on cervical T-cells and dendritic cells was associated with high-risk HPV positivity and increased in parallel with cervical neoplastic progression. In a tissue microarray of a group of 115 patients with FIGO stage IB1–II cervical carcinoma, Karim et al. [125] observed that PD-1 was expressed by >50% of both the infiltrating CD8+ T-cells and the CD4+ Foxp3+ T-cells, irrespective of PD-L1 or PD-L2 expression by the tumors. Finally, Heeren et al. [126] have recently studied PD-1 in tumor-draining lymph nodes in patients with lymph node-positive cervical cancer and found significantly lower CD4+ and higher CD8+ T-cells accompanied by increased surface levels of PD1 and CTLA-4. These studies indicate that the PD-1/PD-L1 axis may be a relevant therapeutic target to address microenvironmental immune suppression in cervical cancer, a disease which remains unclaimed in the PD-1/PD-L1 therapeutic arena. Potential randomized phase II study designs using PD-1/PD-L1 blockade in the recurrent/metastatic cervical cancer population include:

1. First-line metastatic/recurrent disease: Chemotherapy plus bevacizumab (i.e., GOG 240) vs chemotherapy plus PD-1 (or PD-L1) with cross-over to bevacizumab upon progression.

2. Second-line metastatic/recurrent disease (i.e., progression on prior chemotherapy plus bevacizumab): Physician's choice single-agent chemotherapy (e.g., paclitaxel, pemetrexed, or gemcitabine) vs PD-1 (or PD-L1).

At the 2016 ASCO, Frenel presented data from the cervical cancer cohort of the phase 1b KEYNOTE-028 study [127]. Twenty-four patients with inoperable recurrent or metastatic PD-L1-positive cervical cancer received **pembrolizumab** 10 mg/kg IV q2wks. All but one patient had received prior radiotherapy and prior platinum. Nine women (38%) had received three or more prior lines of therapy for advanced disease. Ten women (42%) had received prior bevacizumab. There were five grade 3 AEs, no grade 4 AEs, and no treatment-related mortality. One case each of grade 3 colitis and grade 3 Guillian-Barre syndrome led to treatment discontinuation. In this biomarker-enriched population, the ORR was 17% ($n = 4$, PRs all), with stable disease detected in another three women (13%). The median time to response was 8 weeks and the median duration of response was 26 weeks. Twelve-month OS was 33% [127]. It is difficult to place these results in context because there are no effective therapies used in the second line (and beyond) setting for advanced cervical cancer. The 2014 US FDA approval of chemotherapy plus bevacizumab satisfied a high, unmet clinical need for first-line therapy of recurrence, but in effect created a new population in need of novel therapeutic options in the second line setting. The 17% ORR attributed to pembrolizumab is notable due to the relative lack of activity (10% or less) of other chemotherapeutic agents studied by the Gynecologic Oncology Group (GOG) in this setting. Although the population studied was enriched (i.e., PD-L1 positive), it is not clear that PD-L1 expression is predictive in solid tumors.

Bispecific T-Cell Engagers and Chimeric T-Cell Receptor Antigens

With adoptive T-cell therapy, E6- and E7-based therapeutic vaccination, and anti-PD-1/PD-L1 dominating current immunologic approaches to obtain durable sustained responses that confer a survival advantage in advanced cervical cancer, bispecific T-cell engagers (BiTE) and chimeric antigen receptors (CARs) represent additional immunologic considerations. BiTE antibodies induce a transient cytolytic synapse between a cytotoxic T-cell and the cancer target cell. This results in discharge of cytotoxic T-cell contents and direct tumor cell lysis [93,128].

CARs are engineered autologous T-cells composed of an antigen-binding moiety derived from the variable region of a monoclonal antibody, and linked via a transmembrane motif to a lymphocyte-signaling moiety located in the cytoplasm [93,129]. Variable, extracellular binding motifs allow for recognition of tumor-associated antigens, including cell-surface-specific molecules. The principle benefit of CAR T-cell-based therapy is the elimination of the need to identify and harvest tumor-specific lymphocytes from patients [93,129]. Exploration of CAR-based therapies in advanced cervical cancer is warranted and will require the identification of appropriate ligand-binding domains, transmembrane linkers, and intracellular-signaling elements to optimize tumor cell recognition and limit off-target therapy.

OTHER PROMISING THERAPEUTIC STRATEGIES

Several other novel strategies to treat recurrent and metastatic cervical cancer are listed in Table 14.1 and include other immunologic platforms to manipulate the tumor microenvironment (e.g., dendritic-cell-based vaccines [130] and Newcastle disease virus [131]) as well as agents that enhance drug exposure to the tumor cell (e.g., nanoparticle drug delivery systems [132]) and molecules that directly interfere with critical cellular signal transduction pathways detected through genome-wide analysis of HPV integration [133], genomic alterations [134,135], and elucidation of subtle epigenetic phenomena [136]. Candidates include the PI3K/Akt/mTOR pathway, tumors harboring homologous recombination deficiency, and Notch signaling [88]. Interfering RNAs, cervical cancer stem cell identification and targeting, and gene therapy also represent promising avenues of treatment [137–145].

PI3K/AKT/mTOR Pathway Inhibitors

The mammalian target of rapamycin (mTOR) plays an integral role in angiogenesis, cell growth, proliferation, and survival (Fig. 14.10). In the absence of PTEN inhibition, AKT phosphorylates and inhibits the tuberous sclerosis complex (TSC) which leads to mTOR activation and formation of two different multiprotein complexes, mTOR complex 1 (mTORC1) and mTOR complex 2 (mTORC2) [88]. High-risk HPV-related E6 expression leads to rapid degradation of TSC2, resulting in TORC1 activation and downstream *mTOR signaling* [88,146–148]. The high prevalence of PI3KCA mutations in cervical cancer suggests that mTOR-targeted agents should be explored in this disease [149]. Activation of this pathway in HPV- infected cells occurs through both mutation of the pathway components and activation of upstream-signaling molecules and contributes to genetic instability, deregulation of proliferation, resistance to apoptosis, and changes in cellular metabolism [148]. Fourteen cervical cancer tumors from a phase I program were analyzed for PIK3CA, KRAS, NRAS, and BRAF mutations [150]. The five patients found to have PIK3CA mutations were treated with agents targeting the PI3K/AKT/mTOR pathway and two had a partial response [150]. In an unenriched

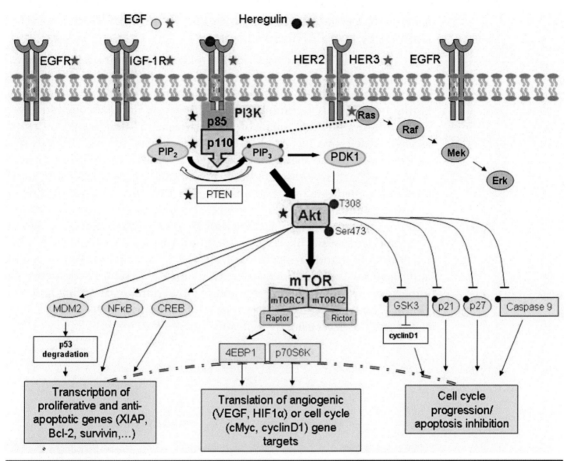

FIGURE 14.10

The PI$_3$K/Akt/mTOR pathway is upregulated via direct upstream stimulation by ligand binding by growth factor receptors, indirect activation via cross-talk with the Ras pathway, or intrinsically via activating genetic alterations in PI$_3$K or Akt, or via loss of function in the tumor suppressor gene product, PTEN. *From www.intechopen.com (open access).*

cervical cancer population ($n = 33$), a phase II trial of single-agent temsirolimus (25 mg IV qwk q28 days) yielded on partial response (3.0%) and stable disease in 19 patients (57.6%) [151]. Six-month PFS was 28% (95% CI: 14–43%) and the median PFS was 3.52 months (95% CI: 1.81–4.7) [151]. No grade 4–5 adverse events were observed.

Synthetic Lethality via Homologous Recombination Deficiency and PARP Inhibition

Poly(ADP-ribose) polymerases (PARPs) play an important role in base excision repair pathway in the setting of single-strand DNA breaks [88]. Through synthetic lethality, PARP inhibitors have demonstrable activity in tumors harboring homologous recombination deficiency, as simultaneous blockage of both pathways prevents repair of DNA damage. In addition, PARP trapping as well as defective BRCA1 and POLQ recruitment to sites of DNA repair may represent additional mechanisms [152]. Selection of enriched populations among patients with advanced cervical cancer for whom treatment with PARPi(s) is predicated on examining tumor DNA for evidence of a homologous recombination defect. However, because the genomic landscape of a tumor can be shaped by a myriad of factors, only a minority of observed DNA aberrations are likely to result from a tumor's inability to faithfully repair double-strand breaks in the DNA [152]. The total burden of genomic abnormalities can be approximated by methods that capture telomeric allelic imbalance, loss of heterozygosity, and the total number of coding mutations. *Genomic scars* may represent a surrogate for homologous recombination deficiency and predict sensitivity to treatment with PARPi(s) [152].

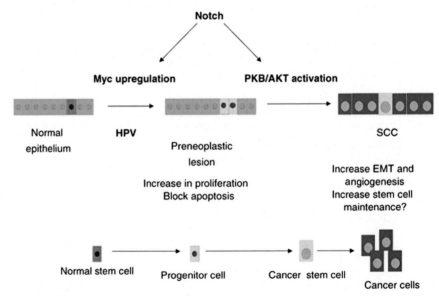

FIGURE 14.11

The Notch binary cell fate signaling pathway promotes malignant conversion by upregulating Myc or by activating the PKB/Akt pathway, resulting in increased proliferation, blocking apoptosis, increasing epithelial–mesenchymal transition, and angiogenesis, and possibly by favoring self-renewal of progenitor or the cancer stem cell pool. *From Maliekal TT, Bajaj J, Giri V, et al. The role of Notch signaling in human cervical cancer: implications for solid tumors. Oncogene 2008;27:5110–14.*

Notch Signaling

The Notch signal transduction pathway is an evolutionarily conserved binary switch important for cell fate decision-making [4] (Fig. 14.11). Notch expression is associated with cell populations undergoing cell fate changes, suggesting that the pathway is active in the cervical cancer transformation zone where cells typically undergo physiologic metaplasia with replacement of columnar cells in the endocervical canal with mature squamous cells which ectocervix is comprised of [4]. It has been postulated that HPV infection may modulate Notch activity, possibly through upregulation of *Jagged* and/or downregulation of the Notch pathway inhibitors, *Manic Fringe* and *Numb* [4,153,154]. In addition, the detection of intracellular forms of Notch1 in human cervical cancers indicate a complex interplay between Notch signaling and papillomaviruses in the context of tumorigenesis, possibly resulting from activation of the PI3K/Akt/mTOR pathway and upregulation of c-Myc [155]. Ligand engagement of the Notch type I transmembrane precursor heterodimer via a canonical pathway induces at least two subsequent cleavages that ultimately releases the intracellular form that functions as a transcriptional activator. Truncated Notch1 alleles complement HPV E6/E7 function in transforming epithelial cells [156]. Prevention of cleavage at the cell surface with *gamma secretase inhibitors* can effectively block Notch signaling [156].

References

[1] Jemal A, Bray F, Cener MM, et al. Global cancer statistics. CA Cancer J Clin 2011;61:69–90.

[2] Siegel RL, Miller KD, Jemal A. Cancer statistics, 2016. CA Cancer J Clin 2016;66:7–30.

[3] Tewari KS, Monk BJ. Invasive cervical cancer DiSaia PJ, Creasman WT, editors. Clinical gynecologic oncology (8th ed.). Philadelphia, PA: Mosby; 2012.

[4] Tewari KS, Taylor JA, Liao SY, et al. Development and assessment of a general theory of cervical carcinogenesis utilizing a severe combined immunodeficiency murine-human xenograft model. Gynecol Oncol 2000;77:137–48.

[5] Morris M, Eifel PJ, Lu J, et al. Pelvic radiation with concurrent chemotherapy compared with pelvic and para-aortic radiation for high-risk cervical cancer. N Engl J Med 1999;340:1137–43.

[6] Rose PG, Bundy BN, Watkins EB, et al. Concurrent cisplatin-based radiotherapy and chemotherapy for locally advanced cervical cancer. N Engl J Med 1999;340:1144–53.

[7] Keys HM, Bundy BN, Stehman FB, et al. Cisplatin, radiation, and adjuvant hysterectomy compared with radiation and adjuvant hysterectomy for bulky stage IB cervical carcinoma. N Engl J Med 1999;340:1154–61.

[8] Whitney CW, Sause W, Bundy BN, et al. Randomized comparison of fluorouracil plus cisplatin versus hydroxyurea as an adjunct to radiation therapy in stage IIB–IVA carcinoma of the cervix with negative para-aortic lymph nodes: a Gynecologic Oncology Group and Southwest Oncology Group study. J Clin Oncol 1999;17:1339–48.

[9] Peters III WA, Liu PY, Barrett II RJ, et al. Concurrent chemotherapy and pelvic radiation therapy compared with pelvic radiation therapy alone as adjuvant therapy after radical surgery in high-risk early-stage cancer of the cervix. J Clin Oncol 2000;18:1606–13.

[10] NIH News Advisory. NCI issues Clinical Announcement on cervical cancer: chemotherapy plus radiation improves survival. Embargoed for release. Monday, February 22, 1999, 10 a.m. EST.

[11] Monk BJ, Tewari KS, Koh WJ. Multi-modality therapy for locally advanced cervical carcinoma: state of the art and future directions. J Clin Oncol 2007;25:2952–65.

[12] Bonomi P, Blessing JA, Stehman FB, et al. Randomized trial of three cisplatin dose schedules in squamous-cell carcinoma of the cervix: a Gynecologic Oncology Group study. J Clin Oncol 1985;3:1970–85.

[13] Thigpen JT, Blessing JA, DiSaia PJ, et al. A randomized comparison of a rapid versus prolonged (24 hr) infusion of cisplatin in therapy of squamous cell carcinoma of the uterine cervix: a Gynecologic Oncology Group study. Gynecol Oncol 1989;32:198–202.

[14] McGuire III WP, Arseneau J, Blessing JA, et al. A randomized comparative trial of carboplatin and iproplatin in advanced squamous carcinoma of the uterine cervix: a Gynecologic Oncology Group study. J Clin Oncol 1989;7:1462–8.

[15] Omura GA, Blessing JA, Vaccarello L, et al. Randomized trial of cisplatin versus cisplatin plus mitolactol versus cisplatin plus ifosfamide in advanced squamous carcinoma of the cervix: a Gynecologic Oncology Group study. J Clin Oncol 1997;15:165–71.

[16] Bloss JD, Blessing JA, Behrens BC, et al. Randomized trial of cisplatin and ifosfamide with or without bleomycin in squamous carcinoma of the cervix: a Gynecologic Oncology Group study. J Clin Oncol 2002;20:1832–7.

[17] Moore DH, Blessing JA, McQuellon RP, et al. Phase III study of cisplatin with or without paclitaxel in stage IVB, recurrent, or persistent squamous cell carcinoma of the cervix: a Gynecologic Oncology Group study. J Clin Oncol 2004;22:3113–19.

[18] Long III HJ, Bundy BN, Grendys Jr EC, et al. Randomized phase III trial of cisplatin with or without topotecan in carcinoma of the uterine cervix: a Gynecologic Oncology Group study. J Clin Oncol 2005;23:4626–33.

[19] Monk BJ, Sill MW, McMeekin DS, et al. Phase III trial of four cisplatin-containing doublet combinations in stage IVB, recurrent, or persistent cervical carcinoma: a Gynecologic Oncology Group study. J Clin Oncol 2009;27:4649–55.

[20] Rose PG, Blessing JA, Gershenson DM, McGhee R. Paclitaxel and cisplatin as first-line therapy in recurrent or advanced squamous cell carcinoma of the cervix: a Gynecologic Oncology Group study. J Clin Oncol 1999;17:2676–80.

[21] Fiorica J, Holloway R, Ndubisi B, et al. Phase II trial of topotecan and cisplatin in persistent or recurrent squamous and nonsquamous carcinomas of the cervix. Gynecol Oncol 2002;85:89–94.

[22] Morris M, Blessing JA, Monk BJ, et al. Phase II study of cisplatin and vinorelbine in squamous cell carcinoma of the cervix: a Gynecologic Oncology Group study. J Clin Oncol 2004;22:3340–4.

[23] Brewer CA, Blessing JA, Nagourney RA, McMeekin DS, Lele S, Zweizig SL. Cisplatin plus gemcitabine in previously treated squamous cell carcinoma of the cervix: a phase II study of the Gynecologic Oncology Group. Gynecol Oncol 2006;100:385–8.

[24] Tewari KS, Monk BJ. Gynecologic Oncology Group trials of chemotherapy for metastatic and recurrent cervical carcinoma. Curr Oncol Rep 2005;7:419–34.

[25] Tewari KS, Monk BJ. Recent achievements and future developments in advanced and recurrent cervical cancer: trials of the Gynecologic Oncology Group. Semin Oncol 2009;36:170–80.

[26] Tewari KS. Patients with metastatic/recurrent cervical cancer should be treated with cisplatin plus paclitaxel. Expert Panel Clin Ovarian Cancer 2011;4:90–3.

[27] Tewari KS. A critical need for reappraisal of therapeutic options for women with metastatic and recurrent cervical carcinoma: commentary on Gynecologic Oncology Group protocol 204. Am J Hematol Oncol 2010;9:31–4.

[28] Tewari KS, Monk BJ. Beyond platinum for metastatic and recurrent carcinoma of the cervix. Onkologie 2009;32:552–4.

[29] Tewari KS, Monk BJ. The rationale for the use of non-platinum chemotherapy doublets for metastatic and recurrent cervical carcinoma. Clin Adv Hematol Oncol 2010;8:108–15.

[30] Neufeld G, Cohen T, Gengrinovitch S, Poltorak Z. Vascular endothelial growth factor (VEGF) and its receptors. FASEB J 1999;13:9–22.

[31] Koch S, Claesson-Welsh L. Signal transduction by vascular endothelial growth factor receptors. Cold Spring Harb Perspect Med 2012;2:a006502.

[32] Goldman EE. Studien zur Biologie der bosartigen Neublidungen. Bruns Beitrage Klimsche Chirurgie 1911;72:1.

[33] Lewis WH. The vascular patterns of tumors. Bulletin of the Johns Hopkins Hospital 1927;41:156.

[34] Koller O. The vascular patterns of cervical cancer. Acta Unio Internationale Contre Cancrum 1959;15:375.

[35] Kolstad P. Vascular changes in cervical intraepithelial neoplasia and invasive cervical carcinoma. Clin Obstet Gynecol 1983;4:938–48.

[36] Sillman F, Boyce J, Fruchter R. The significance of atypical vessels and neovascularization in cervical neoplasia. Am J Obstet Gynecol 1981;139:154–9.

[37] Wiggins D, Granai CO, Steinhoff MM, Calabresi P. Tumor angiogenesis as a prognostic factor in cervical carcinoma. Gynecol Oncol 1995;56:353–6.

[38] Schlenger K, Hockel M, Mitze M, et al. Tumor vascularity: a novel prognostic factor in advanced cervical carcinoma. Gynecol Oncol 1995;59:57–65.

[39] Dinh TV, Hannigan EV, Smith ER, et al. Tumor angiogenesis as a predictor of recurrence in stage Ib squamous cell carcinoma of the cervix. Obstet Gynecol 1996;87:751–4.

[40] Obermair A, Bancher-Todesca D, Bilgi S, et al. Correlation of vascular endothelial growth factor expression and microvessel density in cervical intraepithelial neoplasia. J Natl Cancer Inst 1997;89:1212–17.

[41] Obermair A, Wanner C, Bilgi S, et al. Tumor angiogenesis in stage IB cervical cancer: correlation of microvessel density with survival. Am J Obstet Gynecol 1998;178:314–19.

[42] Tang X, Zhang Q, Nishitani J, et al. Overexpression of human papillomavirus type 16 oncoproteins enhances hypoxia-inducible factor 1α protein accumulation and vascular endothelial growth factor expression in human cervical carcinoma cells. Clin Cancer Res 2007;13:2568–76.

[43] Lopez-Ocejo O, Viloria-Petit A, Bequet-Romero M, et al. Oncogenes and tumor angiogenesis: the HPV-16 E6 oncoprotein activates the vascular endothelial growth factor (VEGF) gene promoter in a p53 independent manner. Oncogene 2000;19:4611–20.

[44] Walker J, Smiley LC, Ingram D, Roman A. Expression of human papillomavirus type 16 E7 is sufficient to significantly increase expression of angiogenic factors but is not sufficient to induce endothelial cell migration. Virology 2011;410:283–90.

[45] Killough JH, Magill GB, Smith RC. The treatment of amebiasis with fumagillin. Science (Washington, DC) 1952;15:71–2.

[46] Ingber D, Fujita T, Kishimoto S, et al. Synthetic analogues of fumagillin that inhibit angiogenesis and suppress tumor growth. Nature (Lond.) 1990;348:555–7.

[47] Sipos EP, Tamargo RJ, Weingart JD, Brem H. Inhibition of tumor angiogenesis. Ann NY Acad Sci 1994;732:263–72.

[48] Brem H, Ingber D, Blood CH, et al. Suppression of tumor metastasis by angiogenesis inhibition. Surg Forum 1991;42:439–41.

[49] Kusaka M, Sudo K, Fujita T, et al. Potent anti-angiogenic action of AGM-14709: comparison to the fumagillin parent. Biochem Biophys Res Commun 1991;174:1070–6.

[50] Kudelka AP, Levy T, Verschraegen CF, et al. A phase I study of TNP-470 administered to patients with advanced squamous cell cancer of the cervix. Clin Cancer Res 1997;3:1501–5.

[51] Kudelka AP, Verschraegen CF, Loyer E. Complete remission of metastatic cervical cancer with the angiogenesis inhibitor TNP-470. N Engl J Med 1998;338:991–2.

[52] Wright JD, Viviano D, Powell MA, Gibb RK, Mutch DG, Grigsby PW, et al. Bevacizumab combination therapy in heavily pretreated, recurrent cervical cancer. Gynecol Oncol 2006;103:489–93.

[53] Monk BJ, Sill MW, Burger RA, Gray HJ, Buekers TE, Roman LD. Phase II trial of bevacizumab in the treatment of persistent or recurrent squamous cell carcinoma of the cervix: a Gynecologic Oncology Group study. J Clin Oncol 2009;27:1069–74.

[54] Monk BJ, Lopez LM, Zarba JJ, et al. Phase II, open-label study of pazopanib or lapatinib monotherapy compared with pazopanib plus lapatinib combination therapy in patients with advanced and recurrent cervical cancer. J Clin Oncol 2010;28:3562–9.

[55] Monk BJ, Pandite LN. Survival data from a phase II, open-label study of pazopanib or lapatinib monotherapy in patients with advanced and recurrent cervical cancer. J Clin Oncol 2011;29:4845.

[56] Miller DS, Belssing JA, Bodurka DC, et al. Evaluation of pemetrexed (Alimta, LY231514) as second line chemotherapy in persistent or recurrent carcinoma of the cervix: a phase II study of the Gynecologic Oncology Group. Gynecol Oncol 2008;110:65–70.

[57] Miller DS, Blessing JA, Ramondetta LM, et al. Pemetrexed and cisplatin for the treatment of advanced, persistent, or recurrent carcinoma of the cervix: a limited access phase II trial of the Gynecologic Oncology Group. J Clin Oncol 2014;32:2744–9.

[58] Monk BJ, Miller DS, Tewari KS. Reply to W.A.A. Tjalma and J.S. Grossman. J Clin Oncol 2015;33:966.

[59] Symonds RP, Davidson SE, Chan S, Reed NS, McMahon T, Rai D, et al. Beatson west of Scotland Cancer Centre. SCOTCERV: a phase II trial of docetaxel and gemcitabine as second line chemotherapy in cervical cancer. Gynecol Oncol 2011;123:105–9.

[60] Bahadori HR, Green MR, Catapano CV. Synergistic interaction between topotecan and microtubule-interfering agents. Cancer Chemother Pharmacol 2001;48:186–96.

[61] Tiersten AD, Selleck MJ, Hershman DL, et al. Phase II study of topotecan and paclitaxel for recurrent, persistent, or metastatic cervical carcinoma. Gynecol Oncol 2004;92:635–8.

[62] Tewari KS, Sill MW, Long III HJ, et al. Improved survival with bevacizumab in advanced cervical cancer. N Engl J Med 2014;370:734–43.

[63] Moore DH, Tian C, Monk BJ, et al. Prognostic factors for response to cisplatin-based chemotherapy in advanced cervical carcinoma: a Gynecologic Oncology Group study. Gynecol Oncol 2010;116:44–9.

[64] NCI Press Release. Bevacizumab significantly improves survival for patients with recurrent and metastatic cervical cancer, <www.cancer.gov/newscenter/newsfromnci/2013/GOG240>.

[65] Penson RT, Huang HQ, Wenzel LB, et al. Bevacizumab for advanced cervical cancer: patient-reported outcomes of a randomised, phase 3 trial (NRG Oncology-Gynecologic Oncology Group protocol 240). Lancet Oncol 2015;16:301–11.

[66] NHS England. National Cancer Drug Fund Prioritisation Scores—Update. Drug: Bevacziumab. Indication: 1st line treatment recurrent or metastatic cervical cancer in combination with carboplatin and paclitaxel. February 2014 Decision Summary.

[67] Campbell D. Women with advanced cervical cancer in England to get Avastin, <www.the-guardian.com>, March 5, 2014.

[68] Genentech Press Release. FDA grants Genentech's Avastin priority review for certain types of cervical cancer, <www.gene.com>, July 14, 2014.

[69] FDA News Release. FDA approves Avastin to treat patients with aggressive and late- stage cervical cancer. August 14, 2014.

[70] NCCN Clinical Practice Guidelines in Oncology (NCCN Guidelines). Cervical Cancer Version 1.2014 NCCN.org.

[71] Tewari KS, Sill MW, Long HJ III, et al. Final protocol-specified overall survival and updated toxicity analysis in the phase III randomized trial of chemotherapy with and without bevacizumab for advanced cervical cancer. European Society of Medical Oncology, Annual Congress, Madrid, Spain, September 26–30, 2014, LBA #26.

[72] Roche Media Release. Roche drug Avastin approved in Switzerland for treatment of advanced cervical cancer. Basel, December 22, 2014.

[73] European Medicines Agency. Committee for Medicinal Products for Human Use (CHMP) Extension of indication variation assessment report. Invented name: Avastin. International non-proprietary name: bevacizumab. February 26, 2015.

[74] Roche Media Release. Roche's Avastin plus chemotherapy receives positive recommendation from CHMP for EU approval in advanced cervical cancer. Basel, February 27, 2015.

[75] Roche Media Release. EU approves Roche's Avastin plus chemotherapy for women with advanced cervical cancer. April 8, 2015.

[76] Minion LE, Bai J, Monk BJ, et al. A Markov model to evaluate cost-effectiveness of antiangiogenesis therapy using bevacizumab in advanced cervical cancer. Gynecol Oncol 2015;137:490–6.

[77] Tewari KS, Sill MW, Monk BJ, et al. Prospective validation of pooled prognostic factors in women with advanced cervical cancer treated with chemotherapy with and without bevacizumab: a NRG Oncology—Gynecologic Oncology Group study. Clin Cancer Res 2015;21:5480–7.

[78] Waggoner Se, Darcy KM, Fuhrman B, et al. Association between cigarette smoking and prognosis in locally advanced cervical carcinoma treated with chemoradiation: a Gynecologic Oncology Group study. Gynecol Oncol 2006;103:853–8.

[79] Waggoner SE, Java JJ, Monk BJ, et al. Impact of smoking on survival among women treated with and without bevacizumab for advanced cervical cancer: a NRG Oncology—Gynecologic Oncology Group study. Gynecol Oncol 2015;137(Suppl. 1):143–4.

[80] Tewari KS, Sill MW, Monk BJ, et al. Impact of circulating tumor cells on overall survival among patients treated with chemotherapy plus bevacizumab for advanced cervical cancer: a NRG Oncology—Gynecologic Oncology Group study. Society of Gynecologic Oncology, Annual Meeting on Women's Cancer, March 28–31, 2015, Chicago, IL, abstract #24.

[81] Willmott L, Java JJ, Monk BJ, et al. Fistulae in women treated with chemotherapy with and without bevacizumab for persistent, recurrent, or metastatic cervical cancer: a Gynecologic Oncology Group study. 15th Biennial Meeting of the International Gynecologic Cancer Society, Melbourne, Australia, November 8–11, 2014, Game Changer's Opening Plenary, abstract# 3.

[82] Seamon LG, Java JJ, Monk BJ, Tewari KS. Prognostic impact of histology in recurrent and metastatic cervical carcinoma: a Gynecologic Oncology Group study. 15th Biennial Meeting of the International Gynecologic Cancer Society, Melbourne, Australia, November 8–11, 2014, Cervix Plenary Session.

[83] Eskander RN, Java JJ, Monk BJ, Tewari KS. Complete responses in the irradiated field following treatment with chemotherapy with and without bevacizumab in advanced cervical cancer: a NRG Oncology—Gynecologic Oncology Group study. Gynecol Oncol 2015;137(Suppl. 1):28.

[84] Eskander RN, Tewari KS. Chemotherapy in the treatment of metastatic, persistent, and recurrent cervical cancer. Curr Opin Obstet Gynecol 2014;26:314–21.

[85] Eskander RN, Tewari KS. Targeting angiogenesis in advanced cervical cancer. Ther Adv Med Oncol 2014;6:280–92.

[86] Eskander RN, Tewari KS. Beyond angiogenesis blockade: targeted therapy for advanced cervical cancer. J Gynecol Oncol 2014;25:249–59.

[87] Krill LS, Adelson JW, Randall LM, Bristow RE. Clinical commentary: medical ethics and the ramifications of equipoise in clinical research. Is a confirmatory trial using a non-bevacizumab containing arm feasible in patients with recurrent cervical cancer. Gynecol Oncol 2014;134:447–9.

[88] Tewari KS, Monk BJ. New strategies in advanced cervical cancer: from angiogenesis blockade to immunotherapy. Clin Cancer Res 2014;20:4349–58.

[89] Tewari KS, Monk BJ. Development of a platform for systemic antiangiogenesis therapy for advanced cervical cancer. Clin Adv Hematol Oncol 2014;12:737–48.

[90] Monk BJ, Tewari KS. Evidence-based therapy for recurrent cervical cancer. J Clin Oncol 2014;32:2687–90.

[91] Krill LS, Tewari KS. Integration of bevacizumab with chemotherapy doublets for advanced cervical cancer. Expert Opin Pharmacother 2015;16:675–83.

[92] Krill LS, Tewari KS. Exploring the therapeutic rationale for angiogenesis blockade in cervical cancer. Clin Ther 2015;37:9–19.

[93] Eskander RN, Tewari KS. Immunotherapy: an evolving paradigm in the treatment of advanced cervical cancer. Clin Ther 2015;37:20–38.

[94] Eskander RN, Tewari KS. Development of bevacizumab in advanced cervical cancer: pharmacodynamic modeling, survival impact and toxicology. Future Oncol 2015;11:909–22.

[95] Liu FW, Cripe J, Tewari KS. Anti-angiogenesis therapy in gynecologic malignancies. Oncology 2015;29:350–60.

[96] Pfaendler KS, Tewari KS. Changing paradigms in the systemic treatment of advanced cervical cancer. Am J Obstet Gynecol 2015 July 26 [Epub ahead of print].

[97] Longoria TC, Tewari KS. Pharmacologic management of advanced cervical cancer: anti-angiogenesis therapy and immunotherapeutic considerations. Drugs 2015;75:1853–65.

[98] Alldredge JK, Tewari KS. Clinical trials of anti-angiogenesis therapy in recurrent/persistent and metastatic cervical cancer. Oncologist 2016;21:576–85.

[99] Symonds RP, Gourley C, Davidson S, et al. Cedirinib combined with carboplatin and paclitaxel in patients with metastatic or recurrent cervix cancer (CIRCCa): a randomised, double blind, placebo-controlled phase 2 trial. Lancet Oncol 2015;16:1515–24.

[100] Tewari KS. Clinical implications for cediranib in advanced cervical cancer. Lancet Oncol 2015 October 13 [Epub ahead of print].

[101] Arseneau J, Blessing JA, Stehman FB, McGehee R. A phase II study of carboplatin in advanced squamous cell carcinoma of the cervix (a Gynecologic Oncology Group study). Invest New Drugs 1986;4:187–91.

[102] Weiss GR, Green S, Hannigan EV, et al. A phase II trial of carboplatin for recurrent or metastatic squamous carcinoma of the uterine cervix: a Southwest Oncology Group study. Gynecol Oncol 1990;39:332–6.

[103] Kitagawa R, Katsumata N, Yamanaka Y, et al. Phase II trial of paclitaxel and carboplatin in patients with recurrent or metastatic cervical carcinoma. Proc Am Soc Clin Oncol 2004;23:5048.

[104] Kitagawa R, Katsumata N, Ando M, et al. A multi-institutional phase II trial of paclitaxel and carboplatin in the treatment of advanced or recurrent cervical cancer. Gynecol Oncol 2012;125:307–11.

[105] Takekuma M, Hirashima Y, Ito K, et al. Phase II trial of paclitaxel and nedaplatin in patients with advanced/recurrent uterine cervical cancer: a Kansai Clinical Oncology Group study. Gynecol Oncol 2012;126:341–5.

[106] Kitagawa R, Katsumata N, Shibata T, et al. Paclitaxel plus carboplatin versus paclitaxel plus cisplatin in metastatic or recurrent cervical cancer: the open-label randomized phase III trial JCOG 0505. J Clin Oncol 2015;33:2129–35.

[107] Cascone T, Heymach JV. Targeting the angiopoietin/Tie2 pathway: cutting tumor vessels with a double-edged sword? J Clin Oncol 2012;30(4):441.

[108] McKeage MJ, Baguley BC. Disrupting established tumor blood vessels. An emerging therapeutic strategy for cancer. Cancer 2010;116:1859–71.

[109] Stevanovic S, Draper LM, Langhan MM, et al. Complete regression of metastatic cervical cancer after treatment with human papillomavirus-targeted tumor-infiltrating T cells. J Clin Oncol 2015;33:1543–50.

[110] Wallecha A, Maciag PC, Rivera S, et al. Construction and characterization of an attenuated Listeria monocytogenes strain for clinical use in cancer immunotherapy. Clin Vaccine Immunol 2009;16:96–103.

[111] Wallecha A, French C, Petit R, et al. Lm-LLO-based immunotherapies and HPV-associated disease. J Oncol 2012;2012:542851.

[112] Basu P, Mehta AO, Jain MM, et al. ADXS11-001 immunotherapy targeting HPV-E7: final results from a phase 2 study in Indian women with recurrent cervical cancer. J Clin Oncol 2014;32:52. [Suppl.; abstract 5610].

[113] Huh WK, Dizon DS, Powell MA, et al. ADXS11-001 immunotherapy in squamous or non-squamous persistent/recurrent metastatic cervical cancer: Results from stage 1 of the phase II GOG/NRG0265 study. J Clin Oncol 2016;34 (suppl; abstr 5516).

[114] Hodi FS, O'Day SJ, McDermott DF, et al. Improved survival with ipilimumab in patients with metastatic melanoma. N Engl J Med 2010;363:711–23.

[115] Topalian SL, Hodi FS, Brahmer JR, et al. Safety, activity, and immune correlates of anti-PD-1 antibody in cancer. N Engl J Med 2012;366:2443–54.

[116] Brahmer JR, Tykodi SS, Chow LQ, et al. Safety and activity of anti-PD-L1 antibody in patients with advanced cancer. N Engl J Med 2012;366:2455–65.

[117] Wolchok JD, Kluger H, Callahan MK, et al. Nivolumab plus ipilimumab in advanced melanoma. N Engl J Med 2013;369:122–33.

[118] Robert C, Long GV, Brady B, et al. Nivolumab in previously untreated melanoma without BRAF mutation. N Engl J Med 2015;372:320–30.

[119] Postow MA, Chesney J, Pavlick AC, et al. Nivolumab and ipilimumab versus ipilimumab in untreated melanoma. N Engl J Med 2015;372:2006–17.

[120] Larkin J, Chiarion-Sileni V, Gonzalez R, et al. Combined nivolumab and ipilimumab or monotherapy in untreated melanoma. N Engl J Med 2015;373:23–34.

[121] Robert C, Schachter J, Long GV, et al. Pembrolizumab versus ipilimumab in advanced melanoma. N Engl J Med 2015;372:2521–32.

[122] Zheng P, Zhou Z. Human cancer immunotherapy with PD-1/PD-L1 blockade. Biomark Cancer 2015;7(Suppl. 2):15–18.

[123] Rice AE, Latchman YE, Balint JP, et al. An HPV-E6/E7 immunotherapy plus PD-1 checkpoint inhibition results in tumor regression and reduction in PD-L1 expression. Cancer Gene Ther 2015;22:454–62.

[124] Yang W, Song Y, Lu YL, et al. Increased expression of programmed death (PD)-1 and its ligand PD-L1 correlates with impaired cell-mediated immunity in high-risk human papillomavirus-related cervical intraepithelial neoplasia. Immunology 2013;139:513–22.

[125] Karim R, Jordanova ES, Piersma SJ, et al. Tumor-expressed B7-H1 and B7-DC in relation to PD-1+ T-cell infiltration and survival of patients with cervical carcinoma. Clin Cancer Res 2009;15:6341–7.

[126] Hereen AM, Koster BD, Samuels S, et al. High and interrelated rates of PD-L1+ CD14+ antigen-presenting cells and regulatory T cells make the microenvironment of metastatic lymph nodes from patients with cervical cancer.. Cancer Immunol Res 2015;3:48–58.

[127] Frenel J-S, Le Tourneau C, ÓNeil B, et al. Pembrolizumab in patients with advanced cervical cancer: Preliminary results from the phase 1b KEYNOTE-028 study. J Clin Oncol 2016;34(suppl; abstr).

[128] Huehls AM, Coupet TA, Stentman CL. Bispecific T-cell engagers for cancer immunotherapy. Immunol Cell Biol 2015;93:290–6.

[129] Feldman SA, Assadipour Y, Kriley I, et al. Adoptive cell therapy—tumor-infiltrating lymphocytes, T-cell receptors, and chimeric antigen receptors. Semin Oncol 2015;42:626–39.

[130] Constantino J, Gomes C, Falcao A, et al. Antitumor dendritic cell-based vaccines: lessons from 20 years of clinical trials and future perspectives. Transl Res 2015 August 3 [Epub ahead of print].

[131] Schirrmacher V. Oncolytic Newcastle disease virus as a prospective anti-cancer therapy. A biologic agent with potential to break therapy resistance. Expert Opin Biol Ther 2015;5:1–15.

[132] Fan Y, Moon JJ. Nanoparticle drug delivery systems designed to improve cancer vaccines and immunotherapy. Vaccines 2015;3:662–85.

[133] Akagi K, Li J, Broutian TR, et al. Genome-wide analysis of HPV integration in human cancers reveals recurrent, focal genomic instability. Genome Res 2014;24:185–99.

[134] Ojesina AI, Lichtenstein L, Freeman SS, et al. Landscape of genomic alterations in cervical carcinomas. Nature 2014;506:371–5.

[135] Rusan M, Li YY, Hammerman PS. Genomic landscape of human papillomavirus-associated cancers. Clin Cancer Res 2015;21:1–9.

[136] Esteller M. Epigenetics in cancer. N Engl J Med 2008;358:1148–59.

[137] Vici P, Mariani L, Pizzuti L, et al. Emerging biological treatments for uterine cervical carcinoma. J Cancer 2014;5:86–9.

[138] Clarke MA, Wenzensen N, Mirabello L, et al. Human papillomavirus DNA methylation as a potential biomarker for cervical cancer. Cancer Epidemiol Biomarkers Prev 2012;21:2125–37.

[139] Dawson MA, Kouozarides T, Huntly BJP. Targeting epigenetic readers in cancer. N Engl J Med 2012;367:647–57.

[140] Appleton K, Mackay HJ, Judson I, et al. Phase I and pharmacodynamic trial of the DNA methyltransferase inhibitor decitabine and carboplatin in solid tumors. J Clin Oncol 2007;25:4603–9.

[141] Bohonowych JE, Gopal U, Isaacs JS. Hsp90 as a gatekeeper of tumor angiogenesis: clinical promise and potential pitfalls. J Oncol 2010;2010:412985.

[142] Diaz-Gonzalez Sd M, Deas J, Benitez-Boijseauneau O, et al. Utility of microRNAs and siRNAs in cervical carcinogenesis. Biomed Res Int 2015 2015:374924.

[143] Gu W, McMillan N, Yu C. Silencing of E6/E7 expression in cervical cancer stem-like cells. Methods Mol Biol 2015;1249:173–82.

[144] Osaki M, Okada F, Ochiya T. miRNA therapy targeting cancer stem cells: a new paradigm for cancer treatment and prevention of tumor recurrence. Ther Deliv 2015;6:323–37.

[145] Das S, Somasundarum K. Therapeutic potential of an adenovirus expressing p73 beta, a p53 homologue, against human papilloma virus positive cervical cancer in vitro and in vivo. Cancer Biol Ther 2006;5:210–17.

[146] Molinoio AA, Marsh C, El Dinali M, et al. mTOR as a molecular target in HPV-associated oral and cervical squamous carcinomas. Clin Cancer Res 2012;18:2558–68.

[147] Schwarz JK, Payton JE, Rashmi R, et al. Pathway-specific analysis of gene expression data identifies the PI3K/Akt pathway as a novel therapeutic target in cervical cancer. Clin Cancer Res 2012;18:1464–71.

[148] Zhang L, Wu J, Ling MT, et al. The role of the PI3K/Akt/mTOR signaling pathway in human cancers induced by infection with human papillomaviruses. Mol Cancer 2015;14:1–13.

[149] Lou H, Villagran G, Boland JF, et al. Genome analysis of Latin American cervical cancer: frequent activation of the PIK3CA pathway. Clin Cancer Res 2015 June 16 [Epub ahead of print].

[150] Janku F, Wheler JJ, Westin SN, et al. PI3K/AKT/mTOR inhibitors in patients with breast and gynecologic malignancies harboring PIK3CA mutations. J Clin Oncol 2012;30:777–82.

[151] Tinker AV, Ellard S, Welch S, et al. Phase II study of temsirolimus (CCI-779) in women with recurrent, unresectable, locally advanced or metastatic carcinoma of the cervix. A trial of the NCIC Clinical Trials Group (NCIC CTG IND 199). Gynecol Oncol 2013;180:269–74.

[152] Watkins JA, Irshad S, Grigoriadis A, Tutt ANJ. Genomic scars as biomarkers of homologous recombination deficiency and drug response in breast and ovarian cancers. Breast Cancer Res 2014;16:1–11.

[153] Zagouras P, Stifani S, Blaumueller CM, et al. Alterations in Notch signaling in neoplastic lesions of the human cervix. Proc Natl Acad Sci USA 1995;92:6414–18.

[154] Veeraraghavalu K, Pett M, Kumar RV, et al. Papillomavirus-mediated neoplastic progression is associated with reciprocal changes in Jagged1 and Manic Fringe expression linked to Notch activation. J Virol 2004;78:8687–700.

[155] Maliekal TT, Bajaj J, Giri V, et al. The role of Notch signaling in human cervical cancer: implications for solid tumors. Oncogene 2008;27:5110–14.

[156] Olsauskas-Kuprys R, Zlobin A, Osipo C. Gamma secretase inhibitors of Notch signaling. Onco Targets Ther 2013;6:943–55.

IV

Vulvar Cancer

CHAPTER 15

New Therapies in Vulvar Cancer

M.H.M. Oonk and A.G.J. van der Zee
University of Groningen, Groningen, The Netherlands

CONTENTS

INTRODUCTION

In vulvar cancer treatment, some major advances have been made during the last decades. Where standard treatment consisted of radical vulvectomy with inguinofemoral lymphadenectomy "en bloc" until late in the 20th century, nowadays treatment is far less radical. Although cure rates were high, postoperative physical and psychological morbidities were very frequently reported [1]. More conservative approaches were suggested in the hope of reducing morbidity without compromising survival. Important steps forward were wide local excision replacing radical vulvectomy [2] and the omission of inguinofemoral lymphadenectomy in patients with microinvasive disease (depth of invasion <1 mm) [3]. Also inguinofemoral lymphadenectomy through separate incisions replaced the "en bloc" procedure [4,5] and patients with well-lateralized vulvar lesions no longer had an indication for bilateral groin treatment [6]. Furthermore, radiotherapy replaced pelvic node dissection in patients with inguinofemoral lymph node involvement [7]. Although all these steps toward less radical treatment made that morbidity decreased, it remained very high in many patients and was predominantly related to groin treatment [8]. Since only about 30% of the patients have lymph node metastases, the remaining 70% of the patients are unlikely to benefit from elective inguinofemoral lymphadenectomy but will be at risk of its significant morbidity. Therefore, a technique was required which could predict lymph node metastases accurately, and that could select patients in whom an inguinofemoral lymphadenectomy could be safely omitted. Such a technique needed to have a very high negative predictive value, since missing lymph node metastases can result in groin recurrences, which are nearly always fatal [9]. Imaging techniques like ultrasound, CT, MRI, and PET did not have a high enough negative predictive value to identify those patients in whom an inguinofemoral lymphadenectomy could be safely omitted [10].

303

Translational Advances in Gynecologic Cancers. DOI: http://dx.doi.org/10.1016/B978-0-12-803741-6.00015-X

One of the most important steps forward in reducing morbidity was the introduction of the sentinel lymph node procedure in well-selected vulvar cancer patients. In this chapter we will describe the advances in vulvar cancer treatment that have been made recently and further efforts to reduce morbidity in vulvar cancer treatment.

SENTINEL LYMPH NODE PROCEDURE

The sentinel lymph node is defined as the first draining lymph node in a lymphatic basin that receives primary lymph flow from a tumor (Fig. 15.1). In 1994, the first results on the application of the sentinel lymph node procedure in vulvar cancer patients were reported by Levenback et al. [11]. Afterward many small single center accuracy studies showed feasibility and accuracy of the sentinel lymph node procedure, especially when performed with the combined technique (use of a radioactive tracer and blue dye) [12]. In 2008, the results of the Groningen International Study on Sentinel nodes in Vulvar cancer (GROINSS-V) were reported. This large international multicenter observational study showed the safety of omitting inguinofemoral lymphadenectomy in patients with a negative sentinel lymph node [13]. Shortly thereafter, GOG-173, the largest accuracy study on application of the sentinel lymph node procedure in vulvar cancer, showed very similar results [14]. Recent studies also showed that the sentinel lymph node procedure is more cost-effective than inguinofemoral lymphadenectomy in early-stage vulvar cancer patients. The total costs were lower because of cheaper initial treatment and less complications, especially lymphedema [15,16]. Currently, the sentinel lymph node

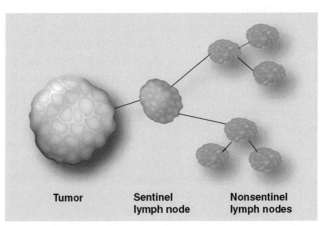

| Tumor | Sentinel lymph node | Nonsentinel lymph nodes |

FIGURE 15.1
Concept of sentinel node procedure.

procedure, performed with the combined technique, is part of standard treatment in well-selected patients with early-stage vulvar cancer (unifocal T1 tumors, smaller than 4 cm, no suspicious groin nodes) [12].

Since all sentinel lymph nodes that are negative on routine hematoxylin–eosin staining are submitted to ultrastaging (multiple sectioning and immunohistochemistry), more and smaller metastasis are detected in these sentinel lymph nodes. In GROINSS-V I 41% of all sentinel lymph node metastases were detected at ultrastaging [13]. The risk of nonsentinel lymph node metastases increased with the size of sentinel-node metastasis: the risk increases from 4.3% for sentinel lymph nodes with isolated tumor cells, up to 62.5% for sentinel lymph nodes metastases >10 mm. No size cutoff seemed to exist below which chances on nonsentinel lymph node metastases were close to zero. Therefore, groin treatment is advised in all patients with metastatic disease in the sentinel node, regardless of the size of metastasis. Furthermore, prognosis of patients with sentinel lymph node metastasis >2 mm is poor compared to those with metastasis ≤2 mm (94.4% vs 69.5%) [17]. Long-term follow-up results of GROINSS-V I (median follow-up 105 months) showed that survival was extremely good for patients with a negative sentinel lymph node, with a 10-year disease-specific survival of 91%, compared to 65% for patients with a positive sentinel lymph node. However, the overall local recurrence rate was very high with 39.5% after 10 years. Although these local recurrences were treated with curative intent, disease-specific survival decreased significantly after local recurrence occurred [18]. Future studies should focus on exploring novel treatment regimens for patients with macrometastatic sentinel-node metastasis (>2 mm) and prevention of local recurrences after primary treatment.

The major disadvantage of the combined technique of the sentinel node procedure is the preoperative injection of the radioactive tracer, which, despite the use of local anesthesia, is a burden for many patients. In 2011, a feasibility study was undertaken with intraoperative near-infrared fluorescence (NIRF) imaging [19]. The great advantage of this technique for sentinel node identification consists of the fact that the optical contrast agent is injected during surgery, under anesthesia. With this technique, injection and lymphoscintigraphy on the day prior to surgery will not be necessary anymore, which hopefully will reduce the psychological stress of these patients. The first clinical results indicated that intraoperative transcutaneous lymphatic mapping using fluorescence is technically feasible in a subgroup of lean vulvar cancer patients. The limited penetration depth of indocyanin green signal made the technique in its current form only feasible in patients with a BMI <25. A more recent study in 12 vulvar cancer patients confirmed feasibility and showed NIRF-guided sentinel-node detection outperforms blue dye staining [20]. In this study, however, indocyanin green was premixed with a radioactive tracer, and thereby passing by the advantage of no preoperative injection.

Recently the use of SPECT/CT was studied in vulvar cancer patients who underwent sentinel lymph node mapping. It was thought that due to its higher spatial resolution and three-dimensional anatomical localization of sentinel lymph nodes, this technique might provide the surgeon important additional information and facilitate intraoperative sentinel node detection. SPECT/CT was found to provide detailed information about the number and anatomical location of sentinel nodes [21,22]. However, more data are needed to confirm its value in the sentinel node procedure in vulvar cancer. It will be hard to demonstrate benefit of the addition of SPECT/CT, since the identification rate of the sentinel node in vulvar cancer patients with the current method is already very high.

RADIOTHERAPY IN VULVAR CANCER PATIENTS WITH A POSITIVE SENTINEL NODE

The GROINSS-V II is currently investigating the safety of radiotherapy to the groins instead of an inguinofemoral lymphadenectomy in patients with a metastatic sentinel node. In this multicenter observational study, patients with a metastatic sentinel node no longer undergo inguinofemoral lymphadenectomy, but are treated with radiotherapy to the groins in a dose of 50 Gy. Stopping rules for the number of groin recurrences watch over the safety in this study.

Earlier studies demonstrated radiotherapy has its value in vulvar cancer treatment. Patients with nodal metastases identified after inguinofemoral lymphadenectomy have better survival when treated with adjuvant radiotherapy to the groins [7,23,24]. Whether patients with only one intranodal metastasis without extracapsular growth also benefit from adjuvant radiotherapy is unclear [24–26]. On the other hand, primary radiotherapy to the groins as an alternative for inguinofemoral lymphadenectomy is controversial. A randomized study by Stehman et al. [27] comparing surgery to radiotherapy in a group of patients with T1–T3 tumors without clinically suspicious nodes, was prematurely stopped on the basis of very adverse findings in the radiotherapy group. Criticism has been made on the radiotherapy dose, the distributions, and techniques in this study [28]. Recently a large retrospective analysis of National Cancer Data Base data showed that the addition of chemotherapy results in a decreased mortality for vulvar cancer patients with metastatic lymph nodes who received radiotherapy [29]. No prospective studies on this subject are available yet.

GROINSS-V II included over 1500 patients at this moment. Final inclusion in this study is expected in the summer of 2016. The final results will be available when all patients completed 2 years of follow-up.

TARGETED THERAPIES

Vulvar cancer originates following two independent pathways [30,31]. The first and most common type occurs in older women and is related to the presence of lichen sclerosus and/or differentiated vulvar intraepithelial neoplasia (dVIN). This type accounts for approximately 80% of all vulvar cancers. Pinto et al. [32] found that dVIN lesions and vulvar squamous cell carcinomas share identical Tp53 mutations, supporting a pathogenetic connection between them. The other type primarily affects younger women. An infection with high-risk HPV, predominantly HPV 16, is involved. This type of carcinoma is associated with usual VIN. The reported proportion of vulvar cancers associated with HPV varies widely, ranging from 15% to 79% [30]. The two different types have different epidemiological, pathological, and molecular features, and should therefore be considered as two separate entities.

Studies investigating the influence of the presence of HPV on the prognosis for the patients are conflicting [33–35]. In head-and-neck cancer it has been shown that HPV-related tumors have a much more favorable prognosis compared to the non-HPV-related tumors [36]. Although mechanisms are not fully understood yet, it seems these HPV-related tumors have a higher response to radiotherapy. Potential underlying mechanisms include impaired DNA repair abilities, differences in activated repopulation-signaling pathways, and cell cycle control mechanisms [37]. Recent studies in oropharyngeal tumors also suggest that increased cytotoxic T-cell-based antitumor response is involved in improved prognosis of patients with HPV-positive tumors [38].

Studies on targeted therapies in vulvar cancer are scarce. Since most data are available on epidermal growth factor receptor (EGFR) and vascular endothelial growth factor (VEGF), we will limit us in this chapter to these two targets.

The EGFR is a transmembrane protein belonging to the family of receptor tyrosine kinases. The binding of a ligand leads to signaling within the cell by activating receptor autophosphorylation through tyrosine kinase activity, triggering a series of intracellular pathways that may result in cancer-cell proliferation, invasion, metastasis, and stimulation of tumor-induced neovascularization [39]. Two classes of EGFR antagonists are available: one that binds to the extracellular domain of the receptor (the anti-EGFR monoclonal antibodies, e.g., cetuximab) and one binding to the intracellular kinase domain of the receptor (selective EGFR tyrosine kinase inhibitors, e.g., erlotinib). Several therapeutic agents against EGFR are now available and have been extensively studied in a large variety of tumors, e.g., non-small-cell lung cancer, colorectal cancer, and head-and-neck cancer [39].

In vulvar cancer, increased expression of EGFR has been observed in 46–72% of the patients [26,40–44]. EGFR overexpression was related to the presence of lymph node metastasis and survival [40,41]. In a study of 183 vulvar cancer patients, Woelber et al. showed in 39% an increase in EGFR gene copy number. This increase was associated with high tumor stage, the number of positive regional lymph nodes, and adverse prognosis in survival analysis. They also observed a reverse relationship between HPV positivity and EGFR amplification/high polysomy, suggesting an HPV-independent role of EGFR copy number increases in vulvar carcinogenesis [26]. Comparable results were found by Dong et al. They found p16 expression was inversely associated with p53 and EGFR expression, supporting HPV-associated and HPV-independent pathogenesis of vulvar cancer and supporting p16 and p53 immunohistochemistry as markers of disease biology [44]. EGFR could be a promising target in selected patients with worse prognosis and HPV-unrelated disease.

Preclinical studies have shown that inhibiting EGFR-mediated signal transduction cascades with ZD1839 (a selective EGFR-tyrosine kinase inhibitor) potentiates the antitumor effect of single and multiple fractions of radiation [45]. Case reports suggest effectiveness of EGFR inhibitors (erlotinib and cetuximab) in patients with advanced stage vulvar cancer or extensive locoregional recurrent disease [46–49].

Only one phase II trial with EGFR inhibition in vulvar cancer patients is available in literature. Horowitz et al. evaluated the efficacy and toxicity of erlotinib, an orally available selective inhibitor of the EGFR tyrosine kinase, in the management of 41 patients with squamous cell carcinoma of the vulva. Two cohorts of patients were treated with erlotinib (150 mg daily); first patients with vulvar lesions amenable to surgery or chemoradiation, and second those with metastatic measurable disease. They observed an overall clinical benefit rate of 67.5% with 27.5% partial response, 40% stable disease, and 17.5% progressive disease. Toxicities were acceptable given the lack of treatment options in these patients [50].

Angiogenesis is a fundamental event in the process of tumor growth and metastatic dissemination. The VEGF pathway is well established as one of the key regulators of this process. The well-established role of VEGF in promoting tumor angiogenesis and the pathogenesis of human cancers has led to the rational design and development of agents that selectively target this pathway [51]. VEGF overexpression has been associated with tumor progression and survival in several tumor types. Recently an anti-VEGF antibody, bevacizumab, was shown to significantly improve survival and response rates in patients with metastatic colorectal cancer [52]. High VEGF expression in vulvar cancer was found to be associated with worse survival [53] and poor tumor differentiation [54]. High expression was also observed in VIN lesions, with VIN3 expressing a significantly higher amount than VIN1 and VIN2

[55]. Also, microvessel density was found to be valuable in identifying the VIN3 lesions at the greatest risk of progression to invasive disease [56]. These data support the idea that VEGF and angiogenesis may play an important (early) role in the development of vulvar cancer. To our knowledge, no data are available on the application of anti-VEGF agents in vulvar cancer patients.

Future studies are needed that investigate the use of targeted therapies in advanced vulvar cancer patients and as radiosensitizers in primary or adjuvant setting.

CONCLUSION

The introduction of the sentinel lymph node procedure was one of the major advances from the last decade in the treatment of early-stage vulvar cancer patients. Long-term survival is very good for patients with a negative sentinel lymph node, but is compromised in case of macrometastasis and locally recurrent disease. Studies are ongoing investigating other techniques for sentinel lymph node detection, like the use of NIRF imaging. Targeted therapy in vulvar cancer patients is still in its infancy. Promising results have been booked with EGFR inhibitors, but large prospective studies are lacking.

RESEARCH AGENDA

Future research should explore novel treatment regimens for patients with macrometastasis in their sentinel lymph node, since their prognosis is significantly worse. Local recurrence is a very frequent event, and these patients have significantly worse prognosis. Therefore, strategies for prevention of locally recurrent disease should be explored. Since vulvar cancer is such a rare disease, studies with new targeted drugs are hard to perform. Basket studies might provide an important opportunity to test therapies for rare cancers, like vulvar cancer.

References

[1] Wills A, Obermair A. A review of complications associated with the surgical treatment of vulvar cancer. Gynecol Oncol 2013;131(2):467–79.

[2] Hacker NF, van der Velden J. Conservative management of early vulvar cancer. Cancer 1993;71(Suppl. 4):1673–7.

[3] Hacker NF, Berek JS, Lagasse LD, Nieberg RK, Leuchter RS. Individualization of treatment for stage I squamous cell vulvar carcinoma. Obstet Gynecol 1984;63(2):155–62.

[4] Hacker NF, Leuchter RS, Berek JS, Castaldo TW, Lagasse LD. Radical vulvectomy and bilateral inguinal lymphadenectomy through separate groin incisions. Obstet Gynecol 1981;58(5):574–9.

[5] Hopkins MP, Reid GC, Morley GW. Radical vulvectomy. The decision for the incision. Cancer 1993;72(3):799–803.

[6] Burger MPM, Hollema H, Bouma J. The side of groin node metastases in unilateral vulvar carcinoma. Int J Gynecol Cancer 1996;6:318–22.

[7] Homesley HD, Bundy BN, Sedlis A, Adcock L. Radiation therapy versus pelvic node resection for carcinoma of the vulva with positive groin nodes. Obstet Gynecol 1986;68(6):733–40.

[8] Gaarenstroom KN, Kenter GG, Trimbos JB, et al. Postoperative complications after vulvectomy and inguinofemoral lymphadenectomy using separate groin incisions. Int J Gynecol Cancer 2003;13(4):522–7.

[9] Cormio G, Loizzi V, Carriero C, Cazzolla A, Putignano G, Selvaggi L. Groin recurrence in carcinoma of the vulva: management and outcome. Eur J Cancer Care 2010;19(3):302–7.

[10] Oonk MH, Hollema H, de Hullu JA, van der Zee AG. Prediction of lymph node metastases in vulvar cancer: a review. Int J Gynecol Cancer 2006;16(3):963–71.

[11] Levenback C, Burke TW, Gershenson DM, Morris M, Malpica A, Ross MI. Intraoperative lymphatic mapping for vulvar cancer. Obstet Gynecol 1994;84(2):163–7.

[12] Covens A, Vella ET, Kennedy EB, Reade CJ, Jimenez W, Le T. Sentinel lymph node biospy in vulvar cancer : systematic review, meta-analysis and guideline recommendations. Gynecol Oncol 2015;137(2):351–61.

[13] Van der Zee AG, Oonk MH, de Hullu JA, et al. Sentinel node dissection is safe in the treatment of early-stage vulvar cancer. J Clin Oncol 2008;26(6):884–9.

[14] Levenback CF, Ali S, Coleman RL, et al. Lymphatic mapping and sentinel lymph node biopsy in women with squamous cell carcinoma of the vulva: a gynecologic oncology group study. J Clin Oncol 2012;30(31):3786–91.

[15] Erickson BK, Divine LM, Leath III CA, Straughn Jr JM. Cost-effectiveness analysis of sentinel lymph node biopsy in the treatment of early-stage vulvar cancer. Int J Gynecol Cancer 2014;24(8):1480–5.

[16] McCann GA, Cohn DE, Jewell EL, Havrilesky LJ. Lymphatic mapping and sentinel lymph node dissection compared to complete lymphadenectomy in the management of early-stage vulvar cancer: a cost–utility analysis. Gynecol Oncol 2015;136(2):300–4.

[17] Oonk MH, van Hemel BM, Hollema H, et al. Size of sentinel-node metastasis and chances of non-sentinel-node involvement and survival in early stage vulvar cancer: results from GROINSS-V, a multicentre observational study. Lancet Oncol 2010;11(7):646–52.

[18] Te Grootenhuis NC, van der Zee AG, van Doorn HC, et al. Sentinel nodes in vulvar cancer: long-term follow-up of the Groningen International Study on Sentinel nodes in Vulvar cancer (GROINSS-V) I. Gynecol Oncol 2016;140(1):8–14.

[19] Crane LM, Themelis G, Arts HJ, et al. Intraoperative near-infrared fluorescence imaging for sentinel lymph node detection in vulvar cancer: first clinical results. Gynecol Oncol 2011;120(2):291–5.

[20] Verbeek FP, Tummers QR, Rietbergen DD, et al. Sentinel lymph node biopsy in vulvar cancer using combined radioactive and fluorescence guidance. Int J Gynecol Cancer 2015;25(6):1086–93.

[21] Klapdor R, Länger F, Gratz KF, Hillemans P, Hertel H. SPECT/CT for SLN dissection in vulvar cancer: improved SLN detection and dissection by preoperative three-dimensional anatomical localisation. Gynecol Oncol 2015;138(3):590–6.

[22] Collarino A, Donswijk ML, van Driel WJ, Stokkel MP, Valdés Olmos RA. The use of SPECT/CT for anatomical mapping of lymphatic drainage in vulvar cancer: possible implications for the extent of inguinal lymph node dissection. Eur J Nucl Med Mol Imag 2015;42(13):2064–71.

[23] Kunos C, Simpkins F, Gibbons H, Tian C, Homesley H. Radiation therapy compared with pelvic node resection for node-positive vulvar cancer: a randomized controlled trial. Obstet Gynecol 2009;114(3):537–46.

[24] Mahner S, Jueckstock J, Hilpert F, et al. Adjuvant therapy in lymph node-positive vulvar cancer: the AGO-CaRE-1 study. J Natl Cancer Inst 2015;107(3) pii:dju426.

[25] Fons G, Groenen SM, Oonk MH, et al. Adjuvant radiotherapy in patients with vulvar cancer and one intra capsular lymph node metastasis is not beneficial. Gynecol Oncol 2009;114(2):343–5.

[26] Woelber L, Hess S, Bohlken H, et al. EGFR gene copy number increase in vulvar carcinomas is linked with poor clinical outcome. J Clin Pathol 2012;65(2):133–9.

[27] Stehman FB, Bundy BN, Thomas G, et al. Groin dissection versus groin radiation in carcinoma of the vulva: a Gynecologic Oncology Group study. Int J Radiat Oncol Biol Phys 1992;24(2):389–96.

[28] McCall AR, Olson MC, Potkul RK. The variation of inguinal lymph node depth in adult women and its importance in planning elective irradiation for vulvar cancer. Cancer 1995;75(9):2286–8.

[29] Gill BS, Bernard ME, Lin JF, et al. Impact of adjuvant chemotherapy with radiation for node-positive vulvar cancer: A National Cancer Data Baser (NCDB) analysis. Gynecol Oncol 2015;137(3):365–72.

[30] Del Pino M, Rodriquez-Carunchio L, Ordi J. Pathways of vulvar intraepithelial neoplasia and squamous cell carcinoma. Histopathology 2013;62(1):161–75.

[31] Van der Avoort IA, Shirango H, Hoevenaars BM, et al. Vulvar squamous cell carcinoma is a multifactorial disease following two separate and independent pathways. Int J Gynecol Pathol 2006;25(1):22–9.

[32] Pinto AP, Miron A, Yassin Y, et al. Differentiated vulvar intraepithelial neoplasia contains Tp53 mutations and is genetically linked to vulvar squamous cell carcinoma. Mod Pathol 2010;23(3):404–12.

[33] Monk BJ, Burger RA, Parham G, Vasilev SA, Wilczynski SP. Prognositic significance of human papillomavirus DNA in vulvar carcinoma. Obstet Gynecol 1995;85(5 Part 1):709–15.

[34] Pinto AP, Schlecht NF, Pintos J, et al. Prognostic significance of lymph node variables and human papillomavirus DNA in invasive vulvar carcinoma. Gynecol Oncol 2004;92(3):856–65.

[35] Alonso I, Fusté V, del Pino M, et al. Does human papillomavirus infection imply a different prognosis in vulvar squamous cell carcinoma? Gynecol Oncol 2011;122(3):509–14.

[36] Chaturvedi AK. Epidemiology and clinical aspects of HPV in head and neck cancers. Head Neck Pathol 2012(6 Suppl. 1):S16–24.

[37] Mirghani H, Amen F, Tao Y, Deutsch E, Levy A. Increased radiosensitivity of HPV-positive head and neck cancers: molecular basis and therapeutic perspectives. Cancer Treat Rev 2015 Oct 13. pii: S0305-7372(15)00172-3. doi:10.1016/j.ctrv.2015.10.001. [Epub ahead of print].

[38] Ward MJ, Thirdborough SM, Mellows T, et al. Tumour-infiltrating lymphocytes predict for outcome in HPV-positive oropharyngeal cancer. Br J Cancer 2014;110(2):489–500.

[39] Ciardiello F, Tortora G. EGFR antagonists in cancer treatment. N Eng J Med 2008;358:1160–74.

[40] Johnson GA, Mannel R, Khalifa M, et al. Epidermal growth factor receptor in vulvar malignancies and its relationship to metastasis and patient survival. Gynecol Oncol 1997;65(3):425–9.

[41] Oonk MH, de Bock GH, van der Veen DJ, et al. EGFR expression is associated with groin node metastases in vulvar cancer, but does not improve their prediction. Gynecol Oncol 2007;104(1):109–13.

[42] Growdon WB, Boisvert SL, Akhavanfard S, et al. Decreased survival in EGFR gene amplified vulvar carcinoma. Gynecol Oncol 2008;111(2):289–97.

[43] De Melo Maia B, Fontes AM, Lavorato-Rocha AM, et al. EGFR expression in vulvar cancer: clinical implications and tumor heterogeneity. Hum Pathol 2014;45(5):917–25.

[44] Dong F, Kojiro S, Borger DR, Growdon WB, Oliva E. Squamous cell carcinoma of the vulva: a subclassification of 97 cases by clinicopathologic, immunohistochemical, and molecular features (p16, p53, and EGFR). Am J Surg Pathol 2015;39(8):1045–53.

[45] Solomon B, Hagekyriakou J, Trivett MK, Stacker SA, McArthur GA, Cullinane C. EGFR blockade with ZD1839 ("Iressa") potentiates the antitumor effects of single and multiple fractions of ionizing radiation in human A431 squamous cell carcinoma. Int J Radiat Oncol Phys 2003;55(3):713–23.

[46] Olawaiye A, Lee LM, Krasner C, Horowitz H. Treatment of squamous cell vulvar cancer with the anti-EGFR tyrosine kinase inhibitor Tarceva. Gynecol Oncol 2007;106(3):628–30.

[47] Bacha OM, Levesque E, Renaud MC, Lalancette M. A case of recurrent vulvar carcinoma treated with erlotinib, an EGFR inhibitor. Eur J Gynaecol Oncol 2011;32(4):423–4.

[48] Bergstrom J, Bidus M, Miles E, Allard J. Use of Cetuximab in combination with cisplatin and adjuvant pelvic radiation for stage IIIB vulvar carcinoma. Case Rep Obstet Gynecol 2015. [Epub 2015 Jul 29].

[49] Richard SD, Krivak TC, Beriwal S, Zorn KK. Recurrent metastatic vulvar carcinoma treated with cisplatin plus cetuximab. Int J Gynecol Cancer 2008;18(5):1132–5.

[50] Horowitz NS, Olawaiye AB, Borger DR, et al. Phase II trial of erlotinib in women with squamous cell carcinoma of the vulva. Gynecol Oncol 2012;127(1):141–6.

[51] Hicklin DJ, Ellis LM. Role of the vascular endothelial growth factor pathway in tumor growth and angiogenesis. J Clin Oncol 2005;23(5):1011–27.

[52] Saltz LB, Clarke S, Diaz-Rubio E, et al. Bevacizumab in combination with oxaliplatin-based chemotherapy as first-line therapy in metastatic colorectal cancer: a randomized phase III study. J Clin Oncol 2008;26(12):2013–19.

[53] Obermair A, Kohlberger P, Bancher-Todesca D, et al. Influence of microvessel density and vascular permeability factor/vascular endothelial growth factor expression on prognosis in vulvar cancer. Gynecol Oncol 1996;63(2):204–9.

[54] Dhakal HP, Nesland JM, Forsund M, Trope CG, Holm R. Primary tumor vascularity, HIF-1α and VEGF expression in vulvar squamous cell carcinomas: their relationships with clinico-pathological characteristics and prognostic impact. BMC Cancer 2013;13:506. http://dx.doi.org/10.1186/1471-2407-13-506.

[55] Bancher-Todesca D, Obermair A, Bilgi S, et al. Angiogenesis in vulvar intraepithelial neoplasia. Gynecol Oncol 1997;64(3):496–500.

[56] Saravanamuthu J, Reid WM, George DS, et al. The role of angiogenesis in vulvar cancer, vulvar intraepithelial neoplasia, and vulvar lichen sclerosus as determined by microvessel density analysis. Gynecol Oncol 2003;89(2):251–8.

V

Fertility and Cancer

CHAPTER 16

Fertility and Cancer: Fertility Preservation and Fertility Sparing Surgery

M.E. Sabatini[1,2]

[1]Harvard Medical School, Boston, MA, United States [2]Massachusetts General Hospital, Boston, MA, United States

CONTENTS

The American Cancer Society estimates that there will be 1.6 million new cases of cancer diagnosed in the United States in 2015, with more than 86,000 of those being in women under the age of 45 years old [1]. At the same time, disease specific 5-year survival has increased from 49% in the 1970s to 68% in the mid-2000s. Concurrent with this trend, the median age of first birth is rising in developed countries [2], thus more and more women will be diagnosed with cancer prior to completing their families. As patients live longer and are starting families later, the consequences of cancer therapy are increasingly important. Infertility has been found to be the most important consequence of cancer treatment among female survivors [3]. The American Society of Clinical Oncology recommends that as part of education and informed consent before cancer therapy, health care providers address the risk of infertility with patients treated during their reproductive years and be prepared to discuss fertility preservation options and/or to refer all patients to reproductive specialists [4].

Normal female reproductive function requires the interaction of a number of systems including the hypothalamus, pituitary, ovaries, uterus, and cervix. Thus damage to any one of these structures can disrupt the recurring cycles of follicular growth, ovulation, and endometrial development that are required for conception to occur. Systemic and or even localized oncologic treatments can impact fertility; however, the management of gynecological cancers can be the most devastating because of the potential acute effects of gynecologic surgery, radiation, and chemotherapy can have on the reproductive organs themselves.

Fertility declines as a woman matures with rapidly declining fertility between 35 and 40 years old. With few exceptions there are few live births in healthy

315

Translational Advances in Gynecologic Cancers. DOI: http://dx.doi.org/10.1016/B978-0-12-803741-6.00016-1

women over 44 years old even with aggressive infertility treatments excepting oocyte or embryo donation. When considering fertility in cancer patients, a woman's age and available family building options must be taken into consideration along with goals for treatment.

While previous sections of the book focus on the primary sites of disease, this chapter will concentrate more on the tissue damage incurred by these three major therapies and possible ways to prevent or mitigate that damage. The fact that cancer therapies are used in combinations makes is difficult to assign toxicity to any particular agent [5]. Additionally because of the rapid change in treatments, we are unable to observe the effects of any agent or combinations of treatments in large populations. As such determining the efficacy of strategies to preserve fertility will also be challenging as they are assessed in the years to come.

SURGERY

Ovarian, Fallopian Tube, and Peritoneal Cancers

Ovarian, fallopian, and peritoneal cancers are among the most feared of gynecological cancer because of the lack of screening and detection methods available and because of the poor associated survival for those diagnosed with late stage disease [6]. Epithelial ovarian cancer represents 95% of ovarian malignancies. Complete or optimal cytoreduction which may include pelvic washings, hysterectomy, bilateral salpingo-oophorectomy cytology of diaphragm, omentectomy, pelvic and para-aortic lymph node dissection, and resection of visible metastases remains the standard of care and has been shown to improve survival [7,8].

Attitudes regarding surgical management of ovarian cancer has shifted somewhat as data have begun to suggest that it may be safe to perform less radical and fertility sparing surgery (FSS). In 1999 Benjamin et al. described 118 cases of stage 1 ovarian cancer with a grossly appearing contralateral ovary. By final pathology, only 2.5% of these cases had microscopic disease on contralateral ovary [9]. This finding suggested that leaving the contralateral ovary could present the opportunity for future fertility in certain women.

If FSS is being considered, surgical staging including washings, unilateral salpingo-oophorectomy, omentectomy, appendectomy, and node biopsies, abdominal exploration, biopsies of suspicious areas, and endometrial biopsy should be done and should be negative. Several retrospective studies describe outcomes of women undergoing FSS for ovarian cancer [10–21] with groups of patients ranging from 11 to 545 patients. Most of these studies included all histological types and grades of cancers. Some were limited to stage 1A and 1C while others included up to stage 3C. With regard to pregnancy and obstetric outcomes data are largely encouraging with conception rates of 38–100% reported [22,23]; however a recent study suggests that FSS is associated with diminished reproductive potential than reported in the general population [24]. Counterintuitively, FSS for early ovarian cancer may not confer a benefit in terms of sexual function or quality of life [25].

FSS must be approached with caution. Based on cumulative data the National Comprehensive Cancer Network 2015 guidelines suggest that FSS may be appropriate for 1A and 1C (but not 1B) and low grade tumors of epithelial, malignant germ cell tumors and malignant sex-cord stromal tumors [26]. The discussions must be balanced in terms of possible need for further surgery, need for chemotherapy, and expectations about fertility outcomes.

Borderline ovarian tumors (BOTs) represent 14–15% of all ovarian neoplasias and the two major subtypes are serous (approximately 70%) and mucinous (approximately 10%). One-third of cases will be diagnosed in women less than 40 years old, and the overwhelming majority of these are stage 1 when they are diagnosed. BOTs are often staged at the time of initial surgery primarily in case the final histologic evaluation shows an invasive carcinoma. Most of these tumors however can be treated with a more conservative surgical approach. Conservative surgery can consist of pelvic washings, omental biopsy, and biopsy of any peritoneal lesions along with either unilateral salpingo-oophorectomy or even cystectomy with frozen section. Appendectomy is performed for mucinous tumors. Because there are discrepancies between frozen and final pathology, patients should be counseled about the possibility of a second more extensive surgery if malignancy is found [27]. In general FSS for BOT is associated with a favorable prognosis from disease and fertility standpoints [28–30].

Endometrial Cancer

Endometrial cancer is the most common gynecologic cancer in the United States with 54,870 new cases and 10,170 death projected for the year 2015 [31]. The median age at diagnosis of endometrial cancer is 62 years; however, 7.1% of women newly diagnosed with endometrial cancer will be under the age of 44 years old. Thus for those women preservation of fertility may be important.

The standard treatment for endometrial cancer is hysterectomy, bilateral salpingo-oophorectomy with or without pelvic and para-aortic lymph node dissection [32]. For women with low risk disease (grade 1 or 2, limited to the endometrium and not high risk histologic type), this treatment confers a high 5 years disease-specific survival [33]. Standard treatment however can impact both fertility and quality of life particularly for premenopausal women, thus more conservative nonsurgical treatments have been reported.

The first case reports of the use of progestin therapy for patients with endometrial hyperplasia or carcinoma in situ was by Kristner in 1959 [34]. Subsequently several other groups reported similar treatment for early stage endometrial cancer [35–38]. Because of the rarity of endometrial cancer in young women, there are no large studies to provide guidelines on fertility sparing treatments for endometrial cancer. Most of the studies are small retrospective studies only reporting on hundreds of cases total which have been reviewed [39–45]. Thus determining who would be an appropriate candidate is still challenging. Based on the current literature the European Society of Gynecological Oncology Task Force for Fertility Preservation as well as others have put for clinical recommendations for fertility sparing management for young women with endometrial cancer [46]. The general consensus is that women offered medical therapy for endometrial cancer should have low risk disease, and specifically grade 1, well-differentiated endometrioid tumors, confined to the endometrium by magnetic resonance imaging (MRI), and after endometrial sampling by dilatation and curettage (D&C) to confirm the tumor grade and histology.

Unfortunately the only way to formally establish the stage and grade of an endometrial cancer is by surgery. In young women grade 1 was the only factor which was a significant predictor for stage 1A disease [38], and thus their recommendation was that these are the only women who should be counseled about the availability of conservative treatment.

Both endometrial biopsy and D&C are possible methods for sampling the endometrium. To date, there are no available reports on prospective studies comparing the two methods. There are retrospective data that suggest that for grade 1 D&C was associated with upgrading the lesion on final hysterectomy specimen of 8.7% compared to 17.4% for those sampled by office pipelle, a finding that was statistically significant [47]. This suggests that for initial diagnostic and management phase D&C would be preferred method. For continued surveillance there is little guidance as to how often one should monitor efficacy of treatment or how often this should be done although a prospective trial has been started to assess for endometrial hyperplasia [48]. Because of the consequences of up or down grading a lesion by pathology, all slides should be examined by at least two pathologists [46].

Myometrial invasion is also important in determining the appropriateness of fertility sparing treatment. Again definitive determination can be made only histologically so patients must be counseled accordingly. Multiple imaging modalities have been studied and analyzed including ultrasound, CT (computed tomography), and MRI with MRI appearing to provide the best accuracy. However in centers with experienced ultrasonographers, efficiency of detection appears to approach that of MRI [46,49].

Unfortunately, there is very little clear guidance as to which progestin treatment is most effective. Thus far medroxyprogesterone acetate (MPA) and megestrol acetate (MA) have been the most widely reported with dosages ranging from 100 to 1200 mg/day for MPA and 40 to 600 mg/day for MA [41,44]. While there is less experience with the levonorgesterol intrauterine device, it is used by some and studies [50] suggest it too can be efficacious. Other medical therapies that will counter the effects of endogenous estrogen such as GnRH analogs, letrozole, and others have also been trialed. Unfortunately, because there is no single agent that has been established, length of treatment is also undetermined; however, studies show 3–4 months and if sampling is not negative by 6 months, late response is not expected and persistent disease may suggest myometrial invasion [46,49].

Several small studies have been done and have suggested that for women only interested in ovarian function and with early stage disease; a hysterectomy can be performed leaving the ovaries in place [51–53].

Cervical Cancer

Traditional treatment for invasive cervical cancer is radical hysterectomy or radiation. By the early 1990s, therapeutic conization was considered appropriate therapy for those with stage 1A1 cancer who wished to preserve fertility [54]. Also in the early 1990s a technique was described for those with early stage invasive cervical cancer. This included laparoscopic pelvic lymphadenectomy followed by radical vaginal trachelectomy [55]. It was modified by others [56,57]. In 1998, Roy and Plante also reported obstetrical outcomes a series of 30 patients who had undergone radical trachelectomy. Median follow up time was 25 months (range 1–79 months). Of the six who had attempted conception there were seven pregnancies, including two term deliveries, two preterm deliveries, two ongoing pregnancies, and one early miscarriage.

This largely established the standard for patient selection that exists today. Appropriate patients must have desire for fertility and be of reproductive age (typically age 40 or less). Lesions should be less than 2 cm and either squamous cell or adenocarcinoma, not high risk lesions such as small cell, and lesions should have limited endocervical extension by colposcopy, and MRI without lymph node metastases. Lymph vascular space invasion within the tumor is a risk factor for recurrence within the lymph nodes but as an isolated finding is not a contraindication for fertility sparing surgery. If positive lymph nodes are identified, the lesion is more extensive than initially assessed or if the endocervical margin is positive by frozen section and radical hysterectomy should be performed [58,59].

Since the initial descriptions of the vaginal procedure, several other surgical approaches have been described including laparoscopic assisted vaginal [60], abdominal [61], and robotic [62]. It is difficult to discern which approach is superior as comparing outcomes because of the variations of techniques, skill levels of surgeons, variations of pathology, etc. There have been a variety of original papers and reviews on outcomes including obstetrical outcomes and recurrence rates. While follow up times vary, recurrence rates have been reported to be between 0% and 5%. Overall pregnancy rates are reported to be from 38% to 76% with premature rupture of membranes and preterm delivery being the most common complications of pregnancy [58,62].

FSS SUMMARY

No matter what disease site, patients who have undergone FSS or conservative surgery should be counseled about risk of recurrence and possible lack of success for fertility pursuits, and need for further surgery. Definitive therapy or surgery once family building is complete should be discussed to avoid risk of recurrence or development of worsening disease.

RADIATION

Radiation has been used in the treatment of tumors and other medical conditions for more than a century. It results in damage to cellular components by ionizing the atoms that make up the cell. The lethality results primarily from DNA damage. Because cancer cells are not as well able to repair DNA damage as healthy cells, they are more subject to cell death from radiation damage. Unfortunately however, despite attempts to reduce the exposure of normal tissues by a variety of methods, some damage to normal tissues is inevitable [64–66].

While the effects of gynecologic surgery are obvious and immediate, the effects of radiation can be equally profound and in some ways more insidious as they can take weeks to years to manifest. Exposure of the cervix and uterus to radiation can have very serious reproductive consequences.

Radiation exposure can result in cervical stenosis and scarring, damage to the endometrium and surrounding myometrium causing impaired implantation, poor myometrial hypertrophy and growth and vascular insufficiency. In general traditional pelvic radiation is considered to be a contraindication to carrying a pregnancy in the future. The degree of damage to the uterus, however, is dependent on both the dose of radiation and the age/developmental stage of the patient [67,68]. In prepubertal girls, doses of radiation of as low as 8.5 Gy can cause damage to uterine growth and blood flow [69,70].

In adults exposure to radiation is associated with increased risk of low birth weight infants and preterm birth and perinatal infant death [71,72], with a positive correlation of poor outcome and total dose of radiation. When stratified by age women treated with radiation premenarchy fared more poorly.

Ovarian transposition can be performed to maintain ovarian function and thus hormonal function and perhaps fertility in appropriate patients. Moving ovaries as high out of field as possible has been described with tagging with clips that can be found by ultrasound [73,74]. In general, most studies have good outcomes in terms of hormonal function and low likelihood of residual cancer [75–78]. Care must be taken to choose correct patient to avoid risk of cancer recurrence [79].

Women anticipating pelvic radiation or who have undergone ovarian transposition can subsequently undergo controlled ovarian hyperstimulation with in vitro fertilization (IVF) and use of a gestational carrier. Typically IVF is done using a vaginal ultrasound probe so this can only be done in centers with experience with transabdominal retrieval. It is also important to note that ovarian reserve may be more limited to both removal of part of the blood supply and distant radiation effects.

CHEMOTHERAPY

Traditional chemotherapy is cytotoxic and kills cells that are growing and dividing. As such they are indiscriminant and destroy any cell that is in the cell cycle while it is being given. Thus chemotherapy results in common side effects such as alopecia, mucositis, and myelosuppression. Newer agents of chemotherapeutics (e.g., monoclonal antibodies) target proteins that are made by cancer cells and result in fewer side effects in general.

Given that the human oocyte population is thought to be nondividing and thus largely quiescent, it is unclear how chemotherapy affects fertility. Perhaps it is acting at an unknown population of germ cells, support cells, or has some other mechanism by which ovarian reserve is diminished [80]. However, there is substantial evidence that traditional chemotherapies are damaging to the ovary and future reproduction [81]. How much damage is done is dependent on the class of medication (with alkylating agents causing the most destruction and plant derivative, antibiotics, and antimetabolites presenting the least), the dose of chemotherapy, the duration of treatment, and a variety of host factors such as age, prior fertility, and starting ovarian reserve which is often unknown.

Targeted chemotherapies represent a largely unknown risk as little data has been accumulated. Bevacizumab has been shown to be associated with a higher rate of primary ovarian insufficiency (POI) in premenopausal women with colon cancer when used in combination with FOLFOX (34%) whereas the rate of POI in women taking FOLFOX alone is much lower (2%). Of those who did experience POI in the combination therapy, 22% had recovery of ovarian function after stopping bevacizumab [82]. The generalizability of this finding to other targeted therapies and other cancers is unknown. Fortunately it appears that women treated with chemotherapy who maintain fertility do not subsequently have increased risk of birth defects or offspring with increased risk of cancers [83].

FERTILITY PRESERVATION EFFORTS AND TECHNIQUES

What fertility options are available to a patient is dependent on a number of factors including age, disease site, and severity, cancer treatments anticipated and even unfortunately insurance and financial resources. The potential impact of cancer therapies on future fertility should be addressed with all women of reproductive age and families of children not yet of reproductive age [4]. Algorithms for approaching these topics have been proposed [84] as seen in Figs. 16.1 and 16.2.

Theoretically, taking ovarian cells out of the cell cycle could provide some protection to ovarian function. Accordingly, some medications such as gonadotropin releasing agonists (GnRHa) are thought to potentially have a protective effect for ovaries during chemotherapy, however the data are mixed [85–88] and GnRHa can have serious side effects [89–91]. Other medications are currently under investigation such as spingosine 1 phosphate [92,93], targeted therapies [94], and selective estrogen receptor modulators [95].

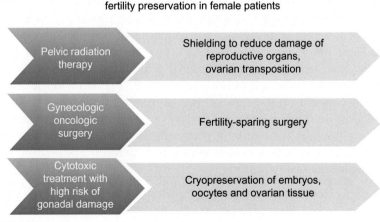

FIGURE 16.1

Fertility-threatening therapies and strategies.

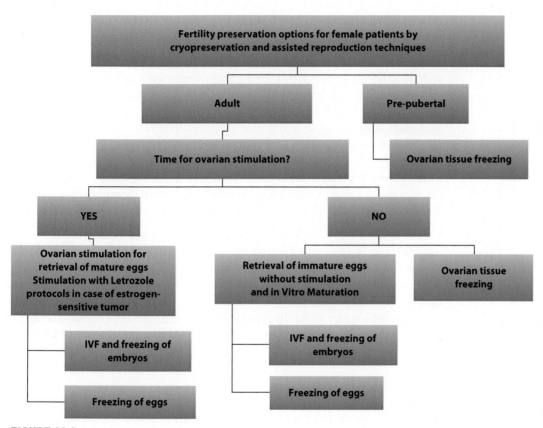

FIGURE 16.2
Fertility preservation options for female patients.

IVF can be employed to freeze either oocytes or embryos for future use. While embryo cryopreservation is the gold standard resulting in an approximately 50% live birth rate [96] it too can be problematic. IVF can only be performed on postpubertal females. Additionally IVF can take 2–6 weeks to complete and can have significant side effects and risks. Oocyte cryopreservation can be used for women who are not partnered, but also requires the process of controlled ovarian hyperstimulation, and outcomes from frozen oocytes are not as well established [97]. For these reasons novel methods need to be explored to help female patients preserve fertility [98,99].

There are several programs worldwide that have reported progress on ovarian tissue freezing. Ovarian tissue freezing can be done both in pre- and post-pubertal females. It is generally a simple laparoscopic procedure that can be scheduled and accomplished quickly. However, in patients with cancer it is important to remember that they can be compromised by a variety of disease-related problems, so careful planning is essential.

Tissue thus obtained can be used in two ways theoretically. First it can be reimplanted into the patient after cancer treatments are over. A number of groups have reported on reimplantation of tissue with resumption of hormonal function and live births [100–104]. Most of these groups have also shown that such surgeries have a very low complication rate; however, it is important to remember that this is still an emerging technology so late term effects, particularly, from reimplantation are not well understood. Additionally, most of these patients had ovarian tissue into which the cryopreserved/thawed tissue was placed so it is difficult to know if they would have conceived without transplantation. American Society for Reproductive Medicine (ASRM) considers ovarian tissue freezing as experimental [105].

Another use of frozen tissue may be the extraction of immature oocytes for in vitro maturation and subsequent fertilization. While still theoretical this may be the only option for those with hematologic malignancies and other cancers that may metastasize to the ovary in which reimplantation may be considered too risky [106]. In general in vitro maturation is considered experimental and currently gives lower pregnancy rates than traditional IVF [107].

ADDITIONAL TOPICS

When considering any conservative therapy for cancer (FSS, ovarian transposition) it is important to have thoroughly evaluated the patient for all risks. For example, prior to conservative medical treatment of endometrial cancer in a young woman, thought should be given to the possibility of Lynch Syndrome and other concurrent cancers. Additionally one must contemplate the patient's predisposition for redeveloping a cancer. For example, if endometrial cancer arose in the setting of an ovulation and obesity a patient should be counseled appropriately about future risks and ways to lower risk.

Those with genetic predisposition to gynecologic cancers (e.g., BRCA mutation carriers) can choose preimplantation genetic diagnosis (PGD) to select embryos that do not carry the mutation. This does result in discarding embryos in a cohort that may otherwise be normal. There are a variety of ethical positions regarding testing the embryos to avoid future disease; however, most patients with hereditary cancer syndromes seem to think it is appropriate to be offered, even if some of them would choose not to use PGD [108,109].

Third-party reproduction (meaning conception with use of gamete, embryo donation, or use of a gestational carrier) is complex socially, legally, ethically, and financially. ASRM has published guidelines for patients to help navigate a process that involves medical evaluation, social and legal counsel, and Food and Drug Administration (FDA) compliance [110]. For those who are unable to preserve ovarian tissue, oocytes or embryos, oocyte donation is an option. The first pregnancy from oocyte donation was reported by Trounson et al. in 1983 [111]. This pregnancy resulted in a miscarriage. The first live birth was reported in 1984 [112]. The legality of oocyte donation varies by country. In the United States, it is regulated by the FDA in the same manner as tissue donation. Excepting in cases of related donors, there is not a genetic link to a child born of oocyte donation, however, in the United States, the intended parent assumes all legal responsibility for the child. Embryo donation is also a possibility. According to the Society of Assisted Reproductive Technology there were more than 20,000 IVF cycles involving donor oocytes or donor embryos in the United States in 2013 [113]. Adoption has been a mainstay of those who are unable to pursue any of the above options who would no less like to pursue family building.

For those who have had hysterectomy, the most widely used method to have a genetic child is to employ a gestational carrier. This was first described in 1985 [114] and involves creating embryos by IVF to obtain oocytes from the intended mother and placing the embryo(s) made into the uterus of another woman. This is a common technique in many places; however, it is banned in some states and countries. The possibility of uterine transplant may be available in the future. This technique was first described in humans in 2002 [115]. The first clinical pregnancy in a patient with a transplanted uterus was described in 2013 [116] and the first live birth occurred on September 4, 2014, to a woman with congenital absence of the uterus and unilateral renal agenesis. The baby was delivered by Cesarean section at 31 weeks gestation for preeclampsia [117]. To date four live births have been reported [118].

The field now known as oncofertility is rapidly growing. There is much hope to those who may want to consider building a family after they have completed cancer treatments. Obviously, however, as the field expands quickly there is a need to provide more educational resources for patients and their providers. Decisions can be liberating but they can also cause distress and anxiety without appropriate understanding of options [119]. Along these lines there are also studies underway to help understand how to best provide patients and families with information that can aid in decision making [120,121]. The hope is that with more breakthroughs and better understanding of both cancer and reproductive physiology the choices will become easier.

References

[1] American Cancer Society. Cancer Facts & Figures 2015. Atlanta (GA): American Cancer Society, <http://www.cancer.org/research/cancerfactsstatistics/cancerfactsfigures2015>; 2015 [accessed 22.08.15].

[2] Finer LB, Philbin JM. Trends in ages at key reproductive transitions in the United States, 1951-2010. Women's Health Issues 2014:e271–9.

[3] Zeltzer EK. Cancer in adolescents and young adults psychosocial aspects. Long-term survivors. Cancer 1993;71(10 Suppl.):3463–8.

[4] Loren AW, Mangu PB, Beck LN, Brennan L, Magdalinski AJ, Partridge AH, et al. Fertility preservation for patients with cancer: American Society of Clinical Oncology clinical practice guideline update. J Clin Oncol 2013;31:2500–10.

[5] Greene MF, Longo DL. Cautious optimism for offspring of women with cancer during pregnancy. N Engl J Med 2015;373:1875–6. [Epub ahead of print] PMID: 26415086.

[6] Heintz AP, Odicino F, Maisonneuve P, Quinn MA, Benedet JL, Creasman WT, et al. Carcinoma of the ovary. Int J Gynaecol Obstet 2006;95:S161–92.

[7] Le T, Adolph A, Krepart GV, Lotocki R, Heywood MS. The benefits of comprehensive surgical staging in the management of early-stage epithelial ovarian carcinoma. Gynecol Oncol 2002;85:351–5.

[8] Elattar A, Bryant A, Winter-Roach BA, Hatem M, Naik R. Optimal primary surgical treatment for advanced epithelial ovarian cancer. Cochrane Database Syst Rev 2011;8:CD007565. http://dx.doi.org/10.1002/14651858.CD007565.pub2.

[9] Benjamin I, Morgan MA, Rubin SC. Occult bilateral involvement in stage I epithelial ovarian cancer. Gynecol Oncol 1999;72:288–91.

[10] Zanetta G, Chiari S, Rota S, Bratina G, Maneo A, Torri V, et al. Conservative surgery for stage I ovarian carcinoma in women of childbearing age. Br J Obstet Gynaecol 1997;104(9):1030–5.

[11] Schilder JM, Thompson AM, DePriest PD, Ueland FR, Cibull ML, Kryscio RJ, et al. Outcome of reproductive age women with stage IA or IC invasive epithelial ovarian cancer treated with fertility-sparing therapy. Gynecol Oncol 2002;87(1):1–7.

[12] Morice P, Leblanc E, Rey A, Baron M, Querleu D, Blanchot J, et al. Conservative treatment in epithelial ovarian cancer: results of a multicentre study of the GCCLCC (Groupe des Chirurgiens de Centre de Lutte Contre le Cancer) and SFOG (Société Francaise d'Oncologie Gynécologique). Hum Reprod 2005;20(5):1379–85.

[13] Borgfeldt C, Iosif C, Måsbäck A. Fertility-sparing surgery and outcome in fertile women with ovarian borderline tumors and epithelial invasive ovarian cancer. Eur J Obstet Gynecol Reprod Biol 2007;134(1):110–14.

[14] Park JY, Kim DY, Suh DS, Kim JH, Kim YM, Kim YT, et al. Outcomes of fertility-sparing surgery for invasive epithelial ovarian cancer: oncologic safety and reproductive outcomes. Gynecol Oncol 2008;110(3):345–53.

[15] Anchezar JP, Sardi J, Soderini A. Long-term follow-up results of fertility sparing surgery in patients with epithelial ovarian cancer. J Surg Oncol 2009;100(1):55–8.

[16] Kwon YS, Hahn HS, Kim TJ, Lee IH, Lim KT, Lee KH, et al. Fertility preservation in patients with early epithelial ovarian cancer. J Gynecol Oncol 2009;20(1):44–7.

[17] Schlaerth AC, Chi DS, Poynor EA, Barakat RR, Brown CL. Long-term survival after fertility-sparing surgery for epithelial ovarian cancer. Int J Gynecol Cancer 2009;19(7):1199–204.

[18] Kajiyama H, Shibata K, Suzuki S, Ino K, Nawa A, Kawai M, et al. Fertility-sparing surgery in young women with invasive epithelial ovarian cancer. Eur J Surg Oncol 2010;36(4):404–8.

[19] Satoh T, Hatae M, Watanabe Y, Yaegashi N, Ishiko O, Kodama S, et al. Outcomes of fertility-sparing surgery for stage I epithelial ovarian cancer: a proposal for patient selection. J Clin Oncol 2010;28(10):1727–32.

[20] Fruscio R, Corso S, Ceppi L, Garavaglia D, Garbi A, Floriani I, et al. Conservative management of early-stage epithelial ovarian cancer: results of a large retrospective series. Ann Oncol 2013;24(1):138–44.

[21] Bentivegna E, Fruscio R, Roussin S, Ceppi L, Satoh T, Kajiyama H, et al. Long-term follow-up of patients with an isolated ovarian recurrence after conservative treatment of epithelial ovarian cancer: review of the results of an international multicenter study comprising 545 patients. Fertil Steril 2015;104(5):1319–24.

[22] Zapardiel I, Diestro MD, Aletti G. Conservative treatment of early stage ovarian cancer: oncological and fertility outcomes. Eur J Surg Oncol 2014;40(4):387–93.

[23] Ditto A, Martinelli F, Lorusso D, Haeusler E, Carcangiu M, Raspagliesi F. Fertility sparing surgery in early stage epithelial ovarian cancer. J Gynecol Oncol 2014;25(4):320–7.

[24] Letourneau J, Chan J, Salem W, Chan SW, Shah M, Ebbel E, et al. Fertility sparing surgery for localized ovarian cancers maintains an ability to conceive, but is associated with diminished reproductive potential. J. Surg Oncol 2015;112(1):26–30.

[25] Chan JL, Letourneau J, Salem W, Cil AP, Chan SW, Chen LM, et al. Sexual satisfaction and quality of life in survivors of localized cervical and ovarian cancers following fertility-sparing surgery. Gynecol Oncol 2015;139(1):141–7.

[26] NCCN Clinical Practice Guidelines in Oncology. Ovarian cancer including fallopian tube cancer and primary peritoneal cancer, Version 2.2015. <http://www.nccn.org/professionals/physician_gls/pdf/ovarian.pdf>; [accessed 10.10.15].

[27] Geomini P, Bremer G, Kruitwagen R, Mol BW. Diagnostic accuracy of frozen section diagnosis of the adnexal mass: a metaanalysis. Gynecol Oncol 2005;96(1):1–9.

[28] Zanetta G, Rota S, Chiari S, Bonazzi C, Bratina G, Mangioni C. Behavior of borderline tumors with particular interest to persistence, recurrence, and progression to invasive carcinoma: a prospective study. J Clin Oncol 2001;19(10):2658–64.

[29] Daraï E, Fauvet R, Uzan C, Gouy S, Duvillard P, Morice P. Fertility and borderline ovarian tumor: a systematic review of conservative management, risk of recurrence and alternative options. Hum Reprod Update 2013;19(2):151–66.

[30] Helpman L, Beiner ME, Aviel-Ronen S, Perri T, Hogen L, Jakobson-Setton A, et al. Safety of ovarian conservation and fertility preservation in advanced borderline ovarian tumors. Fertil Steril 2015;104(1):138–44.

[31] National Cancer Institute Surveillance, Epidemiology and End Results Program. <http://seer.cancer.gov/statistics/summaries.html>; [accessed 15.11.15].

[32] Pecorelli S. Revised FIGO staging for carcinoma of the vulva, cervix, and endometrium. Int J Gynaecol Obstet 2009;105(2):103–4.

[33] Lewin SN, Herzog TJ, Barrena Medel NI, Deutsch I, Burke WM, Sun X, et al. Comparative performance of the 2009 international Federation of gynecology and obstetrics' staging system for uterine corpus cancer1. Obstet Gynecol 2010;116(5):1141.

[34] Kistner RW. Histological effects of progestins on hyperplasia and carcinoma in situ of the endometrium. Cancer 1959;12:1106–22.

[35] Kimmig R, Strowitzki T, Müller-Höcker J, Kürzl R, Korell M, Hepp H. Conservative treatment of endometrial cancer permitting subsequent triplet pregnancy. Gynecol Oncol 1995;58(2):255–7.

[36] Kim YB, Holschneider CH, Ghosh K, Nieberg RK, Montz FJ. Progestin alone as primary treatment of endometrial carcinoma in premenopausal women. Report of seven cases and review of the literature. Cancer 1997;79:320–7.

[37] Randall TC, Kurman RJ. Progestin treatment of atypical hyperplasia and well-differentiated carcinoma of the endometrium in women under age 40. Obstet Gynecol 1997;90:434–40.

[38] Duska LR, Garrett A, Rueda BR, Haas J, Chang Y, Fuller AF. Endometrial cancer in women 40 years old or younger. Gynecol Oncol 2001;83(2):388–93.

[39] Tangjitgamol S, Manusirivithaya S, Hanprasertpong J. Fertility-sparing in endometrial cancer. Gynecol Obstet Invest 2009;67(4):250–68.

[40] Erkanli S, Ayhan A. Fertility-sparing therapy in young women with EC: 2010 update. Int J Gynecol Cancer 2010;20:1170–87.

[41] Gallos ID, Yap J, Rajkhowa M, et al. Regression, relapse, and live birth rates with fertility-sparing therapy for EC and atypical complex endometrial hyperplasia: a systematic review and metaanalysis. Am J Obstet Gynecol 2012;207(4) 266.e1–12.

[42] Gunderson CC, Fader AN, Carson KA, Bristow RE. Oncologic and reproductive outcomes with progestin therapy in women with endometrial hyperplasia and grade 1 adenocarcinoma: a systematic review. Gynecol Oncol 2012;125:477.

[43] Bovicelli A, D'Andrilli G, Giordano A, De Iaco P. Conservative treatment of early endometrial cancer. J Cell Physiol 2013;228(6):1154–8.

[44] Park JY, Kim DY, Kim TJ, Kim JW, Kim JH, Kim YM, et al. Hormonal therapy for women with stage IA endometrial cancer of all grades. Obstet Gynecol 2013;122(1):7–14.

[45] Koskas M, Uzan J, Luton D, Rouzier R, Daraï E. Prognostic factors of oncologic and reproductive outcomes in fertility-sparing management of endometrial atypical hyperplasia and adenocarcinoma: systematic review and meta-analysis. Fertil Steril 2014;101(3):785–94.

[46] Rodolakis A, Biliatis I, Morice P, Reed N, Mangler M, Kesic V, et al. European Society of Gynecological Oncology Task Force for fertility preservation: clinical recommendations for fertility-sparing management in young endometrial cancer patients. Int J Gynecol Cancer 2015;25(7):1258–65.

[47] Leitao Jr MM, Kehoe S, Barakat RR, Alektiar K, Gattoc LP, Rabbitt C, et al. Comparison of D&C and office endometrial biopsy accuracy in patients with FIGO grade 1 endometrial adenocarcinoma. Gynecol Oncol 2009;113(1):105–8.

[48] Kim MK, Seong SJ, Lee TS, Ki KD, Lim MC, Kim YH, et al. Comparison of diagnostic accuracy between endometrial curettage and pipelle aspiration biopsy in patients treated with progestin for endometrial hyperplasia: a Korean Gynecologic Oncology Group Study (KGOG 2019). Jpn J Clin Oncol 2015;45(10):980–2.

[49] Kalogera E, Dowdy SC, Bakkum-Gamez JN. Preserving fertility in young patients with endometrial cancer: current perspectives. Int J Womens Health 2014;6:691–701.

[50] Cade TJ, Quinn MA, Rome RM, Neesham D. Long-term outcomes after progestogen treatment for early endometrial cancer. Aust N Z J Obstet Gynaecol 2013;53(6):566–70.

[51] Lau HY, Twu NF, Yen MS, Tsai HW, Wang PH, Chuang CM, et al. Impact of ovarian preservation in women with endometrial cancer. J Chin Med Assoc 2014;77(7):379–84.

[52] Kinjyo Y, Kudaka W, Ooyama T, Inamine M, Nagai Y, Aoki Y. Ovarian preservation in young women with endometrial cancer of endometrioid histology. Acta Obstet Gynecol Scand 2015;94(4):430–4.

[53] Lin KY, Miller DS, Bailey AA, Andrews SJ, Kehoe SM, Richardson DL, et al. Ovarian involvement in endometrioid adenocarcinoma of uterus. Gynecol Oncol 2015;138(3):532–5.

[54] Jones WB, Mercer GO, Lewis Jr JL, Rubins SC, Hoskins WJ. Early invasive carcinoma of the cervix. Gynecol Oncol 1993;51:26–32.

[55] Dargent D, Brun JL, Roy M, Mathevet P. Remy I: La trachelectomie elargie une alternativea. Ll'hysterectomie radicale dans le traitement des cancers infiltrants. JobGyn 1994;2:285–92.

[56] Shepherd JH, Crawford RA, Oram DH. Radical trachelectomy: a way to preserve fertility in the treatment of early cervical cancer. Br J Obstet Gynaecol 1998;105(8):912–16.

[57] Roy M, Plante M. Pregnancies after radical vaginal trachelectomy for early-stage cervical cancer. Am J Obstet Gynecol 1998;179(6 Pt 1):1491–6.

[58] Mejia-Gomez J, Feigenberg T, Arbel-Alon S, Kogan L, Benshushan A. Radical trachelectomy: a fertility-sparing option for early invasive cervical cancer. Isr Med Assoc J 2012;14(5):324–8.

[59] NCCC Guidelines Version 2.2015, Cervical Cancer. <http://www.nccn.org/professionals/physician_gls/pdf/cervical.pdf>; [updated 18.09.14, accessed 14.11.15].

[60] Dargent D, Martin X, Sacchetoni A, Mathevet P. Laparoscopic vaginal radical trachelectomy: a treatment to preserve the fertility of cervical carcinoma patients. Cancer 2000;88(8):1877–82.

[61] Smith JR, Boyle DC, Corless DJ, Ungar L, Lawson AD, Del Priore G, et al. Abdominal radical trachelectomy: a new surgical technique for the conservative management of cervical carcinoma. Br J Obstet Gynaecol 1997;104(10):1196–200.

[62] Burnett AF, Stone PJ, Duckworth LA, Roman JJ. Robotic radical trachelectomy for preservation of fertility in early cervical cancer: case series and description of technique. J Minim Invasive Gynecol 2009;16(5):569–72.

[63] Deleted in review.

[64] Meirow D, Biederman H, Anderson RA, Wallace WH. Toxicity of chemotherapy and radiation on female reproduction. Clin Obstet Gynecol 2010;53(4):727–39.

[65] Viswanathan AN, Lee LJ, Eswara JR, Horowitz NS, Konstantinopoulos PA, Mirabeau-Beale KL, et al. Complications of pelvic radiation in patients treated for gynecologic malignancies. Cancer 2014;120(24):3870–83.

[66] Peterson CM, Menias CO, Katz DS. Radiation-induced effects to nontarget abdominal and pelvic viscera. Radiol Clin North Am 2014;52(5):1041–53.

[67] Wo JY, Viswanathan AN. Impact of radiotherapy on fertility, pregnancy, and neonatal outcomes in female cancer patients. Int J Radiat Oncol Biol Phys 2009;73(5):1304–12.

[68] Ghadjar P, Budach V, Köhler C, Jantke A, Marnitz S. Modern radiation therapy and potential fertility preservation strategies in patients with cervical cancer undergoing chemoradiation. Radiat Oncol 2015;10:50.

[69] Holm K, Nysom K, Brocks V, Hertz H, Jacobsen N, Müller J. Ultrasound B-mode changes in the uterus and ovaries and Doppler changes in the uterus after total body irradiation and allogeneic bone marrow transplantation in childhood. Bone Marrow Transplant 1999;23:259–63.

[70] Critchley HO, Wallace WH. Impact of cancer treatment on uterine function. J Natl Cancer Inst Monogr 2005:64–8.

[71] Chiarelli AM, Marrett LD, Darlington GA. Pregnancy outcomes in females after treatment for childhood cancer. Epidemiology 2000;11(2):161–6.

[72] Signorello LB, Cohen SS, Bosetti C, Stovall M, Kasper CE, Weathers RE, et al. Female survivors of childhood cancer: preterm birth and low birth weight among their children. J Natl Cancer Inst 2006;98(20):1453–61.

[73] Hwang JH, Yoo HJ, Park SH, Lim MC, Seo SS, Kang S, et al. Association between the location of transposed ovary and ovarian function in patients with uterine cervical cancer treated with (postoperative or primary) pelvic radiotherapy. Fertil Steril 2012;97(6):1387–93.

[74] Yoon A, Lee YY, Park W, Huh SJ, Choi CH, Kim TJ, et al. Correlation between location of transposed ovary and function in cervical cancer patients who underwent radical hysterectomy. Int J Gynecol Cancer 2015;25(4):688–93.

[75] Barahmeh S, Al Masri M, Badran O, Masarweh M, El-Ghanem M, Jaradat I, et al. Ovarian transposition before pelvic irradiation: indications and functional outcome. J Obstet Gynaecol Res 2013;39(11):1533–7.

[76] Irtan S, Orbach D, Helfre S, Sarnacki S. Ovarian transposition in prepubescent and adolescent girls with cancer. Lancet Oncol 2013;14(13):e601–8.

[77] Gubbala K, Laios A, Gallos I, Pathiraja P, Haldar K, Ind T. Outcomes of ovarian transposition in gynaecological cancers; a systematic review and meta-analysis. J Ovarian Res 2014;7:69.

[78] Mossa B, Schimberni M, Di Benedetto L, Mossa S. Ovarian transposition in young women and fertility sparing. Eur Rev Med Pharmacol Sci 2015;19(18):3418–25.

[79] Zhao C, Wang JL, Wang SJ, Zhao LJ, Wei LH. Analysis of the risk factors for the recurrence of cervical cancer following ovarian transposition. Eur J Gynaecol Oncol 2013;34(2):124–7.

[80] Morgan S, Anderson RA, Gourley C, Wallace WH, Spears N. How do chemotherapeutic agents damage the ovary? Hum Reprod Update 2012;18(5):525–35.

[81] Brydøy M, Fosså SD, Dahl O, Bjøro T. Gonadal dysfunction and fertility problems in cancer survivors. Acta Oncol 2007;46(4):480–9.

[82] http://www.accessdata.fda.gov/drugsatfda_docs/label/2013/125085s263lbl.pdf. [accessed 11.11.15].

[83] Green DM, Whitton JA, Stovall M, Mertens AC, Donaldson SS, Ruymann FB, et al. Pregnancy outcome of female survivors of childhood cancer: a report from the Childhood Cancer Survivor Study. Am J Obstet Gynecol 2002;187(4):1070–80.

[84] Rodriguez-Wallberg KA, Oktay K. Options on fertility preservation in female cancer patients. Cancer Treat Rev 2012;38(5):354–61.

[85] Beck-Fruchter R, Weiss A, Shalev E. GnRH agonist therapy as ovarian protectants in female patients undergoing chemotherapy: a review of the clinical data. Hum Reprod Update 2008;14:553–61.

[86] Munster PN, Moore AP, Ismail-Khan R, Cox CE, Lacevic M, Gross-King M, et al. Randomized trial using gonadotropin-releasing hormone agonist triptorelin for the preservation of ovarian function during (neo) adjuvant chemotherapy for breast cancer. J Clin Oncol 2012;30:533–8.

[87] Moore HC, Unger JM, Phillips KA, Boyle F, Hitre E, Porter D, et al. Goserelin for ovarian protection during breast-cancer adjuvant chemotherapy. N Engl J Med 2015;372:923–32.

[88] Oktay K, Turan V. Failure of ovarian suppression with gonadotropin-releasing hormone analogs to preserve fertility: an assessment based on the quality of evidence. JAMA Oncol 2015:1–2. <http://dx.doi.org/10.1001/jamaoncol.2015.3252>. [Epub ahead of print] PMID: 2642640.

[89] Walker LM, Tran S, Robinson JW. Luteinizing hormone-releasing hormone agonists: a quick reference for prevalence rates of potential adverse effects. Clin Genitourin Cancer 2013;11(4):375–84.

[90] Joffe H, et al. Adverse effects of induced hot flashes on objectively recorded and subjectively reported sleep: results of a gonadotropin-releasing hormone agonist experimental protocol. Menopause 2013;20(9):905–14.

[91] BenDor R, et al. Effects of pharmacologically induced hypogonadism on mood and behavior in healthy young women. Am J Psychiatry 2013;170(4):426–33.

[92] Zelinski MB, Murphy MK, Lawson MS, Jurisicova A, Pau KY, Toscano NP, et al. In vivo delivery of FTY720 prevents radiation-induced ovarian failure and infertility in adult female non-human primates. Fertil Steril 2011;95(4):1440–5. e1–7.

[93] Li F, Turan V, Lierman S, Cuvelier C, De Sutter P, Oktay K. Sphingosine-1-phosphate prevents chemotherapy-induced human primordial follicle death. Hum Reprod 2014;29(1):107–13.

[94] Gonfloni S, Di Tella L, Caldarola S, Cannata SM, Klinger FG, Di Bartolomeo C, et al. Inhibition of the c-Abl-TAp63 pathway protects mouse oocytes from chemotherapy-induced death. Nat Med 2009;15(10):1179–85.

[95] Mahran YF, El-Demerdash E, Nada AS, Ali AA, Abdel-Naim AB. Insights into the protective mechanisms of tamoxifen in radiotherapy-induced ovarian follicular loss: impact on insulin-like growth factor 1. Endocrinology 2013;154(10):3888–99.

[96] Cardozo ER, Thomson AP, Karmon AE, Dickinson KA, Wright DL, Sabatini ME. Ovarian stimulation and in-vitro fertilization outcomes of cancer patients undergoing fertility preservation compared to age matched controls: a 17-year experience. J Assist Reprod Genet 2015;32(4):587–96.

[97] McLaren JF, Bates GW. Fertility preservation in women of reproductive age with cancer. Am J Obstet Gynecol 2012;207(6):455–62.

[98] Rodriguez-Wallberg KA, Oktay K. Fertility preservation during cancer treatment: clinical guidelines. Cancer Manag Res 2014;6:105–17.

[99] Chung K, Donnez J, Ginsburg E, Meirow D. Emergency IVF versus ovarian tissue cryopreservation: decision making in fertility preservation for female cancer patients. Fertil Steril 2013;99(6):1534–42.

[100] Donnez J, Dolmans M-M, Pellicer A, Diaz-Garcia C, Sanchez Serrano M, Schmidt KT, et al. Restoration of ovarian activity and pregnancy after transplantation of cryopreserved ovarian tissue: a review of 60 cases of reimplantation. Fertil Steril 2013;99:1503–13.

[101] Imbert R, Moffa F, Tsepelidis S, Simon P, Delbaere A, Devreker F, et al. Safety and usefulness of cryopreservation of ovarian tissue to preserve fertility: a 12-year retrospective analysis. Hum Reprod 2014;29(9):1931–40.

[102] Dittrich R, Hackl J, Lotz L, Hoffmann I, Beckmann MW. Pregnancies and live births after 20 transplantations of cryopreserved ovarian tissue in a single center. Fertil Steril 2015;103(2):462–8.

[103] Jensen AK, Kristensen SG, Macklon KT, Jeppesen JV, Fedder J, Ernst E, et al. Outcomes of transplantations of cryopreserved ovarian tissue to 41 women in Denmark. Hum Reprod 2015;30(12):2838–45.

[104] Tanbo T, Greggains G, Storeng R, Busund B, Langebrekke A, Fedorcsak P. Autotransplantation of cryopreserved ovarian tissue after treatment for malignant disease - the first Norwegian results. Acta Obstet Gynecol Scand 2015;94(9):937–41.

[105] https://www.asrm.org/uploadedFiles/ASRM_Content/News_and_PublicationsPractice_Guidelines/Committee_Opinions/OvarianTissueCryoprservation2014-noprint.pdf [accessed 16.11.15].

[106] Salama M, Mallmann P. Emergency fertility preservation for female patients with cancer: clinical perspectives. Anticancer Res 2015;35(6):3117–27.

[107] Sauerbrun-Cutler MT, Vega M, Keltz M, McGovern PG. In vitro maturation and its role in clinical assisted reproductive technology. Obstet Gynecol Surv 2015;70(1):45–57.

[108] Rubin LR, Werner-Lin A, Sagi M, Cholst I, Stern R, Lilienthal D, et al. "The BRCA clock is ticking!": negotiating medical concerns and reproductive goals in preimplantation genetic diagnosis. Hum Fertil (Camb) 2014;17(3):159–64.

[109] Woodson AH, Muse KI, Lin H, Jackson M, Mattair DN, Schover L, et al. Breast cancer, BRCA mutations, and attitudes regarding pregnancy and preimplantation genetic diagnosis. Oncologist 2014;19(8):797–804.

[110] ASRM. Third-party reproduction: sperm, egg, and embryo donation and surrogacy: a guide for patients, <https://www.asrm.org/BOOKLET_Third-party_Reproduction/>; 2012 [accessed on 12.11.15].

[111] Trounson A, Leeton J, Besanko M, Wood C, Conti A. Pregnancy established in an infertile patient after transfer of a donated embryo fertilised in vitro. Br Med J (Clin Res Ed) 1983;286(6368):835–8.

[112] Bustillo M, Buster JE, Cohen SW, Hamilton F, Thorneycroft IH, Simon JA, et al. Delivery of a healthy infant following nonsurgical ovum transfer. JAMA 1984;251(7):889.

[113] https://www.sartcorsonline.com/rptCSR_PublicMultYear.aspx?ClinicPKID=0 [accessed 02.11.15].

[114] Utian WH, Sheean L, Goldfarb JM, Kiwi R. Successful pregnancy after in vitro fertilization and embryo transfer from an infertile woman to a surrogate. N Engl J Med 1985;313(21):1351–2.

[115] Fageeh W, Raffa H, Jabbad H, Marzouki A. Transplantation of the human uterus. Int J Gynaecol Obstet 2002;76(3):245–51.

[116] Erman Akar M, Ozkan O, Aydinuraz B, Dirican K, Cincik M, Mendilcioglu I, et al. Clinical pregnancy after uterus transplantation. Fertil Steril 2013;100:1358–63.

[117] Braanstrom M, Johannesson L, Bokstrom H, et al. Live birth after uterus transplantation. Lancet 2015;385:607–16.

[118] Braanstrom M. Uterus transplantation. Curr Opin Organ Transplant 2015;20(6):621–8.

[119] Lawson AK, Klock SC, Pavone ME, Hirshfeld-Cytron J, Smith KN, Kazer RR. Psychological counseling of female fertility preservation patients. J Psychosoc Oncol 2015;33(4):333–53.

[120] Baysal Ö, Bastings L, Beerendonk CC, Postma SA, IntHout J, Verhaak CM, et al. Decision-making in female fertility preservation is balancing the expected burden of fertility preservation treatment and the wish to conceive. Hum Reprod 2015;30(7):1625–34.

[121] Chiavari L, Gandini S, Feroce I, Guerrieri-Gonzaga A, Russell-Edu W, Bonanni B, et al. Difficult choices for young patients with cancer: the supportive role of decisional counseling. Support Care Cancer 2015;23(12):3555–62.

Index

Printed in the United States
By Bookmasters